Basic
Communication
Theory

Basic Communication Theory

A teacher's eye view of the way non-mathematicians see the mathematics and use of the basic ideas of modulation

J. E. Pearson
Electronic and Electrical Engineering Department
King's College, London

PRENTICE HALL
New York London Toronto Sydney Tokyo Singapore

First published 1992 by
Prentice Hall International (UK) Ltd
Campus 400, Maylands Avenue, Hemel Hempstead
Hertfordshire HP2 7EZ
A division of
Simon & Schuster International Group

The author acknowledges the permission of the University of London for the use of problems from their examination papers.

Typeset in 10/12 pt Monotype Times
by P&R Typesetters Ltd, Salisbury, Wilts

Printed and bound in Great Britain by
Redwood Books, Trowbridge, Wiltshire

Library of Congress Cataloging-in-Publication Data

Pearson, J. E. (John E.)
Basic communication theory / J.E. Pearson.
p. cm.
Includes bibliographical references and index.
ISBN 0-13-061078-X (pbk.)
1. Telecommunication. 2. Modulation (Electronics) 3. Signal processing. I. Title
TK5101.P33 1991
621.382—dc20 91-18249
CIP

British Library Cataloging in Publication Data

Pearson J.E.
Basic communication theory: A teacher's eye view of the way non-mathematicians see the mathematics and use of the basic ideas of modulation.
I. Title
621.382

ISBN 0-13-061078-X

4 5 96 95

Contents

Preface

This book is intended to help undergraduate students to understand the basic ideas and methods used for the modulation of signals in communication systems.

The plan has been to present it all – explanations, descriptions and mathematics – as clearly and simply as possible, gradually building up complexity when needed without leaping straight in. The core of the material is the three broad areas of modulation – analogue, keying and digital – including the main methods of modulation and demodulation as well as the basic ideas mentioned at the start. This material seems naturally to require a preliminary treatment of broader ideas of what communication systems are for and how they work. This needs a consideration of the language used, both definitions and mathematics, and especially of signal representation which seems always to confuse students, often terminally. All this naturally leads to the inclusion of analogue Fourier methods and of transfer functions.

A problem throughout has been to decide where to stop in a subject. Keying modulations are a vast and growing topic full of detailed practice and spilling over into digital techniques, especially multilevel PSK. For these the importance of signal degradation, its causes and cures, means the inclusion of ISI and noise for which, in turn, there seemed to be a need for as uncluttered a summary of basic ideas as can be reached.

All of this has arisen from more than 10 years' teaching to undergraduates and MSc students of the University of London, England, first at Chelsea College and now at King's College. Full "notes" were written (and much rewritten) to provide students with a clear accurate record of what the lectures should have included. These have been rewritten again and amplified to produce this book.

As any author must be, this one is very grateful to many people who have helped him, knowingly or not. In particular to the late Eric Houldin for the opportunity to find and develop my own methods of teaching; to Neville Mathews whose "notes" I took over and who may still recognize bits of them; and to students and colleagues who have read all or some of the manuscript. Also to many colleagues over the years who have tolerantly coped with an obsession to sort out particular points I did not fully understand (e.g. negative frequency) and to Hamid Aghvami and the rest of the Communications Research Group at King's for providing an exciting environment in which to work.

Thanks are also due to my wife and family who survived weeks of losing me to a word processor and to my superiors for turning a blind eye to analogous use of College time. But perhaps the most remarkable thing is that I have survived it all myself – it really is unbelievably hard work.

So to you, dear readers, I extend the hope that you will find the book both useful and enjoyable.

JOHN PEARSON

Acronyms and Abbreviations

A/D	Analogue-to-digital converter
ADC	Analogue-to-digital converter
ADPM	Adaptive delta pulse modulation
ADPCM	Adaptive delta pulse code modulation
a.f.	Audio frequency
a.f.s.k.	Audio frequency shift keying (at baseband for VHF)
AM	Amplitude modulation
a.m.	Amplitude-modulated (signal)
AWGN	Additive white gaussian noise
ASK	Amplitude shift keying
BBC	British Broadcasting Corporation
BCD	Binary-coded decimal (signal)
BER	Bit error rate
BJT	Bipolar junction transistor
BP	Bandpass (filter)
BPCM	Binary pulse code modulation
bps	Bits per second
BPSK	Bipolar phase shift keying
BT	British Telecom
B/W	Bandwidth
B/W	Black and white (television)
CB	Citizens' band (radio)
CCIR	Comité Consultatif International des Radiocommunications (International Radio Consultative Committee – of URSI)
CCITT	Comité Consultatif International de Telegraphie et Telephonie (International Telegraph and Telephone Consultative Committee – of URSI)
CCW	Counter clockwise
CDM	Code division multiplexing
CEPT	Conference Européenne des Administrations des Postes et des Telecommunications (European Conference of Postal and Telecommunications Administrations)
CODEC	COder/DECoder

CVSDM	Continuously variable slope delta modulation
c.w.	Continuous wave (signal)
DAC	Digital-to-analogue converter
D/A	Digital-to-analogue converter
DPCM	Delta pulse code modulation
DSBSC	Double sideband suppressed carrier
DSBWC	Double sideband with carrier
d.s.p.	Digital signal processing
FDM	Frequency division multiplexing
FDMA	Frequency division multiple access
FET	Field-effect transistor
FM	Frequency modulation
f.m.	Frequency-modulated (signal)
FSK	Frequency shift keying
f.s.k.	Frequency shift keying (direct to r.f. at HF)
HF	High-frequency (radio broadcast band)
HP	High-pass (filter)
i.c.	Integrated circuit
i.f.	Intermediate frequency
ISI	Intersymbol interference
ITU	International Telecommunications Union – of UNO
LF	Low-frequency (radio broadcast band)
LP	Low-pass (filter)
LPF	Low-pass filter
L–R	Left–right (stereo)
LSB	Lower sideband
M-...	M-level signal (e.g. M-QAM)
M-ary	M level (signal)
M-level	Multilevel (signal)
M-PAM	Multilevel phase and amplitude modulation
M-PSK	Multilevel phase shift keying
M-QAM	Multilevel quadrature amplitude modulation
OOK	On–off keying
NBPM	Narrow-band phase modulation
NF	Nyquist frequency
NICAM	Near instantaneous companding audio multiplex
NRZ	Non return to zero
NTSC	National Television Standards Committee (of USA)
PAL	Phase alternation line
PAM	Phase and amplitude modulation
PAM	Pulse amplitude modulation
PAM	Pulse analogue modulation
PCM	Pulse code modulation
PDH	Plesiochronous digital hierarchy

PDM	Pulse duration modulation
PISO	Parallel in–serial out
PLL	Phase locked loop
PM	Phase modulation
PM	Pulse modulation
p.m.	Phase modulated
PPM	Pulse position modulation
PSK	Phase shift keying
P/S	Parallel in–serial out
PTM	Pulse time modulation (= PDM)
PWM	Pulse width modulation (= PDM)
QAM	Quadrature amplitude modulation
QPSK	Quadrature phase shift keying
r.f.	Radio frequency
RTTY	Radio teletype
Rx	Receiver
SDH	Synchronous digital hierarchy
S/H	Sample and hold
SIPO	Serial in–parallel out
S/N	Signal-to-noise (power) ratio
$(S/N)_Q$	Quantization signal-to-noise (power) ratio
S/P	Serial in–parallel out
SSB	Single sideband
SSBSC	Single sideband suppressed carrier
TDM	Time division multiplexing
TDMA	Time division multiple access
TF	(Voltage) Transfer function
TV	Television
Tx	Transmitter
UHF	Ultra-high-frequency (TV broadcast band)
URSI	Union Radio Scientifique International (International Union of Radio Science – of ITU)
USB	Upper sideband
VCO	Voltage-controlled oscillator
VFCT	Voice frequency carrier telephony
VHF	Very-high-frequency (radio broadcast band)
VSB	Vestigial sideband (TV signal)
VTF	Voltage transfer function
WBFM	Wideband frequency modulation
WBPM	Wideband phase modulation
WDM	Wavelength division multiplexing
ΔM	Delta modulation
ΔPCM	Delta pulse code modulation

9-QPRS	9-level quadrature partial response
16-QAM	16-level quadrature amplitude modulation
64-QAM	64-level quadrature amplitude modulation

• 1 •
Introduction

1.1 • Communication Systems

One paramount characteristic which distinguishes man from other animals is the ability to communicate with his fellow men at a very high level of complexity and speed. David Attenborough entitled the final programme of his *Life On Earth* TV series "Man the Compulsive Communicator" – and that is by no means an exaggeration.

In animals communication is nevertheless an essential part of behaviour. Without it enemies and rivals could not be warned; herds could not show group behaviour; mating could not occur. However, most of it is at a very instinctive level (e.g. scent marking of territory, plumage displays in mating, birdsong, rump marking in deer, etc., etc.); even the limited range of voiced signals in higher mammals comes into this category.

By contrast man's survival and dominance has depended on the ability to recognize individuals and to work together when hunting; to discuss complex ideas and actions; and, above all, to pass on information once gained. These methods became possible once mankind had developed a large versatile brain enabling speech and, eventually, writing as means of instant communication and permanent memory, respectively.

Both these techniques have the three main parts which any communication system must have to be useful. These parts are shown in Figure 1.1.

Speech

This is the simplest communication system man uses, yet it has all the three main parts quite obvious to see.

There is a **sending end** – the person speaking converts thoughts into muscular movements which operate a transmitter (the vocal cords and voicing system), converting them into pressure variations in the **transmission medium** (the air). These pressure variations travel outwards as a **signal** through the **transmission channel** (more air) to reach the **receiving end** (the listener), where a **receiver** (the ear drum, etc.) converts the pressure variations back into first movement, then electrical signals, and

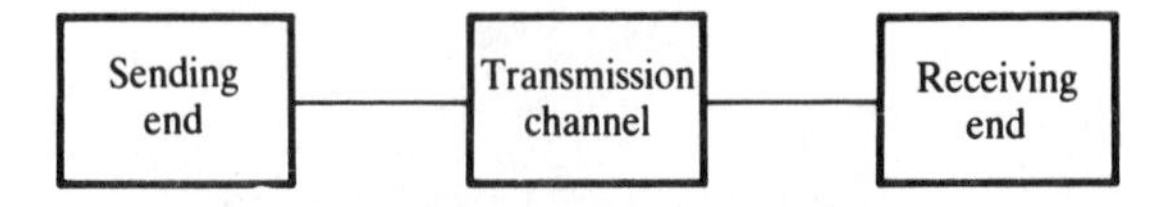

Figure 1.1 Basic parts of a communication system

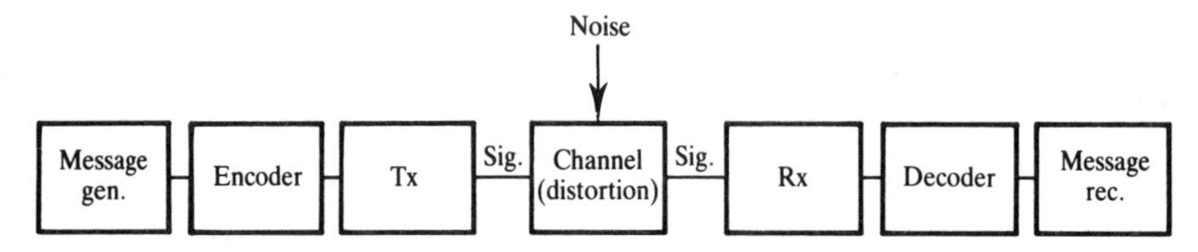

Figure 1.2 Parts of a communication system

finally thoughts. Thus the listener has had planted into his or her brain a replica of the speaker's **message** and **information** has been conveyed.

Some other aspects of communication systems are also illustrated by this trivial example. They are as follows.

Coding

The speaker's message is conveyed in a language the receiver can understand. If it were in, say, Swahili, few of us could decode the message and so would receive little information. Morse code is a simple engineering example.

Noise

Unwanted random signals added to the original message may mean the listener has difficulty decoding it. Try talking across a crowded room at a party and you will get the point. Noise can occur in all three parts of the system. At that party it came in the transmission channel, but it could be generated by the sender (wheezy breathing) or by the receiver (tinnitus).

Distortion

Here the signal is changed but nothing is added. This change may be so great that the receiver cannot decode the message correctly. Try talking across a large empty hall. Some voice frequencies resonate. Reverberation occurs, introducing time delays and attenuation which cause the instantaneous spectrum of the signal to become so distorted as to become unintelligible. Distortion usually occurs in the transmission channel, although it may occur elsewhere (e.g. if the speaker has a strong accent).

Therefore perhaps a more complete general representation of a communication system should be as shown in Figure 1.2.

Now let us concentrate on the sort of system this course is about.

1.2 • Telecommunications

1.2.1 • *Communication at a distance*

First we must restrict ourselves to artificial methods of communication which enable us to send information over longer distances than can be spanned by the human voice. At our noisy party hand signals or lip reading might help for simple ideas. Carrying memorized or written messages has always worked and, for centuries, the greatest speed with which a message could be conveyed was that of the galloping horse (as in famous events like Paul Revere, Ghent-to-Aix and the Pony Express) but usually it was much slower (as at Thermopylae).

Even then there was one way of transmitting messages much more quickly – by sight. For centuries these ways were very simple indeed (e.g. bonfires heralding the approach of the Armada). Native smoke signals were one stage better, as were flags, but the most advanced techniques were those of the heliograph and the semaphore. Actual words could be conveyed, although incredibly slowly by modern standards. The need for sending information quickly over long distances was so great that governments went to the lengths of building large wooden semaphore towers on hilltops along important routes. A well-known English example occurred during the Napoleonic wars when the Admiralty built such a chain from London to Portsmouth. Their sites are often still called Telegraph Hill and there is one on Putney Common near London.

Not until electrical signals were used in the nineteenth century can we say that we were getting anywhere near the type of signal with which this course deals. The spur to it all was the development of the railways which needed to let their operators know, beforehand, that a train was coming down the track. They were able to do this by an electric telegraph which used pulses of current along copper wire with a variety of electromechanical indicators at the terminals. Transmission was virtually instantaneous but only one letter or number could be conveyed at a time, making all but the shortest messages as slow as by semaphore. Of course greater distances could be covered and suitable coding (e.g. Morse) speeded things up.

But the telephone, invented in 1869, marked the real breakthrough. It must have seemed miraculous at the time. You could now talk to someone miles away as easily and as quickly as if you were standing next to them (subject, of course, to noise, distortion, breakdown, etc.). By using repeaters, exchanges and multiplexing, vast complex telephone networks were built up culminating in long submarine cable links. Much of the content of this course applies to these wired systems as well as to the next development.

Wireless communication completed the breakthrough. It all started in 1895 when Hertz carried out some experiments using spark gaps in resonant loops. He transmitted near-microwave frequencies. Many workers (e.g. Lodge, Popov, Kelvin) tried to develop the use of these new electromagnetic waves to provide telecommunications, but eventual success went to Marconi who was initially supported by the British Admiralty in its race to communicate with its fleets at sea. He founded the Marconi

Wireless Telegraphy Company Ltd and, by sheer hard work, developed an enormous spark transmitter which, on 12 December 1901, was able to signal across the Atlantic using very long wavelengths – and without any amplification at all.

After this developments occurred in two main areas – hardware and software. The main hardware changes have been the introduction of amplifiers (vacuum valves for a long time, then semiconductors); the use of higher frequencies; improvements in aerial design (and an understanding of propagation mechanisms); optical fibres and cables; the use of satellites. Software changes have been mainly concerned with developing methods of modulation and coding to reduce bandwidth and power and to improve range, speed, and reliability. That is what this book is about.

1.3 • Design considerations

The basic parts of a communication system are very simple. The complexity comes when trying to create practical systems to do specific tasks to exact specifications. Then many other considerations must be taken into account:

1. Range.
2. Power.
3. Cost.
4. Bandwidth.
5. Speed.
6. Reliability.
7. Convenience.
8. Accuracy/Quality.

Some detailed considerations follow.

1.3.1 • *Range*

The further information has to be transmitted, the more difficult it is to get the message through uncorrupted.

Wired links require repeaters for longer distances but work well at low frequencies. However, high frequencies require special cables – coaxial, waveguide (and now optical fibre).

Terrestrial radio links require different frequencies for different purposes. Microwave for line-of-sight links; HF for long-distance ionospheric communication across the world; VHF for a variety of shorter-range uses; UHF to give large bandwidths for TV; medium wave for local broadcasting; and so on.

Satellite links can be used for both small and large terrestrial distances but always have long path lengths with the consequent problems of attenuation and noise.

1.3.2 • *Power*

The less the power required at the sending end, the simpler and cheaper is the transmitting installation required (but the Rx has to be more sophisticated). Where other considerations allow it, transmitted power is always kept to a minimum. Compare Marconi's megawatts of peak power with the milliwatts used by the early satellites.

Some of the factors involved are that higher frequencies produce a higher proportion of radiated power from aerials ($\propto f^2$); radiated power may need to be kept high to allow the use of cheap receivers (as in radio broadcasting); pulse coding techniques increase signal accuracy and enable very weak signals to be recovered from noise (e.g. space probes); directional aerials increase effective use of radiated power (e.g. on microwave links); multiplexing enables more information to be sent for the same power.

1.3.3 • *Cost*

This obviously has to be kept as low as is compatible with achieving the desired system performance. What is economic depends on the application. For example, it is worth while spending many millions of pounds on the ground installations for transatlantic communication satellite reception (e.g. Goonhilly) but £100 would be too much for a roof-top aerial for direct reception of educational TV from satellites in India. Cost is almost irrelevant for stringent military requirements, such as in the guidance systems of nuclear missiles, whereas it becomes the major factor for mass commercial systems such as CB radio. Much of the ingenuity in developing systems goes into doing the same thing but cheaper (e.g. optical fibres not microwaves).

1.3.4 • *Bandwidth*

No information will be obtained at all unless the signal received contains at least a small range of frequencies – its bandwidth (an unchanging single carrier frequency tells you very little, except that the transmitter is on). But using a larger bandwidth than is essential increases cost and complexity unnecessarily. The narrower the individual channel the more channels can be sent simultaneously over the same link bandwidth. Thus considerable ingenuity goes into reducing channel bandwidth whilst retaining acceptable quality of information. Often a compromise is reached between the two.

For example, in a telephone channel the bandwidth needed is halved at the start by using single sideband (SSB) techniques and then making what is left just wide enough so that voices are recognizable. The result is the standard 4 kHz B/W voice frequency channel for telephones. On the other hand, where virtual 100% accuracy is required (as in many data transmission requirements) digital techniques must be used. These require larger bandwidths than other modulation techniques but this is acceptable because much higher carrier frequencies can be used. This is one reason for the present shift to optical fibre transmission.

1.3.5 • *Speed*

Real-time transmission is very common (e.g. telephones, TV). If you send information more slowly you save bandwidth but take more time. This may be acceptable and cheaper (e.g. teleprinters, facsimile) and so is used. On the other hand, by sending information faster more bandwidth is required but less time is taken. This may be necessary (e.g. high-speed data) or even cheaper (e.g. if hiring time on a BT landline) and is used.

Again the modulation and encoding methods used can be designed to get the best out of the system – digital methods in particular.

1.3.6 • *Reliability*

How much does it matter if your signal arrives corrupted? Most of this is considered above but it is worth summarizing separately here. Obviously reliability of equipment is a factor but not the one meant here. We mean factors causing signal degradation in a working system. The aim is to use the cheapest and simplest system which will give acceptable reproducibility of signal. A telephone channel is a good example of an uncomplicated system which has acceptable quality without achieving complete accuracy. Narrow bandwidths, low frequencies and intense multiplexing can be used.

Digital data streams obviously need much greater accuracy and have to have much larger bandwidths to get it. But even here a very low error rate (e.g. 1 in 10^5) may be achieved by the use of coding techniques. There always seems to be a trade-off between bandwidth and accuracy.

1.3.7 • *Convenience*

This encompasses a multitude of factors of which the most restricting technically is the need for new systems to be compatible with older existing systems. A well-known example occurred when colour was introduced into TV – it had both to keep the same bandwidth and to be compatible with existing B/W sets. It is now occurring again with direct broadcast satellites. A more recent example has been the digitizing of the telephone system.

Other aspects of convenience occur with the growth of multifacility networks (they must be digital); the use of larger and more comprehensive integrated circuits wherever possible (perhaps fixing details of the modulation method); the need for ease of production and cheaper repair (modular design); and so on.

1.3.8 • *Accuracy/Quality*

These considerations have been mentioned in most of the others above. The more accurate the received information signal must be compared with the original, the more complex and expensive the communication system has to be, in general. Thus economies can be made because there is absolutely no need to recover the signal at the receiver more accurately than it needs to be to serve the purpose for which it was sent. Take speech signals as an example. Here accuracy is a fairly subjective matter better described by the term quality and is something we automatically take into account every day when telephoning or listening to the radio knowing that what we hear is understandable and recognizable but not perfect. Qualitative considerations of acceptability of received sound have led to quantitative requirements for system specifications which differ for different purposes. The 4 kHz standard telephone channel bandwidth has been mentioned already, but in "steam" radio 3 kHz is regarded as enough with even less for HF communications. On the other hand a minimum of 15 kHz is considered necessary on VHF radio to reproduce music at acceptable quality. By contrast digital signals need much higher accuracy, partly because a single missed bit can cause a much more serious error than the loss of a short piece of analogue signal, but also because digital data is often of no use unless very accurate indeed (e.g. needs bit error rates of better than 1 in 10^6).

1.4 • Conclusion

But enough of these generalized preliminaries, however interesting they may be. Our aim in the rest of the book is to make a detailed examination of the basics of the various modulation methods used, their theories, and something about implementation and use. Much of this is mathematical, at least at the start, so our first task is to establish the mathematical language used.

•2•
Sinusoids

2.1 • Introduction

This chapter defines the mathematical language used in the book and is largely concerned with sinusoids and the various ways of representing them. There is a choice of several mathematical and pictorial ways of doing this, different ones being chosen for different situations. Thus it is necessary to define each representation clearly and to understand it.

There are six main aspects to consider:

1. Sinusoidal expressions – equations.
2. Time domain pictures – graphs.
3. Frequency domain pictures – spectra.
4. Rotating phasor representation – "real" exponential form.
5. Stationary phasor representation – time removed.
6. Complex exponential form – negative frequency.

These develop one from the other and are best explained in sequence.

2.2 • Sinusoidal expressions

A voltage which is varying in time (written $v(t)$ or just v) sinusoidally at an angular frequency ω_0 ($= 2\pi f_0$) will have an amplitude V_0 and must start at some initial phase θ_0 at $t = 0$. These are related by

$$\boxed{v(t) = V_0 \cos(\omega_0 t + \theta_0)}$$

Note that this could just as correctly be written as $\sin(\omega_0 t + \theta_0)$, since there is only a difference of $\pi/2$ in the value of θ_0 ($\cos\theta = \sin(\theta + \pi/2)$). But I have chosen to use the cosine form as it makes the mathematical analysis a little more straightforward.

2.3 • Time domain graph

A plot of v against t gives an easy-to-understand pictorial representation of the sinusoid above as in Figure 2.1. It is a fairly elementary idea which is now given the rather grandiose name of **time domain representation**.

Note how $+\theta_0$ radians cause a negative shift of $-\theta_0/\omega$ seconds on the time axis.

The usefulness occurs with non-sinusoidal periodic waveforms made up of more than one sinusoid related harmonically. The equation can look quite confusing but the graph can be beautifully simple as shown in the example below (Figure 2.2):

$$v = V_0 \cos \theta_0 - \frac{V_0}{3} \cos 3\omega_0 t$$

The beginnings of a square wave emerge very clearly – try adding $(V_0/5) \cos 5\omega_0 t$ as well. Note that the same function is obtained from $V_0 \sin \omega_0 t + (V_0/3) \sin 3\omega_0 t$ if $t = 0$ occurs at the point marked X.

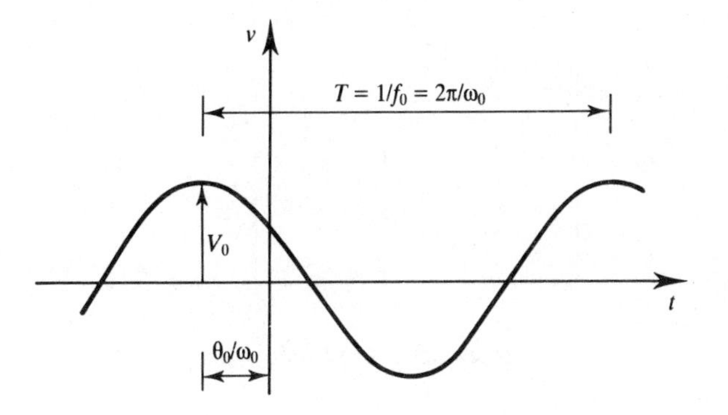

Figure 2.1 Time domain representation

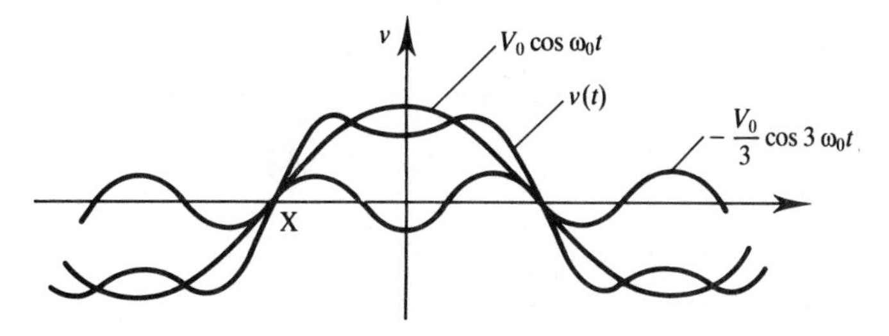

Figure 2.2 Square wave build-up

2.4 • Frequency domain spectra

Each sinusoid occurs at one particular frequency (ω radians per second or f hertz) with a definite amplitude (V) and an initial phase (θ). V and θ can each be plotted separately against ω (or f) and the result is a spectrum – or rather two spectra – amplitude and phase. This is the **frequency domain representation**. Take the voltage

$$v = V_1 \cos(\omega_1 t + \theta_1) + V_2 \cos(\omega_2 t - \theta_2)$$

Its amplitude spectrum is as shown in Figure 2.3.

Its phase spectrum is as shown in Figure 2.4.

Again this is a fairly familiar idea. It is particularly useful when representing a series of harmonically related sinusoids (e.g. a square wave) or a non-periodic function (i.e. a single pulse) whose spectra are continuous (i.e. all frequencies are present) but decreasing in amplitude as in Figure 2.5. However, all sinusoids must be expressed as cosines first to make sure the phase relationships are correctly plotted.

If a sinusoidal expression contains no θ terms it can be shown as a single amplitude spectrum, negative amplitudes being simply drawn negative. For example, the spectrum of Figure 2.2 is shown in Figure 2.6 below.

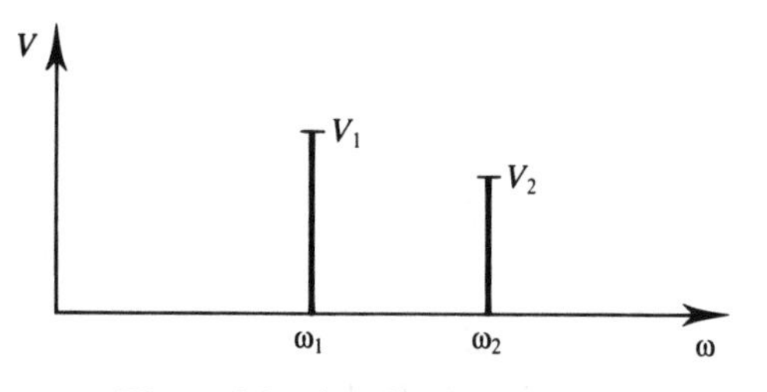

Figure 2.3 Amplitude spectrum

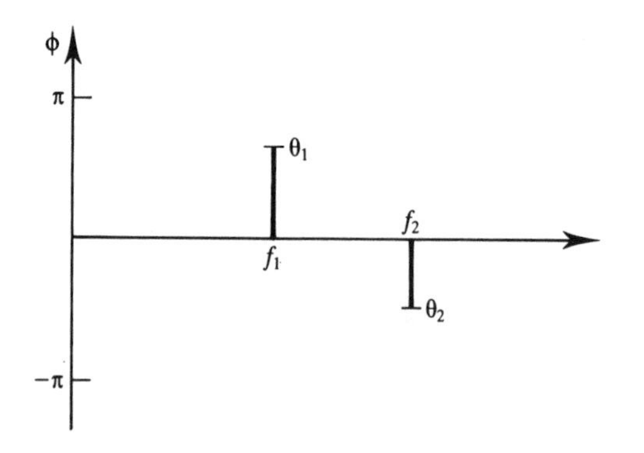

Figure 2.4 Phase spectrum

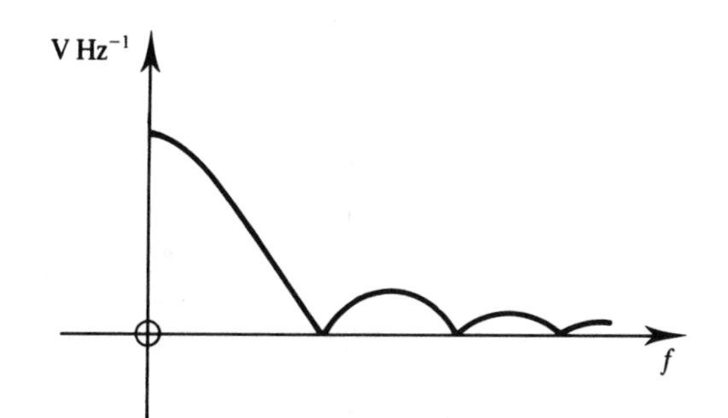

Figure 2.5 Continuous amplitude spectrum for pulse

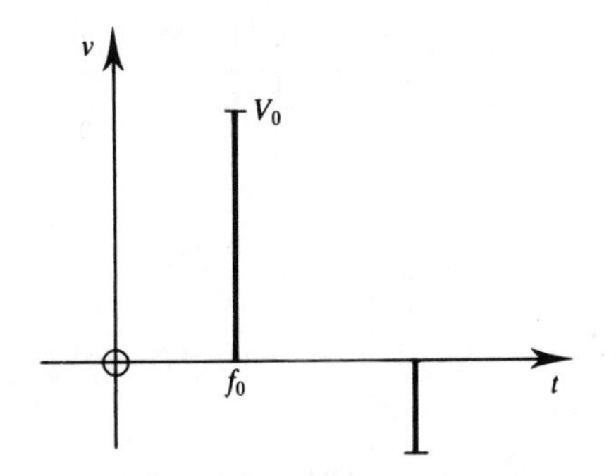

Figure 2.6 Single spectrum when all $\theta = 0$

But beware of two things. If $\theta \neq 0$ and negative amplitudes occur these must be expressed as a phase change of π (strictly $\pm n\pi$ as we shall see later) to get the phase spectrum completely correct. Also, all sine terms must be re-expressed as cosines. For example

$$v = V_1 \cos \omega_0 t - V_2 \cos 2\omega_0 t + V_3 \sin 3\omega_0 t$$

becomes

$$v = V_1 \cos \omega_0 t + V_2 \cos(2\omega_0 t + \pi) + V_3 \cos(3\omega_0 t - \pi/2)$$

using

$$-\cos \theta = \cos(\theta + \pi) \quad \text{and} \quad \sin \theta = \cos(\theta - \pi/2)$$

and the expression now plots as in Figure 2.7.

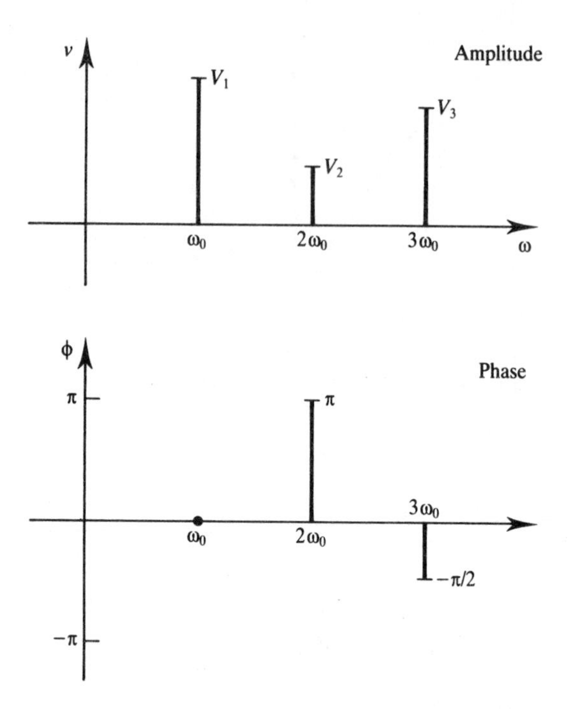

Figure 2.7 Spectra when sines expressed as cosines

2.5 • Rotating phasors – the exponential representation

An elementary way to represent sinusoids pictorially is to treat them as projections on the x or y axis of a line of constant length rotating at a constant angular speed anti-clockwise about one end fixed at the origin. The line length is that of the amplitude of the sinusoid. The x axis projection is the cosine and the y axis projection is the sine as shown in Figure 2.8.

This rotating line is very similar to a vector, in that it has magnitude and direction, but the "direction" is mathematical rather than in space, hence the separate name **phasor**. Also, because the line is moving round, its full name is **rotating phasor**. Note that it is rotating anti-clockwise since angles in the complex plain are measured counter-clockwise (CCW) from the positive real axis. Figure 2.8 shows this representation for the simple expression

$$v = V_0 \cos(\omega_0 t + \theta_0)$$

And as long as it is clearly understood that we mean the cosine projection of the phasor, we can also write v as

$$v = V_0 \angle (\omega_0 t + \theta_0)$$

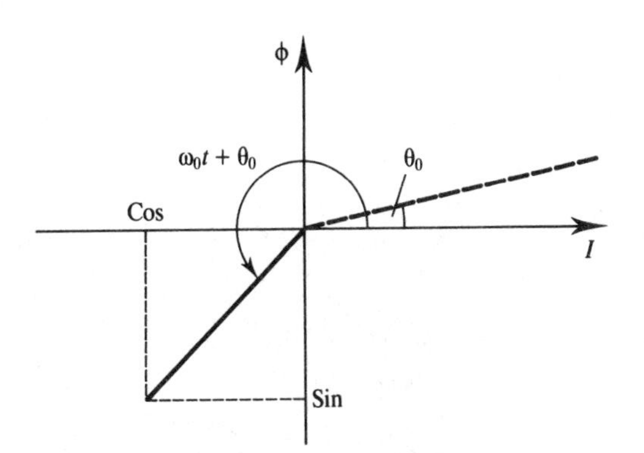

Figure 2.8 Rotating phasor

But that is not all. Mathematically v is now a **vector** sum of its sine and cosine components so it can be written

$$\bar{v} = V_0 \cos(\omega_0 t + \theta_0) + V_0 \sin(\omega_0 t + \theta_0)$$

or

$$v = V_0 \cos(\omega_0 t + \theta_0) + \mathrm{j} V_0 \sin(\omega_0 t + \theta_0)$$

because j causes a phase change of $\pi/2$ ($\mathrm{j}^2 = -1 \equiv \pi$ phase change, so $\mathrm{j} \equiv \pi/2$ phase change). Also, because $\exp \mathrm{j}\theta = \cos\theta + \mathrm{j}\sin\theta$, we can write

$$v = V_0 \operatorname{Re}\{\exp[\mathrm{j}(\omega_0 t + \theta_0)]\}$$

Again if it is clearly understood that we are taking the cosine or *real* part of the exponential, we merely write

$$v = V_0 \exp[\mathrm{j}(\omega_0 t + \theta_0)]$$

[Note that people who write $v = V \sin \omega t$ use the imaginary part of the exponential.] This last equation is of great use in the analysis of communication systems because so many operations involve integrations with other exponentials (e.g. Fourier transforms), but then it usually implies the full complex exponential (as in Section 2.6) assuming ω can be negative as well as positive. It is not much use as the "real part" form above because there are only a limited number of mathematical operations which can be done with it – really only addition and subtraction. You will find that there is much loose and confusing usage of both forms but not, I hope, in this book.

Rotating phasor diagrams themselves are also of only limited use but do have the advantage that they can show phasors of different frequencies. For example, you

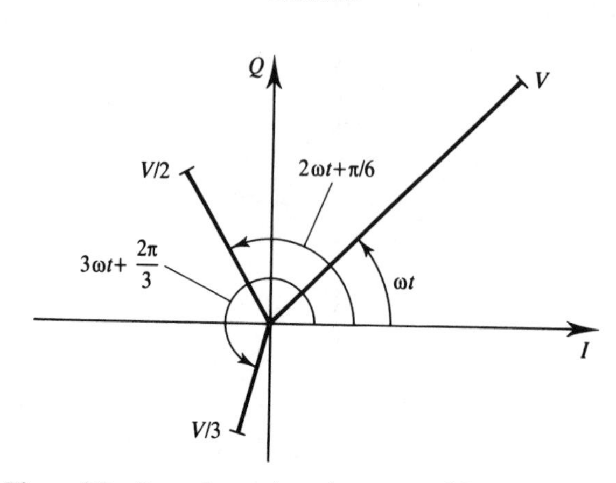

Figure 2.9 Several rotating phasors at different frequencies

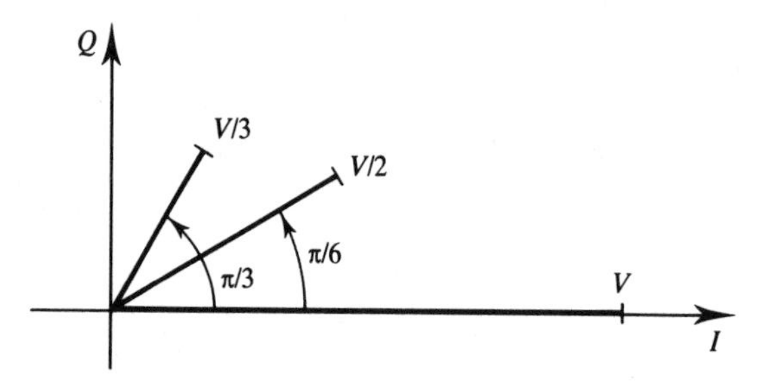

Figure 2.10 Stationary phasor representation

could use them to draw this voltage as in Figure 2.9:

$$v = V\cos\omega t + \frac{V}{2}\cos(2\omega t + \pi/6) + \frac{V}{3}\cos(3\omega t + 2\pi/3)$$

2.6 • Stationary phasors

The special case of rotating phasors at $t = 0$ is of much more interest because it shows the initial phase constant (θ) very clearly and, more important, illustrates the relative values of θ for several phasors. This is particularly useful if they are all of the same frequency because their relative relationship stays constant for all t. They are then known as **stationary phasors** and can be shown clearly on a **stationary phasor diagram** as in Figure 2.10 for the voltage

$$v = V\cos\omega t + \frac{V}{2}\cos(\omega t + \pi/6) + \frac{V}{3}\cos(\omega t + \pi/3)$$

This representation is also very illustrative where there is actually a small frequency difference between the phasors as occurs in AM and FM later on. Then one phasor (the carrier) can be held stationary whilst the others rotate slowly with respect to it (see Figure 7.4).

2.7 • Complex exponential representation – negative frequency

By using the Euler expression

$$\exp \mathrm{j}\theta = \cos\theta + \mathrm{j}\sin\theta$$

you can easily show that

$$V_0\cos(\omega t + \theta) = \frac{V_0}{2}\{\exp[\mathrm{j}(\omega t + \theta)] + \exp[-\mathrm{j}(\omega t + \theta)]\}$$

which is the full **complex exponential representation** for a sinusoidal voltage. This can be rewritten to bring out the need for the idea of negative frequency as follows:

$$V_0\cos(\omega t + \theta) = \frac{V_0}{2}\{\exp[\mathrm{j}((+\omega)t + \theta)] + \exp[\mathrm{j}((-\omega)t - \theta)]\}$$

In words, this means that a cosinusoid can be expressed as a sum of two complex exponentials, one a function of positive frequency and the other of negative frequency. This is a very useful mathematical idea used for the analysis of all communication systems (except the simpler introductory ones we use soon). Like time, frequency becomes a normal double-sided function with both positive and negative values giving a satisfying analytical symmetry about it all. The only trouble is that everybody is so used to frequency being a purely positive quantity (how can it be anything else?) that it is a very difficult idea to accept. After using it a lot one can learn to accept it as a useful convenience but the apparent unreality of it still tends to rankle. One helpful way to overcome this built-in prejudice is to think of it in terms of rotating phasors as shown for the expression above in Figure 2.11.

Positive frequency is the familiar anti-clockwise rotating phasor whereas **negative** frequency is merely the same phasor rotating clockwise. This gives a sort of "physical" meaning to it which most people find helpful.

Once the "existence" of negative frequency is accepted, the two-term expression for $V_0\cos(\omega_0 t + \theta)$ above can be simplified to

$$V_0\cos(\omega_0 t + \theta) \equiv V_0\exp[\mathrm{j}(\omega_0 t + \theta)]$$

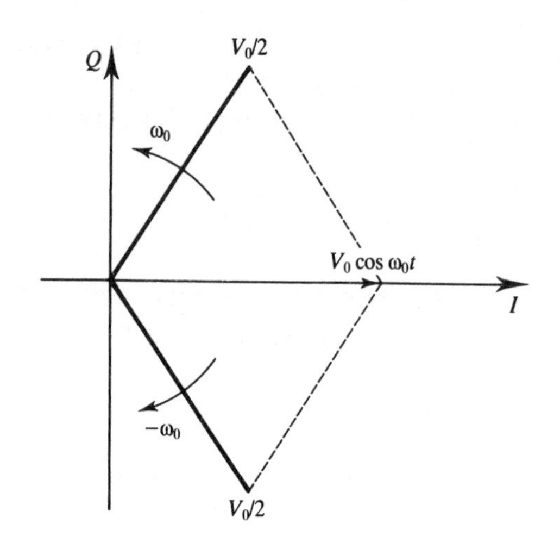

Figure 2.11 A sinusoid as positive and negative phasors

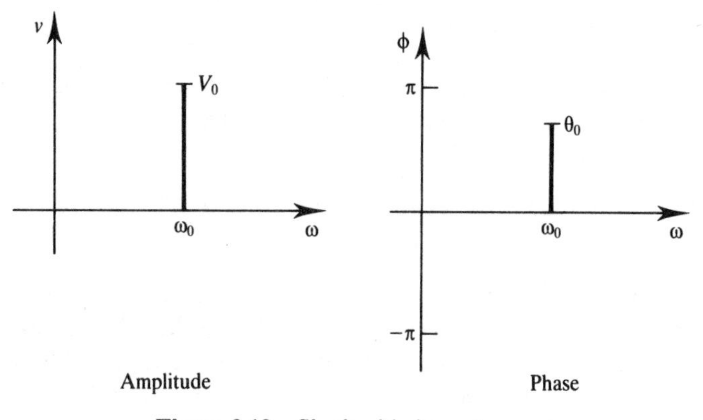

Figure 2.12 Single-sided spectrum

as long as it is clearly understood that both positive and negative frequencies are included in the one value of ω (this is where the $\frac{1}{2}$ has gone). This is now the same expression as was arrived at when discussing rotating phasors in Section 2.5 above, but this time its true origin is revealed although it is often referred to as the **rotating phasor** form.

But note the slight anomaly which must also be understood in the use of negative frequency. This is that θ also changes sign with ω, being positive for $+\omega$ and negative for $-\omega$. This always occurs for real signals and is shown clearly in the spectra below (Figures 2.12 and 2.13). They need to be double sided not single sided as hitherto.

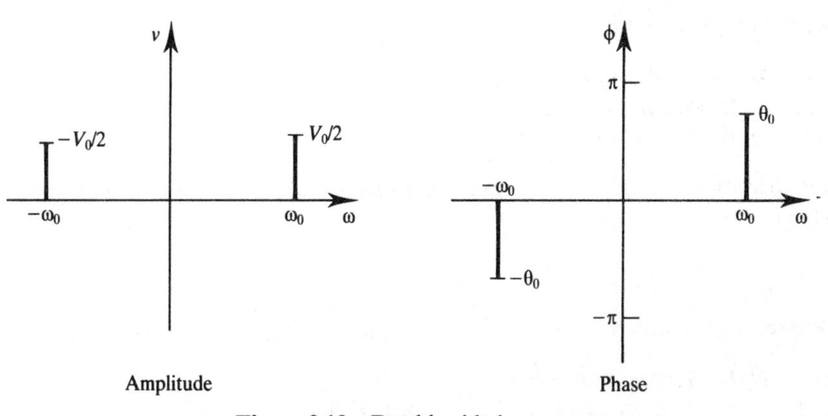

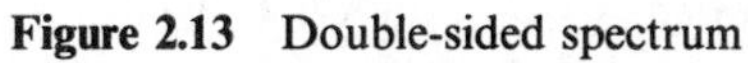
Figure 2.13 Double-sided spectrum

2.8 • Summary of forms

Sinusoidal expression: $v = V_0 \cos(\omega_0 t + \theta_0)$

"Real" exponential: $v = \mathrm{Re}\{V_0 \exp[\mathrm{j}(\omega_0 t + \theta_0)]\}$

"Double-sided" exponential:

$$v = \frac{V_0}{2}\{\exp[\mathrm{j}(\omega_0 t + \theta_0)] + \exp[-\mathrm{j}(\omega_0 t + \theta_0)]\}$$

$$= \frac{V_0}{2}\{\exp[\mathrm{j}(\omega_0 t + \theta_0)] + \exp[\mathrm{j}((-\omega_0)t - \theta_0)]\}$$

Rotating phasor: $v = V_0 \exp[\mathrm{j}(\omega_0 t + \theta_0)] = V_0 \angle (\omega_0 t + \theta_0)$

Stationary phasor: $v = V_0 \exp \mathrm{j}\theta_0 = V_0 \angle \theta_0$

2.9 • Conclusion

Now we know all the various forms in which signals are expressed in this book. The next stage is to see some of these forms in use in descriptions of the types of analysis common in communications. The most common of these are the Fourier techniques examined in the next two chapters.

2.10 • Problems

2.1 A signal is given by the expression

$$v = V\cos(\omega t + \phi) + \frac{V}{2}\cos(2\omega t + \phi/2)$$

Represent this signal

(i) on a rotating phasor diagram
(ii) on a stationary phasor diagram
(iii) on double-sided amplitude and phase spectra.

Put in as many quantities as you can in terms of the constants of the expression above.

2.2 Represent the following sinusoids in as many different ways as you can.

(i) $v(t) = V_0 \cos(\omega_0 t - \theta_0)$
(ii) $v(t) = V_0 \cos \omega_0 (t + T/2)$ [period T]
(iii) $v = V_0 \cos \omega_0 t - (V_0/3) \cos \omega_0 t$
(iv) $v = V_0 \cos(\omega_0 t - \pi/2) + V_0 \cos(2\omega_0 t - \pi)$
(v) $v = (V_0 \cos \omega_0 t)^2$ [use $\cos 2\theta = 2\cos^2 \theta - 1$]
(vi) $v = V_0 \cos(\omega_0 t + \pi/2)$
(vii) $v = V_0 \cos \omega_0 t + V_0 \cos(2\omega_0 t + \pi/2)$
(viii) $v = [V_1 \cos(\omega_1 t + \theta_1)][V_2 \cos(\omega_2 t + \theta_2)]$
(ix) $v = 1.0 \cos(2\pi \times 10^3 t + \pi/2) + 1.0(2\pi \times 10^3 t + \pi/4)$
(x) $2.0 \sin 2\pi \times 10^6 (t + 2.5 \times 10^{-7}) + 3.0 \cos 2\pi \times 10^6 (t + 5.0 \times 10^{-7})$
(xi) $2.0 \sin(\omega_0 t + \pi/8) + 3.0 \sin(\omega_0 t + \pi/4) + 4.0 \sin(\omega_0 t + \pi/2)$

$$+ 5.0 \sin(\omega_0 t + \pi)$$

(xii) $v = \mathrm{Re}[V_0 \exp \mathrm{j}(\omega_0 t + \pi/2)]$
(xiii) $v = V_0\{\exp(\mathrm{j}\omega_0 t + \theta_0) - \exp[-(\mathrm{j}\omega_0 t - \theta_0)]\}$
(xiv) $v = \mathrm{j}V[\exp(-\mathrm{j}\omega t) - \exp(\mathrm{j}\omega t)]$

2.3 Write down expressions, in as many forms as you can, for the sinusoids shown in the diagrams in Figure 2.14.

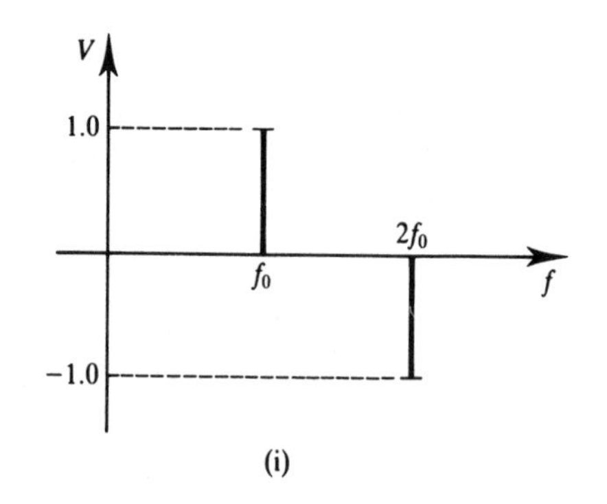

Figure 2.14 Diagrams for Problem 2.3

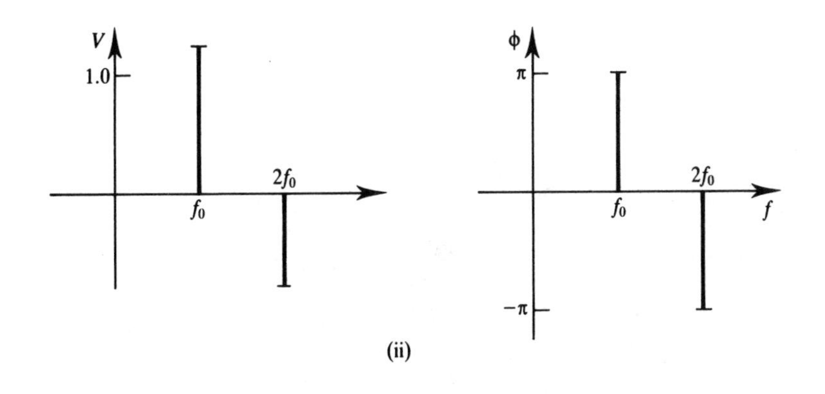

(ii)

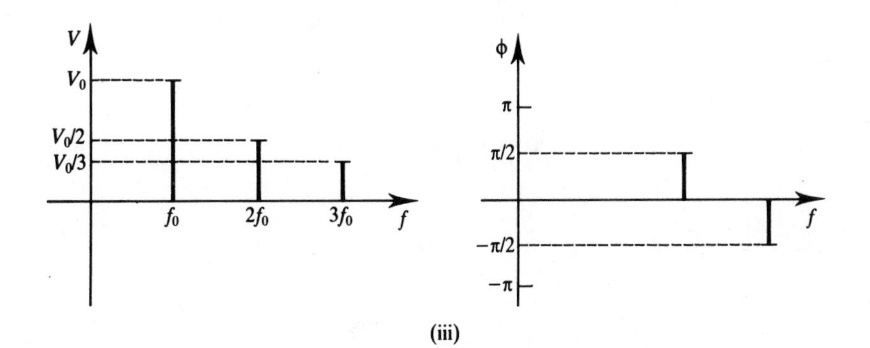

(iii)

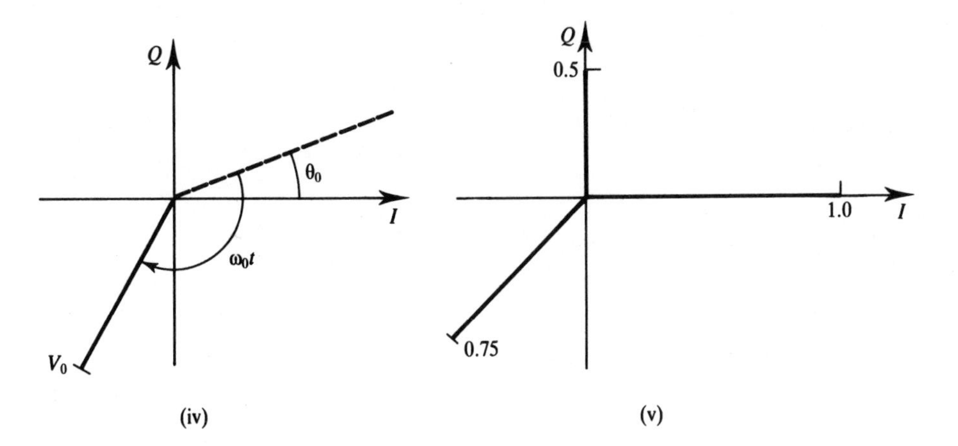

(iv) (v)

Figure 2.14 (*continued*)

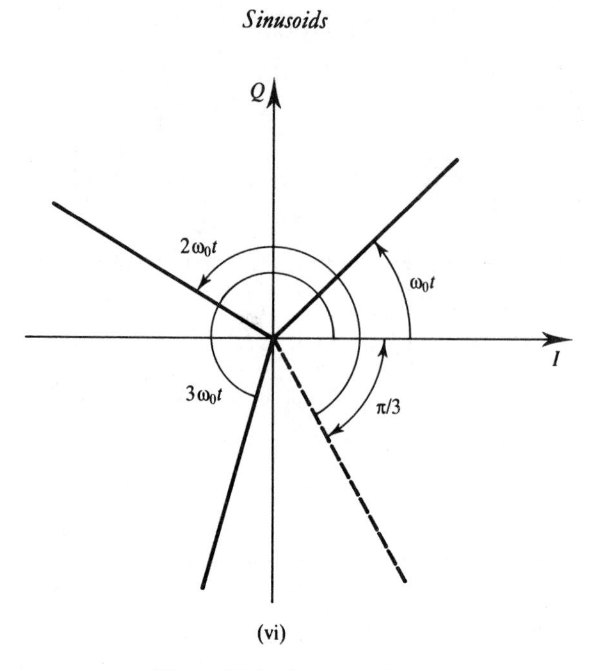

(vi)

Figure 2.14 (*continued*)

• 3 •
Fourier series

3.1 • Introduction

The preceding chapter showed various ways of expressing sinusoidal signals. Some were essentially in the time domain and others were mainly in the frequency domain (sort out which is which for yourself). The question then is: how do you get from one to the other?

For the simple examples used in that chapter it is fairly obvious – but how is it done in general? The answer is provided by the **Fourier techniques** named after the French mathematician who developed them in the nineteenth century. As far as we are concerned, there are two of these.

1. **Fourier series**: how to get the spectrum of a **periodic** signal.
2. **Fourier transforms**: how to get the spectrum of an **aperiodic** signal (or pulse) and the reverse.

The first technique only goes one way but the second is a reversible process (a true transform), so you can go from time to frequency and also from frequency to time if required.

Only the series methods are described in this chapter; transforms are discussed in Chapter 4.

3.2 • Periodic and aperiodic signals

A **periodic** signal is one which keeps repeating itself with the same pattern like that in Figure 3.1. The repetitition occurs at equal intervals of time (the period), T_0, seconds, which is stated formally as

$$v(t) = v(t \pm nT_0)$$

where n is an integer.

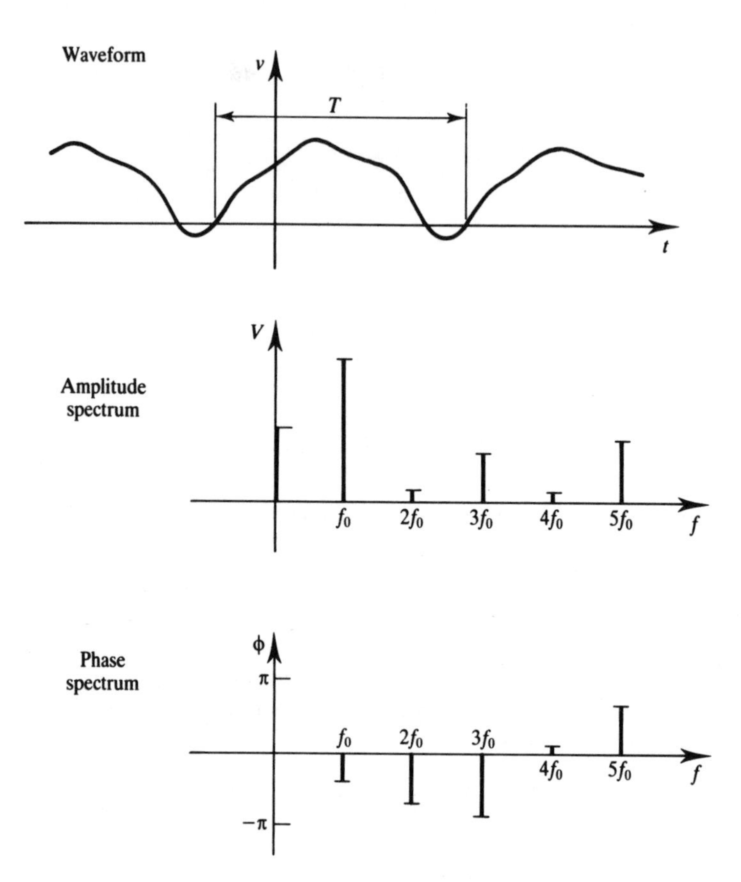

Figure 3.1 A periodic function and its line spectra

The repetition period, T_0, is related to its fundamental frequency, f_0, by

$$f_0 = 1/T_0 \text{ Hz}$$

or

$$\omega_0 = 2\pi/T_0 \text{ rad s}^{-1}$$

The spectrum of a signal like this always consists of a series of spectral lines spaced at equal intervals, f_0 apart.

An **aperiodic** signal does not repeat in time (and in practice always tails off to zero in the past and future) so that it has the appearance of a **pulse** as in Figure 3.2. It is then like a periodic signal with an infinitely long period ($T_0 = \infty$) so that the frequency separation between the lines goes to zero and the spectrum becomes

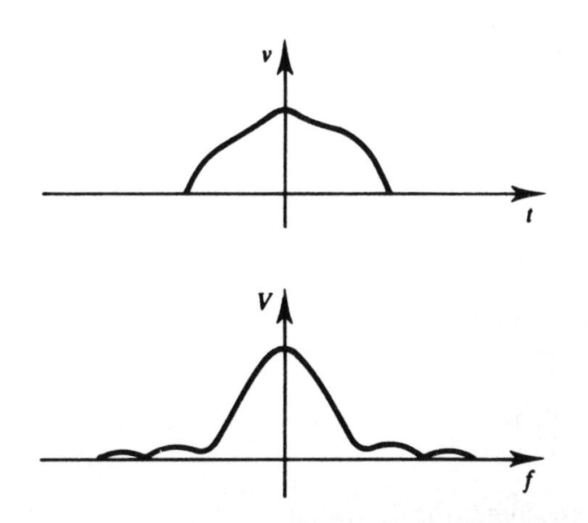

Figure 3.2 An aperiodic signal and its spectral density distribution

continuous. It is then strictly a **spectral density distribution** with axes as also shown in Figure 3.2.

3.3 • General forms of the Fourier series

The basic idea is that any periodic signal can be expressed as an infinite series of sine and cosine functions of the harmonics of the fundamental frequency of the signal, each term having an amplitude coefficient multiplier. This leads to the following forms.

3.3.1 • *General – single sided*

$$\begin{aligned} v(t) = a_0 &+ a_1 \cos \omega_0 t + a_2 \cos 2\omega_0 t + a_3 \cos 3\omega_0 t + \ldots \\ &+ a_n \cos n\omega_0 t + \ldots + b_1 \sin \omega_0 t + b_2 \sin 2\omega_0 t + b_3 \sin 3\omega_0 t + \ldots \\ &+ b_n \sin n\omega_0 t + \ldots \end{aligned}$$

or

$$v(t) = a_0 + \sum_{n=1}^{\infty} a_n \cos n\omega_0 t + \sum_{n=1}^{\infty} b_n \sin n\omega_0 t$$

where a_n and b_n are the **Fourier coefficients**. But the two terms in n can be combined into one sinusoid with a phase term as follows.

3.3.2 • *Single sinusoid form – single sided*

$$v(t) = a_0 + \sum_{n=1}^{\infty} (a_n^2 + b_n^2)^{1/2} \cos(n\omega_0 t + \theta_n) \qquad \theta_n = \tan^{-1}(-b_n/a_n)$$

Show this for yourself.

If we now write

$$\cos(n\omega_0 t) = \tfrac{1}{2}[\exp(jn\omega_0 t) + \exp(-jn\omega_0 t)]$$

$$\sin(n\omega_0 t) = \frac{1}{2j}[\exp(jn\omega_0 t) - \exp(-jn\omega_0 t)]$$

then the basic form becomes the following.

3.3.3 • *Double-sided complex exponential form*

$$v(t) = \sum_{n=-\infty}^{\infty} c_n \exp(jn\omega_0 t) \qquad c_n = \tfrac{1}{2}(a_n - jb_n)$$

Prove this for yourself too. The c_n are also Fourier coefficients.

The coefficients are obtained by means of the following integrals:

$$a_0 = \frac{1}{T}\int_0^T v(t)\,dt$$

$$a_n = \frac{2}{T}\int_0^T v(t)\cos(n\omega_0 t)\,dt$$

$$b_n = \frac{2}{T}\int_0^T v(t)\sin(n\omega_0 t)\,dt$$

$$c_n = \frac{1}{T}\int_0^T v(t)\exp(-jn\omega_0 t)\,dt$$

Each of these can be proven by substituting the relevant series for $v(t)$ and using the fact that all integrals of a product of sinusoids over a complete period are zero except the squared ones (i.e. $\sin^2$ and $\cos^2$). Again, prove these for yourself if you care to.

Note that the limits of integration must cover a whole number of periods of the waveform (usually only one), but need not start at zero. In fact it is very often much more convenient, when evaluating these integrals, to start at a convenient zero crossing of the waveform.

3.4 • Simplification by symmetry

In finding the Fourier series of a periodic voltage, the problem usually comes down to carrying out the integrals to find the expressions (or values) for the coefficients. Often some of these are zero and this can easily be seen from symmetry in the shape of the waveform. This can occur in five distinct ways:

1. a_0 is merely the **time average** of $v(t)$. This average can often be seen quickly because of vertical symmetry. A value of zero is usually easily seen.
2. $b_n = 0$ if $v(t)$ has cosine-like **even** symmetry along the time axis about $t = 0$ (i.e. v has the same value at t and $-t$).
3. $a_n = 0$ if $v(t)$ has sine-like **odd** symmetry along the time axis about $t = 0$ (i.e. v has the same magnitude but opposite sign at t and $-t$).
4. $a_n = b_n = 0$ for all even values of n (except a_0) if $v(t)$ has **skew** (rotational) symmetry (i.e. successive half periods have mirror-image shapes).
5. These symmetries may be **hidden** by a d.c. level (i.e. $a_0 \neq 0$) but are still there and allow valid simplification.

Figure 3.3 gives examples of each of these symmetries.

These symmetries often occur together as in the fourth diagram of Figure 3.3. Proofs of these simplifications are as follows.

1. Even symmetry: $v(t) = v(-t)$

$$
\begin{aligned}
b_n &= \frac{2}{T}\int_{-T/2}^{T/2} v(t)\sin(n\omega_0 t)\,\mathrm{d}t \\
&= \frac{2}{T}\left(\int_{-T/2}^{0} v(t)\sin(n\omega_0 t)\,\mathrm{d}t + \int_{0}^{T/2} v(t)\sin(n\omega_0 t)\,\mathrm{d}t\right) \\
&= \frac{2}{T}\int_{0}^{-T/2} v(-t)\sin[n\omega_0(-t)]\,\mathrm{d}(-t) + \int_{0}^{T/2} v(t)\sin(n\omega_0 t)\,\mathrm{d}t \\
&= \frac{2}{T}\left(\int_{0}^{T/2} v(-t)\sin(-n\omega_0 t)\,\mathrm{d}t + \int_{0}^{T/2} v(t)\sin(n\omega_0 t)\,\mathrm{d}t\right) \\
&= \frac{2}{T}\left(\int_{0}^{T/2} v(t)[-\sin(n\omega_0 t)]\,\mathrm{d}t + \int_{0}^{T/2} v(t)\sin(n\omega_0 t)t\,\mathrm{d}t\right) \\
&= \frac{2}{T}\left(-\int_{0}^{T/2} v(t)\sin(n\omega_0 t)\,\mathrm{d}t + \int_{0}^{T/2} v(t)\sin(n\omega_0 t)\,\mathrm{d}t\right) \\
&= 0
\end{aligned}
$$

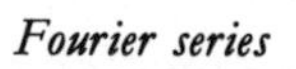

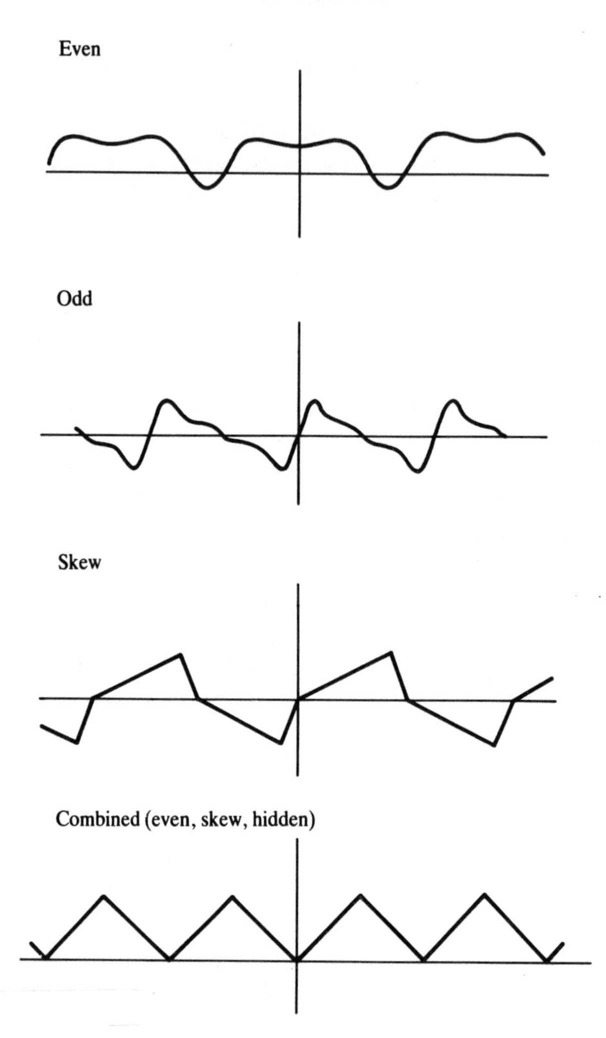

Figure 3.3 Types of symmetry

2. Odd symmetry: $v(t) = -v(-t)$

$$a_n = \frac{2}{T}\int_{-T/2}^{T/2} v(t)\cos(n\omega_0 t)\,\mathrm{d}t$$

$$= \frac{2}{T}\left(\int_{-T/2}^{0} v(t)\cos(n\omega_0 t)\,\mathrm{d}t + \int_{0}^{T/2} v(t)\cos(n\omega_0 t)\,\mathrm{d}t\right)$$

$$= \frac{2}{T}\left(\int_{0}^{-T/2} v(-t)\cos[n\omega_0(-t)]\,\mathrm{d}(-t) + \int_{0}^{T/2} v(t)\cos(n\omega_0 t)\,\mathrm{d}t\right)$$

$$= \frac{2}{T}\left(\int_0^{T/2} v(-t)\cos(-n\omega_0 t)\,dt + \int_0^{T/2} v(t)\cos(n\omega_0 t)\,dt\right)$$

$$= \frac{2}{T}\left(-\int_0^{T/2} v(t)\cos(n\omega_0 t)\,dt + \int_0^{T/2} v(t)\cos(n\omega_0 t)\,dt\right)$$

$$= 0$$

3. Skew symmetry: $v(t) = -v(t \pm T/2)$

$$a_n = \frac{2}{T}\int_0^{T} v(t)\cos(n\omega_0 t)\,dt$$

$$= \frac{2}{T}\left(\int_0^{T/2} v(t)\cos(n\omega_0 t)\,dt + \int_{T/2}^{T} v(t)\cos(n\omega_0 t)\,dt\right)$$

$$= \frac{2}{T}\left(\int_0^{T/2} v(t)\cos(n\omega_0 t)\,dt\right.$$

$$\left. + \int_{T/2-T/2}^{T-T/2} v(t+T/2)\cos[n\omega_0(t+T/2)]\,d(t+T/2)\right)$$

$$= \frac{2}{T}\left(\int_0^{T/2} v(t)\cos(n\omega_0 t)\,dt + \int_0^{T/2} -v(t)\cos[n(\omega_0 t + \pi)]\,dt\right)$$

$$= \frac{2}{T}\left(\int_0^{T/2} v(t)\cos(n\omega_0 t)\,dt - \int_0^{T/2} v(t)\cos(n\omega_0 t)\cos(n\pi)\,dt\right)$$

$$(\sin(n\pi) = 0)$$

$$= \frac{2}{T}[1 - \cos(n\pi)]\int_0^{T/2} v(t)\cos(n\omega_0 t)\,dt$$

$$= \begin{cases} \dfrac{4}{T}\displaystyle\int_0^{T/2} v(t)\cos(n\omega_0 t)\,dt & \text{when } n \text{ odd} (\cos n\pi = -1) \\ 0 & \text{when } n \text{ even} (\cos n\pi = 1) \end{cases}$$

Thus a_n only exists when n is even (2, 4, etc.). This is also true for b_n as can be shown similarly. Note that a_0 is *not* included.

A common example of this type of symmetry is a square pulse train. Note that it has both skew and even (or odd) symmetry.

3.5 • Worked examples

Three worked examples of increasing hardness follow.

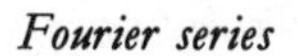

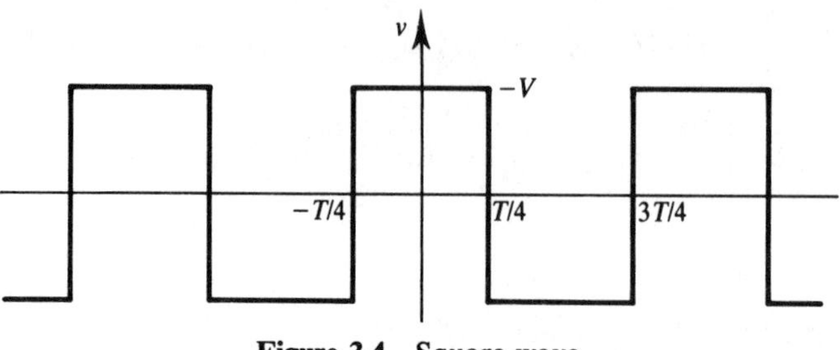

Figure 3.4 Square wave

1. An easy one

This refers to a square wave (Figure 3.4) with period T, amplitude V, and even symmetry. Here

$a_0 = 0$ as average is zero (work it out from the integral)
$b_n = 0$ even symmetry (do this one too, as an exercise)
$a_n = 0$ for even n values – skew symmetry

so we only need to work out a_n for odd n values:

$$
\begin{aligned}
a_n &= \frac{2}{T}\int_{-T/4}^{3T/4} V(t)\cos(n\omega_0 t)\,\mathrm{d}t \\
&= \frac{2}{T}\left(\int_{-T/4}^{T/4} V\cos(n\omega_0 t)\,\mathrm{d}t + \int_{T/4}^{3T/4} (-V)\cos(n\omega_0 t)\,\mathrm{d}t\right) \\
&= \frac{2}{T}\left(\int_{-T/4}^{T/4} V\cos(n\omega_0 t)\,\mathrm{d}t - \int_{T/4}^{3T/4} V\cos(n\omega_0 t)\,\mathrm{d}t\right) \\
&= \frac{2V}{T}\left[\frac{\sin(n\omega_0 t)}{n\omega_0}\right]_{-T/4}^{T/4} - \frac{2V}{T}\left[\frac{\sin(n\omega_0 t)}{n\omega_0}\right]_{T/4}^{3T/4} \\
&= \frac{2V}{nT\omega_0}\left[\sin\left(\frac{n\omega_0 T}{4}\right) - \sin\left(\frac{n\omega_0(-T)}{4}\right) - \sin\left(\frac{3n\omega_0 T}{4}\right) + \sin\left(\frac{n\omega_0 T}{4}\right)\right] \\
&= \frac{4V}{n\pi}\left[\sin\left(\frac{n\pi}{2}\right)\right] \qquad (\omega_0 = 2\pi/T;\ \sin(-\theta) = -\sin\theta) \\
&= \begin{cases} 0 & \text{for even } n \text{ values, of course} \\ 4V/n\pi & \text{for } n = 1, 5, 9, \text{etc.} \\ -4V/n\pi & \text{for } n = 3, 5, 7, \text{etc.} \end{cases}
\end{aligned}
$$

So the complete single-sided Fourier series expression for this square wave (i.e. its

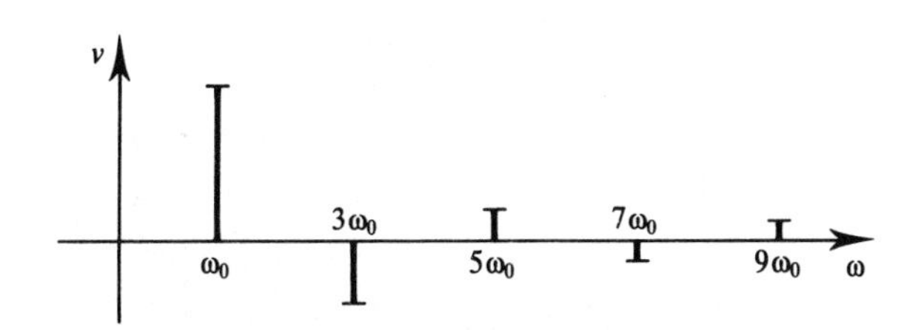

Figure 3.5 Single-sided spectrum of Figure 3.4

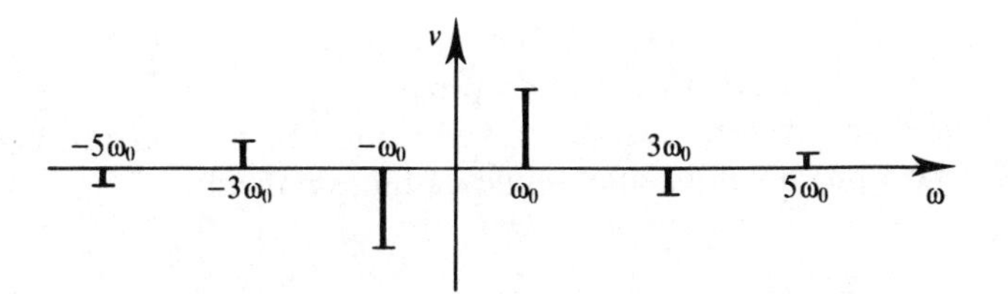

Figure 3.6 Double-sided spectrum of Figure 3.4

spectrum) is given by

$$v(t) = \frac{4V}{\pi}(\cos \omega_0 t - \tfrac{1}{3}\cos 3\omega_0 t + \tfrac{1}{5}\cos 5\omega_0 t - \ldots)$$

This has a spectrum as in Figure 3.5.

To get the double-sided result we can use either $c_n = \frac{1}{2}(a_n - jb_n)$ (i.e. $c_n = \frac{1}{2}a_n$ here) or the integral for c_n direct. You may care to try this to show that

$$c_n = \frac{2V}{n\pi}(-1)^{[1/2(n-1)]}$$

and

$$v(t) = \frac{2V}{\pi}[\ldots - \tfrac{1}{5}\exp(-5j\omega_0 t) + \tfrac{1}{3}\exp(-3j\omega_0 t) - \exp(-j\omega_0 t)$$

$$+ \exp(j\omega_0 t) - \tfrac{1}{3}\exp(3j\omega_0 t) + \tfrac{1}{5}\exp(5j\omega_0 t) - \ldots]$$

which plots as the double-sided spectrum in Figure 3.6.

Note the odd symmetry for phase.

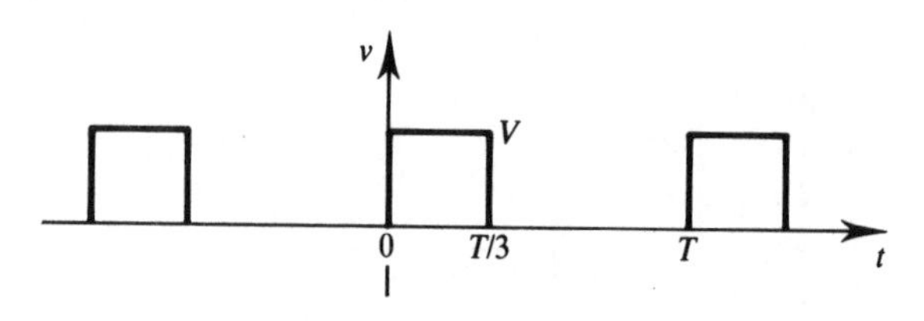

Figure 3.7 Pulse train

2. A harder one

This concerns a pulse train (Figure 3.7) of mark/space ratio 1:2.

Here $a_0 = V/3$ by inspection, but do prove it.

There is no time symmetry, so both a_n and b_n must be worked out. The integration is straightforward but finding actual values is more difficult:

$$\begin{aligned} a_n &= \frac{2}{T}\int_0^T v(t)\cos(n\omega_0 t)\,\mathrm{d}t \\ &= \frac{2}{T}\int_0^{T/3} V\cos(n\omega_0 t)\,\mathrm{d}t + \frac{2}{T}\int_{T/3}^{T} 0\cos(n\omega_0 t)\,\mathrm{d}t \\ &= \frac{2V}{T}\left[\frac{\sin(n\omega_0 t)}{n\omega_0}\right]_0^{T/3} + 0 \\ &= \frac{2V}{n\omega_0 T}\left[\sin\left(\frac{2n\pi}{3}\right) - 0\right] \end{aligned}$$

Thus

$$\boxed{a_n = (V/n\pi)[\sin(2n\pi/3)]}$$

and has values as follows:

$$n = 1 \quad a_1 = \frac{V3^{1/2}}{\pi 2} = 0.87V/\pi = 0.28V$$

$$n = 2 \quad a_2 = -\frac{V3^{1/2}}{2\pi 2} = -0.43V/\pi = -0.14V \quad (= a_1/2)$$

$$n = 3 \quad a_3 = \frac{V}{3\pi}.0 = 0$$

$$n = 4 \quad a_4 = \frac{V3^{1/2}}{4\pi 2} = 0.21V/\pi = 0.07V \qquad (= a_1/4)$$

$$n = 5 \quad a_5 = -\frac{V3^{1/2}}{5\pi 2} = -\frac{0.17V}{\pi} = -0.05V \quad (=a_1/5)$$

$$n = 6 \quad a_6 = \frac{V}{6\pi}.0 = 0$$

and so on

$$b_n = \frac{2}{T}\int_0^T v(t)\sin(n\omega_0)\,dt$$

$$= \frac{2}{T}\int_0^{T/3} V\sin(n\omega_0 t)\,dt$$

$$= \frac{2V}{T}\left[-\frac{\cos(n\omega_0 t)}{n\omega_0}\right]_0^{T/3}$$

$$= \frac{-2V}{nT\omega_0}\left[\cos\left(\frac{2n\pi}{3}\right) - 1\right]$$

Thus

$$\boxed{b_n = (V/n\pi)[1 - \cos(2n\pi/3)]}$$

which has values as follows:

$$n = 1 \quad b_1 = \frac{3V}{2\pi} \qquad n = 4 \quad b_4 = \frac{3V}{8\pi}$$

$$n = 2 \quad b_2 = \frac{3V}{4\pi} \qquad n = 5 \quad b_5 = \frac{3V}{10\pi}$$

$$n = 3 \quad b_3 = 0 \qquad n = 6 \quad b_6 = 0$$

Therefore

$$v(t) = V/3 + (V3^{1/2}/2\pi)[\cos\omega_0 t - \tfrac{1}{2}\cos 2\omega_0 t + \tfrac{1}{4}\cos 4\omega_0 t - \ldots]$$
$$+ (3V/\pi)[\sin\omega_0 t + \tfrac{1}{2}\sin 2\omega_0 t + \tfrac{1}{4}\sin 4\omega_0 t + \ldots]$$

which is rather clumsy, so re-express it as a single sinusoid with phase

$$v(t) = V/3 + (V/2\pi)[(3 + 9)^{1/2}\cos(\omega_0 t - \tan^{-1} 3^{1/2})$$
$$- (0.75 + 2.25)^{1/2}\cos(2\omega_0 t + \tan^{-1} 3^{1/2}) + 0$$
$$+ (3/16 + 9/16)^{1/2}\cos(4\omega_0 t - \tan^{-1} 3^{1/2}) + \ldots]$$

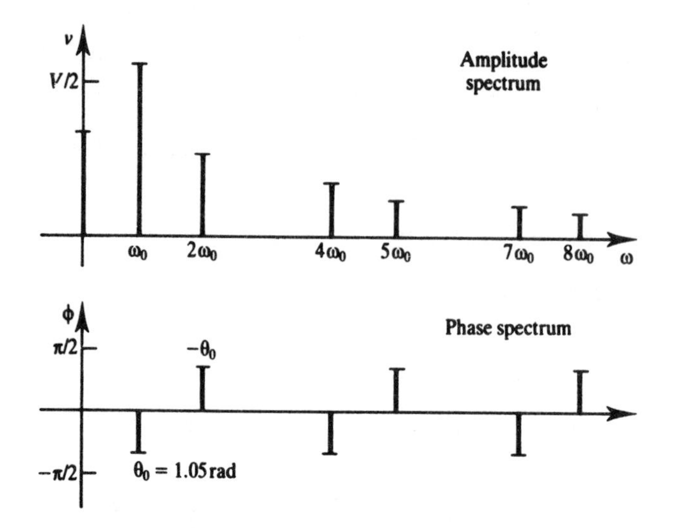

Figure 3.8 Spectra of pulse train

$$v(t) = V/3 + (V3^{1/2}/\pi)[\cos(\omega_0 t - \theta_0) - \tfrac{1}{2}\cos(2\omega_0 t(-) + \theta_0) + \tfrac{1}{4}\cos(4\omega_0 t - \theta_0) + \ldots]$$

where $\theta_0 = \tan^{-1} 3^{1/2} = \pi/3$. This form is useful if you need to sketch spectra. In the first example there was no need for separate amplitude and phase spectra because the only possible phase differences were due to the minus signs on amplitude (i.e. $\pm\pi$), but these are the only phases which can be shown adequately on just the amplitude spectrum. In general, as here, two spectra are needed (Figure 3.8).

Now, as an exercise, re-express $v(t)$ as a double-sided series and sketch its spectra. Compare your results with those above: what do you notice?

3. An even harder one

This concerns a triangular wave above ground (Figure 3.9). Here

$a_0 = V/2$ average height by inspection (work it out too if you like)

$b_n = 0$ even symmetry (again do it in full if you feel the need)

$$a_n = \frac{2}{T}\int_{-T/2}^{T/2} v(t)\cos(n\omega_0 t)\,dt$$

$$= \frac{2}{T}\int_{-T/2}^{0} V(1 + 2t/T)\cos(n\omega_0 t)\,dt + \frac{2}{T}\int_{0}^{T/2} V(1 - 2t/T)\cos(n\omega_0 t)\,dt$$

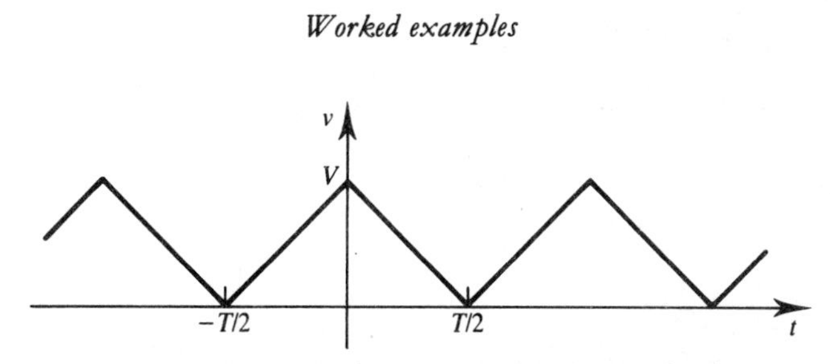

Figure 3.9 Triangular waveform with d.c. level

Note that the integration is over one peak's length and that the voltage variations in each half have been expressed as linear equations. Check these.

Therefore

$$a_n = \frac{2V}{T}\int_{-T/2}^{0} \cos(n\omega_0 t)\,dt + \frac{4V}{T^2}\int_{-T/2}^{0} t\cos(n\omega_0 t)\,dt + \frac{2V}{T}\int_{0}^{T/2} \cos(n\omega_0 t)\,dt$$
$$- \frac{4V}{T^2}\int_{0}^{T/2} t\cos(n\omega_0 t)\,dt$$
$$= \underset{①}{\frac{2V}{T}\int_{-T/2}^{T/2} \cos(n\omega_0 t)\,dt} + \frac{4V}{T^2}\left(\underset{②}{\int_{-T/2}^{0} t\cos(n\omega_0 t)\,dt} - \underset{③}{\int_{0}^{T/2} t\cos(n\omega_0 t)\,dt}\right) = 0$$

$$① = \frac{2V}{T}\left[\frac{\sin(n\omega_0 t)}{n\omega_0}\right]_{-T/2}^{T/2} = \frac{2V}{T}[\sin n\pi - \sin(-n\pi)] = 0$$

The other two integrals require integration by parts ($d(uv) = v\,du + u\,dv$; therefore $v\,du = u\,dv - uv$; for ② $u = t$ and $v = \sin(n\omega_0 t)$):

$$② = \frac{4V}{T^2}\int_{-T/2}^{0} \frac{t\,d[\sin(n\omega_0 t)]}{n\omega_0}$$
$$= \frac{4V}{n\omega_0 T^2}\left([t\sin(n\omega_0 t)]_{-T/2}^{0} - \int_{-T/2}^{0} \sin(n\omega_0 t)\,dt\right)$$
$$= \frac{4V}{n\omega_0 T^2}\left([0 + (T/2)\sin(-n\pi)] + \left[\frac{\cos(n\omega_0 t)}{n\omega_0}\right]_{T/2}^{0}\right)$$
$$= 0 + \frac{4V}{n^2\omega_0^2 T^2}[1 - \cos(-n\pi)]$$
$$= \frac{V}{n^2\pi^2}(1 - \cos n\pi)$$

$$③ = \frac{V}{n^2\pi^2}(1 - \cos n\pi)$$

also (do not forget the minus sign when you prove it). Therefore

$$a_n = \frac{2V}{n^2\pi^2}(1 - \cos n\pi)$$

Thus

$$a_n = 0 \qquad \text{when } n \text{ is even} \quad (\cos n\pi = 1)$$

$$a_n = \frac{4V}{n^2\pi^2} \qquad \text{when } n \text{ is odd} \quad (\cos n\pi = -1)$$

Therefore

$$\boxed{v(t) = \frac{V}{2} + \frac{4V}{\pi^2}\left(\cos \omega_0 t + \frac{1}{9}\cos 3\omega_0 t + \frac{1}{25}\cos 5\omega_0 t + \dots\right)}$$

Now plot its spectrum – only an amplitude one is needed this time.

Integration by parts is tedious so yɔu may care to know of a cunning method. The square waveform in Figure 3.10 is the differential of $v(t)$ from Figure 3.9 so that

$$v(t) = \int v'(t)\,\mathrm{d}t + \text{constant}$$

where $V' = \pm 2V/T$.

Now $v'(t)$ can be worked out much more easily. It is

$$v'(t) = -\frac{8V}{\pi T}(\sin \omega_0 t + \tfrac{1}{3}\sin 3\omega_0 t + \dots)$$

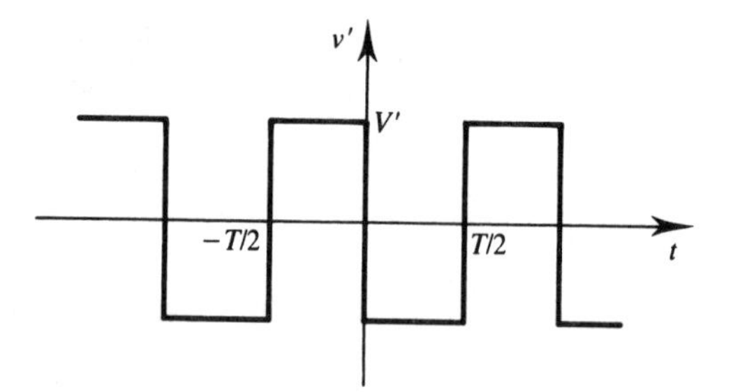

Figure 3.10 Differential waveform of $v(t)$ (i.e. $v'(t)$)

which gives the result above for $v(t)$ when integrated (how can you avoid the difficulty of evaluating the constant?).

This is an example of a general method of especial use for Fourier transforms as we shall see later.

3.6 • Summary of expressions

$$v(t) = a_0 + \sum_{n=1}^{\infty} a_n \cos(n\omega_0 t) + \sum_{n=1}^{\infty} b_n \sin(n\omega_0 t)$$

or

$$v(t) = a_0 + \sum_{n=1}^{\infty} (a_n^2 + b_n^2)^{1/2} \cos(n\omega_0 t - \theta_n) \qquad (\tan\theta_n = b_n/a_n)$$

$$v(t) = \sum_{n=-\infty}^{n=\infty} c_n \exp(\mathrm{j}n\omega_0 t) \qquad (c_n = \tfrac{1}{2}(a_n - \mathrm{j}b_n))$$

$$a_0 = \frac{1}{T}\int_0^T v(t)\,\mathrm{d}t$$

$$a_n = \frac{2}{T}\int_0^T v(t)\cos(n\omega_n t)\,\mathrm{d}t$$

$$b_n = \frac{2}{T}\int_0^T v(t)\sin(n\omega_n t)\,\mathrm{d}t$$

$$c_n = \frac{1}{T}\int_0^T v(t)\exp(-\mathrm{j}n\omega_0 t)\,\mathrm{d}t$$

3.7 • Conclusion

We have now seen the techniques for obtaining for a periodic signal both its single-sided and, more important in the long run, its double-sided spectrum. The basic ideas are fairly straightforward although rather tedious; hence the need to make as much use as possible of simplifying methods such as symmetry.

Periodic signals are obviously very common in communication systems, but of even more importance, especially for digital signals, are the techniques required for finding the spectral characteristics of an aperiodic pulsed signal – and its converse. These are the Fourier transforms dealt with in the next chapter.

3.8 • Problems

3.1 Where appropriate, for the waveforms in Figure 3.11:

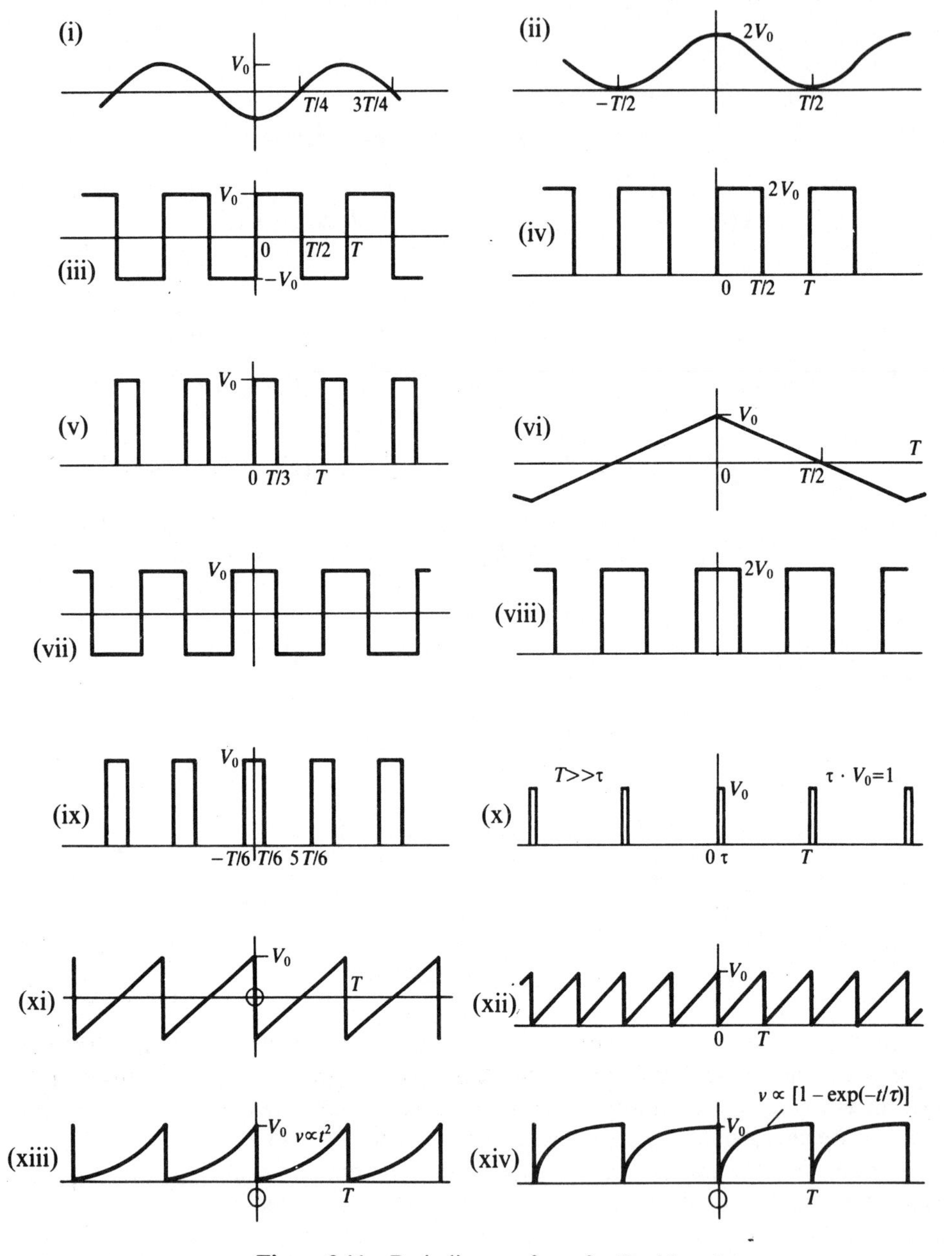

Figure 3.11 Periodic waveform for Problem 3.1

(a) obtain the Fourier coefficients a_n and b_n

(b) express as Fourier series

  (i) with separate sine and cosine series

  (ii) as cosine series with phase

(c) for b(ii) sketch the amplitude and phase spectra

(d) obtain the Fourier coefficients, c_n, making sure you can get them both directly and from a_n and b_n.

(e) express as a complex Fourier series and sketch its amplitude and phase spectra – double-sided now.

Use symmetry wherever possible. Some waveforms may be trivial enough not to need the full treatment.

3.2 Obtain a Fourier series for the periodic waveform in Figure 3.12. Plot its amplitude spectrum and assign a suitable transmission bandwidth giving reasons.

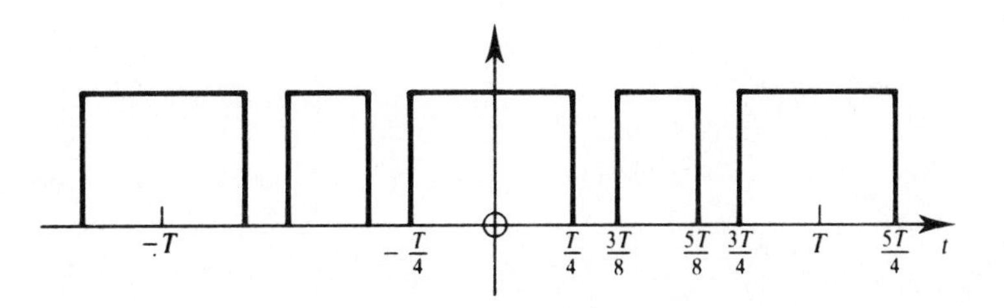

Figure 3.12 Periodic function for Problem 3.2

3.3 Obtain a general expression for the double-sided spectrum of the saw-tooth waveform in Figure 3.13, showing clearly how you obtain it.

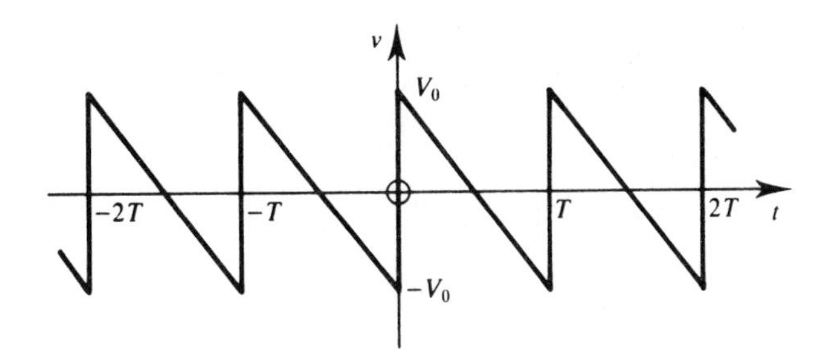

Figure 3.13 Saw-tooth waveform for Problem 3.3

Sketch the amplitude and phase spectra of your result.

Now, without deriving these spectra afresh, convert your expression in the following ways:

(i) Into a double-sided spectrum of the same waveform but with the time zero at point A.

(ii) Into a single-sided spectrum.

Again show clearly how you do these.

• 4 •
Fourier transforms

4.1 • Introduction

A Fourier series expresses a **continuous periodic waveform** as a sum of sinusoids at harmonic frequencies. The effect is to transform the time domain expression for the waveform into a frequency domain one. The frequency domain picture is a series of spectral lines separated by equal intervals of frequency and of different amplitudes and phase. This forms a **line spectrum** as shown in Figure 4.1 for a rectangular pulse train.

Now we are going to consider the closely similar process whereby a completely non-periodic waveform (i.e. a single voltage pulse) can also be expressed as a frequency spectrum leading to diagrams similar to those above. The time domain equation is transformed mathematically into a frequency domain one (and vice versa), leading now to spectral components at all frequencies (not just harmonics) in a **continuous spectral distribution**. Again there are two spectra — amplitude density and phase. They are shown in Figure 4.2 for one pulse of the pulse train above. Note that the vertical axis is now *not* voltage but voltage density in volts per hertz.

To see the similarity to series, imagine the period in the pulse train is getting bigger and bigger. Thus the pulses will get farther and farther apart (until you can only see one of them in the present time) and the spectral lines will come closer and closer together until they coalesce. This similarity extends to the mathematics and, by taking T to infinity for the series, you get the transforms.

But, for now, we will just state the transform equations and see how to use them.

4.2 • The Fourier transforms stated

If you have a voltage expressed as a time function, $v(t)$, it will have a spectral distribution as a function of frequency, $V(\omega)$ (or $V(f)$), such that the expression for one can be obtained from the other. The change is made by means of the Fourier transform equations given below.

$V(\omega)$ is then called the **Fourier transform** of $v(t)$ (also written $F[v(t)]$) and $v(t)$

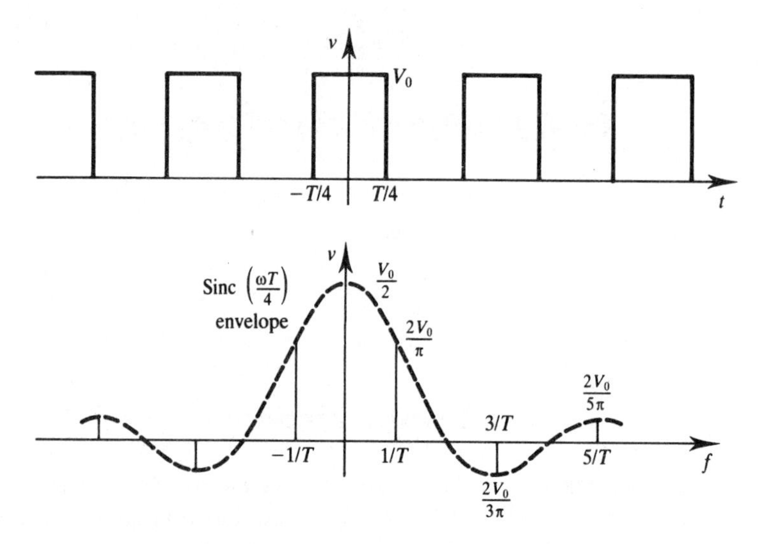

Figure 4.1 A rectangular pulse train and its spectrum

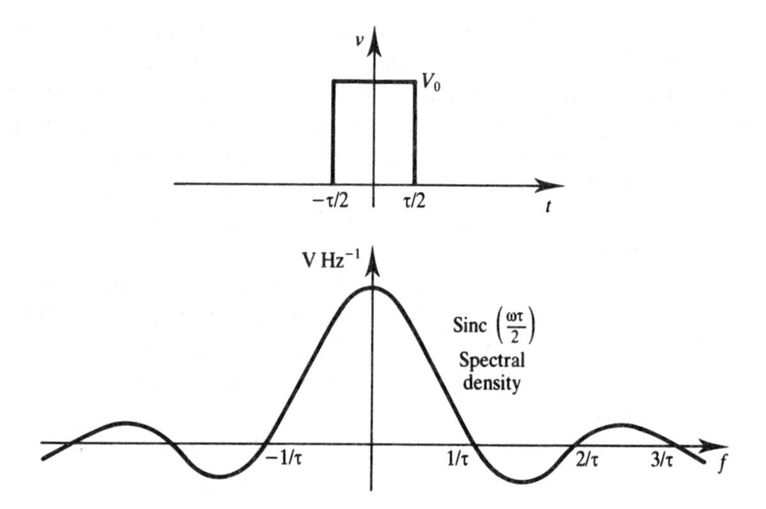

Figure 4.2 Single rectangular pulse and its spectrum

is called the **inverse Fourier transform** of $V(\omega)$ (or $F^{-1}[V(\omega)]$). This is summarized in the following way:

$$v(t) \leftrightarrow V(\omega)$$

The actual transforms are the integral equations

$$V(\omega) = \int_{-\infty}^{\infty} v(t)\exp(-j\omega t)\,dt \qquad \textbf{Fourier transform (FT)}$$

$$v(t) = \frac{1}{2\pi}\int_{-\infty}^{\infty} V(\omega)\exp(j\omega t)\,d\omega \qquad \textbf{Inverse FT}$$

The upper **forward** transform gives $V(\omega)$ from $v(t)$

The lower **inverse** transform gives $v(t)$ from $V(\omega)$

These expressions assume frequency is double sided. The $1/2\pi$ factor appears to give an ugly asymmetry and occurs because of the use of ω not f. It is done because ω is somewhat more fundamental in communications, but many people express the inverse transform in f as

$$v(t) = \int_{-\infty}^{\infty} V(f)\exp(2\pi j f t)\,df$$

4.3 • A straightforward example

Consider a rectangular pulse, $\Pi(t)$, symmetrical in time (see Figure 4.3).

Always start by writing down the general definition of the transform:

$$\begin{aligned}
V(\omega) &= \int_{-\infty}^{\infty} v(t)\exp(-j\omega t) \\
&= \int_{-\infty}^{\infty} \Pi(t)\exp(-j\omega t) \\
&= \int_{-\tau/2}^{\tau/2} V_0\exp(-j\omega t) \\
&= V_0\left[\frac{\exp(-j\omega t)}{-j\omega}\right]_{-\tau/2}^{\tau/2} \\
&= \frac{V_0}{-j\omega}[\exp(-j\omega\tau/2) - \exp(j\omega\tau/2)] \\
&= \frac{2V_0}{\omega}\left[\frac{\exp(j\omega\tau/2) - \exp(-j\omega\tau/2)}{2j}\right] \\
&= \frac{2V_0}{\omega}\sin\left(\frac{\omega\tau}{2}\right)
\end{aligned}$$

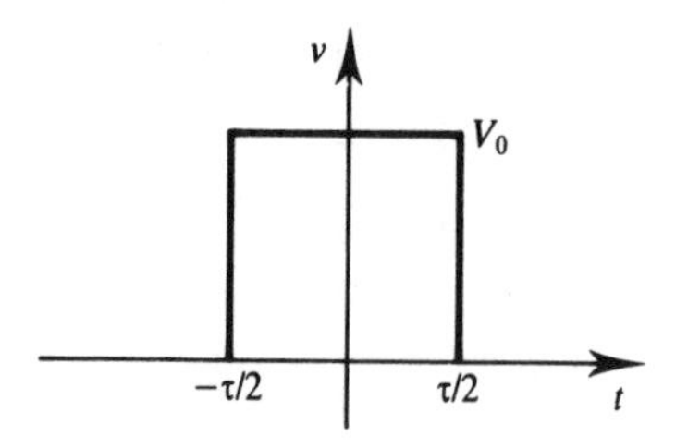

Figure 4.3 Single pulse

This function cannot easily be plotted as it is, so it is developed further:

$$V(\omega) = V_0\tau\left[\frac{\sin(\omega\tau/2)}{\omega\tau/2}\right]$$

or

$$\boxed{V(\omega) = V_0\tau\,\mathrm{sinc}(\omega\tau/2)}$$

where

$$\mathrm{sinc}(\omega\tau/2) = \frac{\sin(\omega\tau/2)}{\omega\tau/2}$$

This **sinc function** is always occurring in Fourier transforms (and in series too – as an envelope; see Figure 4.1). It really is a separate function in its own right with a definite recognizable shape. It has a large central peak with a maximum of one at $\omega = 0$ and tails out to zero as $\omega \to \pm\infty$, like a decaying sine wave with zero crossings at constant intervals of $\omega = 2\pi/\tau$.

The **shape** of this function is something you should learn to recognize readily but, before doing so, note that there is a certain amount of confusion in textbooks about exactly how to name it. Some call it $\mathrm{sinc}(x)$, as is done here, but others call it $\mathrm{Sa}(x)$ and more define $\mathrm{sinc}(x)$ as $\sin(\pi x)/\pi x$. Most avoid the problem by just calling it the $\sin(x)/x$ function. Whatever the name, the important point is its shape and, especially, its zero crossing positions which we now look at in more detail below.

Writing $\mathrm{sinc}\,x$ where $x = \omega\tau/2$, values up to the fourth zero crossing are given in the table below.

x	sinc x	$\text{sinc}^2\, x$	ω	f
0	1	1	0	0
$\pm\pi/4$	0.900	0.810	$\pi/2\tau$	$1/4\tau$
$\pm\pi/2$	0.637	0.405	π/τ	$1/2\tau$
$\pm 3\pi/4$	0.300	0.090	$3\pi/2\tau$	$3/4\tau$
$\pm\pi$	0	0	$2\pi/\tau$	$1/\tau$
$\pm 5\pi/4$	−0.180	0.032	$5\pi/2\tau$	$5/4\tau$
$\pm 3\pi/2$	−0.212	0.045	$3\pi/\tau$	$3/2\tau$
$\pm 7\pi/4$	−0.129	0.017	$7\pi/2\tau$	$7/4\tau$
$\pm 2\pi$	0	0	$4\pi/\tau$	$2/\tau$
$\pm 9\pi/4$	0.100	0.010	$9\pi/2\tau$	$9/4\tau$
$\pm 5\pi/2$	0.127	0.016	$5\pi/\tau$	$5/2\tau$
$\pm 11\pi/4$	0.082	0.007	$11\pi/2\tau$	$11/4\tau$
$\pm 3\pi$	0	0	$6\pi/\tau$	$3/\tau$
$\pm 13\pi/4$	−0.069	0.005	$13\pi/2\tau$	$13/4\tau$
$\pm 7\pi/2$	−0.091	0.008	$7\pi/\tau$	$7/2\tau$
$\pm 15\pi/4$	−0.060	0.004	$15\pi/2\tau$	$15/4\tau$
$\pm 4\pi$	0	0	$8\pi/\tau$	$4/\tau$

Now sinc x will plot in two ways giving three plots in all.

1. As a total spectrum

The plot in Figure 4.4 is the best one to show the transform of the pulse we have just worked on.

2. As amplitude and phase spectra

The phase spectrum has been drawn taking a minus sign to be $+\pi$ or $-\pi$ to give a plot (Figure 4.5) with the necessary odd symmetry, but it could be drawn in an

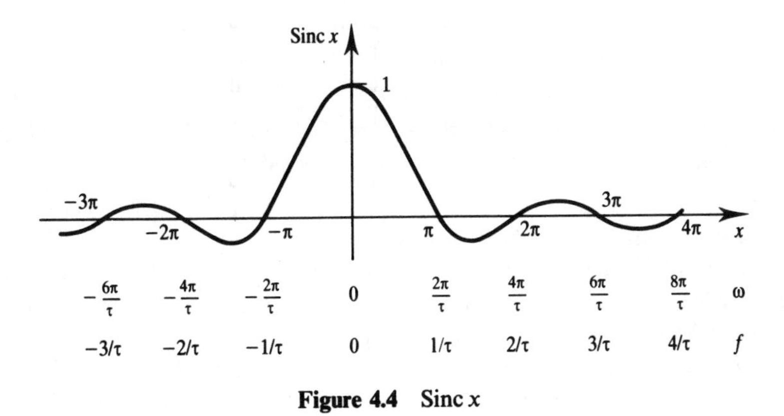

Figure 4.4 Sinc x

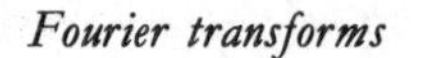

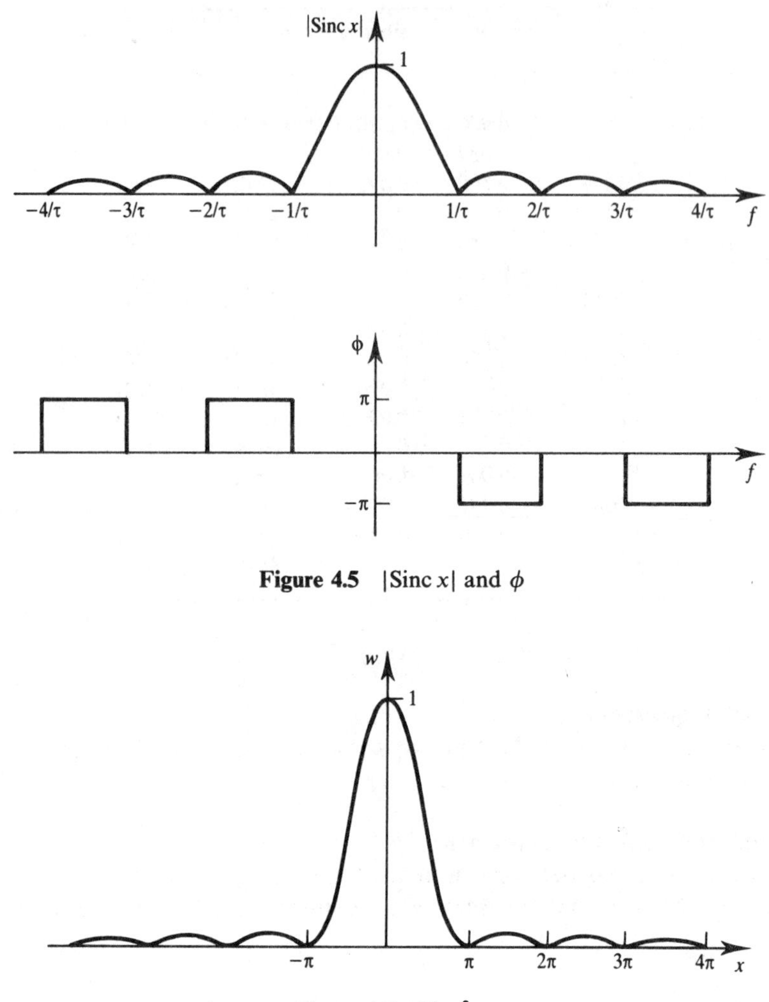

Figure 4.5 $|\text{Sinc}\, x|$ and ϕ

Figure 4.6 $\text{Sinc}^2\, x$

infinity of other ways because, strictly, $-1 \equiv (2n - 1)\pi$ and $+1 \equiv 2n\pi$ where n can have any value. This can be useful as we shall see later.

3. Now $\text{sinc}^2\, x$ can be sketched

Note how the plot in Figure 4.6 shows very clearly that nearly all the power in a square pulse is contained in the central peak of the spectral density distribution within the frequency range $0 \rightarrow 1/\tau$. This is a very relevant result in communications where systems invariably act as filters restricting the transmitted spectrum. If the system acts as a low-pass filter cutting off at $1/\tau$, most of the pulse will be received and the system will be adequate. This is one way of looking at the usual criterion that **bandwidth** equals **bit rate**.

Having done this example we shall now look at ways to simplify the integration.

4.4 • Simplification from symmetry

Most pulses are symmetric in time, which enables the transforms to be simplified. This is also true of filters and the inverse transforms.

The pulse in the left-hand diagram of Figure 4.7 has **even** symmetry (i.e. $v(-t) = v(t)$), while that on the right has **odd** symmetry (i.e. $v(-t) = -v(t)$).

Let us see what happens when we put these conditions into the general transform integrals:

$$V(\omega) = \int_{-\infty}^{\infty} v(t)\exp(-\mathrm{j}\omega t)\,\mathrm{d}t$$

$$= \int_{-\infty}^{0} v(t)\exp(-\mathrm{j}\omega t)\,\mathrm{d}t + \int_{0}^{\infty} v(t)\exp(-\mathrm{j}\omega t)\,\mathrm{d}t$$

$$= \int_{-\infty}^{0} v(t)(\cos\omega t - \mathrm{j}\sin\omega t)\,\mathrm{d}t + \int_{0}^{\infty} v(t)(\cos\omega t - \mathrm{j}\sin\omega t)\,\mathrm{d}t$$

$$= \int_{-\infty}^{0} v(t)\cos\omega t\,\mathrm{d}t + \int_{0}^{\infty} v(t)\cos\omega t\,\mathrm{d}t - \mathrm{j}\int_{-\infty}^{0} v(t)\sin\omega t\,\mathrm{d}t$$

$$-\mathrm{j}\int_{0}^{\infty} v(t)\sin\omega t\,\mathrm{d}t$$

Now cosine is an **even** function of time and sine is an **odd** function. So if $v(t)$ is even

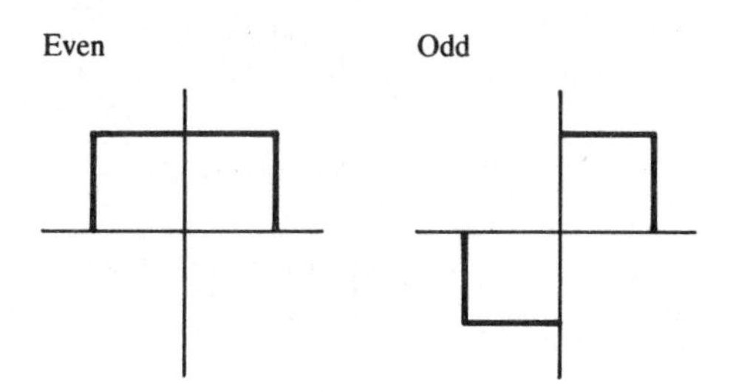

Figure 4.7 Even and odd pulses

then $v(t) \cos \omega t$ is also even, but $v(t) \sin \omega t$ is odd. Therefore

$$\int_{-\infty}^{0} v(t) \cos \omega t \, \mathrm{d}t = \int_{0}^{\infty} v(t) \cos \omega t \, \mathrm{d}t$$

and

$$\int_{-\infty}^{0} v(t) \sin \omega t \, \mathrm{d}t = -\int_{0}^{\infty} v(t) \sin \omega t \, \mathrm{d}t$$

Therefore

$$V(\omega) = 2\int_{0}^{\infty} v(t) \cos \omega t \, \mathrm{d}t + 0$$

that is

$$\boxed{V(\omega) = 2\int_{0}^{\infty} v(t) \cos \omega t \, \mathrm{d}t} \qquad v(t) \text{ even}$$

Similarly you can show that (try it)

$$\boxed{V(\omega) = -2\mathrm{j}\int_{0}^{\infty} v(t) \sin \omega t \, \mathrm{d}t} \qquad v(t) \text{ odd}$$

And there are corresponding expressions for the inverse transforms (try to work them out).

Now apply this to the even pulse in Figure 4.7. It has even symmetry, so

$$\begin{aligned} V(\omega) &= 2\int_{0}^{\infty} v(t) \cos \omega t \, \mathrm{d}t \\ &= 2\int_{0}^{\infty} V_0 \cos \omega t \, \mathrm{d}t \\ &= 2V_0 \left[\frac{\sin \omega t}{\omega}\right]_0^{\tau/2} \\ &= \frac{2V_0 \sin \omega\tau/2}{\omega} \end{aligned}$$

or

$$V(\omega) = V_0\tau \operatorname{sinc} \omega\tau/2$$

Very quick, you will agree.

Try doing the odd pulse in Figure 4.7. You will probably get

$$V(\omega) = \frac{-2jV_0\tau}{\omega}(1 - \cos \omega\tau/2)$$

But this is not a sinc form and so cannot easily be sketched. A more useful form is

$$V(\omega) = -j(V_0/4)\omega\tau^3 \operatorname{sinc}^2(\omega\tau/4)$$

Now draw its spectra – separate amplitude and phase ones this time.

4.5 • Time-shifted pulse

A pulse whose centre is at time t_d has the same transform as the same pulse centred on $t = 0$ but multiplied by $\exp(-j\omega t_d)$ as in Figure 4.8. That is

$$F[v(t - t_d)] = F[v(t)] \exp(-j\omega t_d)$$

Prove this for yourself by working out $V'(\omega)$ directly. Of course the theorem also applies if t_d is negative, that is an early pulse. Again this theorem simplifies the algebra because all you need to do is work out the original $V(\omega)$.

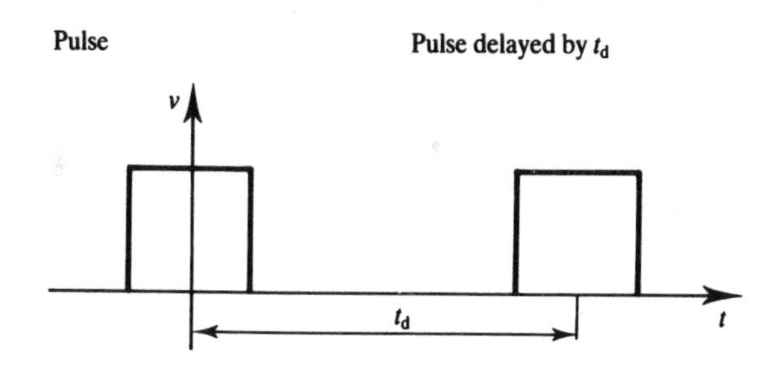

Figure 4.8 A pulse and its delayed twin

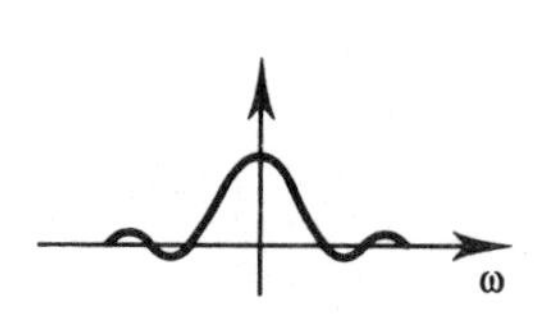

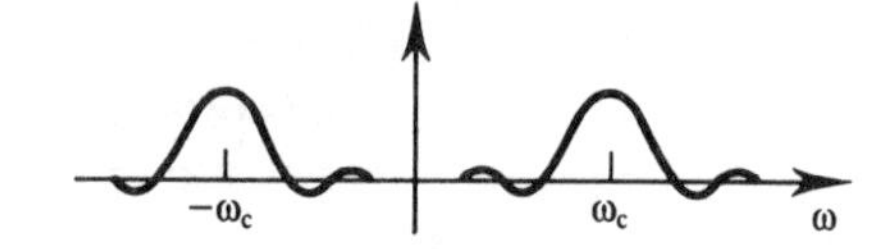

Figure 4.9 Spectra of pulse and modulated pulse

For IFTC – $F^{-1}\{V(\omega - \omega_c)\} = F^{-1}\{V(\omega)\}e^{j\omega_c t}$

4.6 • Modulated pulse

Suppose the pulse is not a d.c. one but a pulse of a single frequency sinusoid (a carrier, f_c) as we shall meet later on in amplitude shift keyed (ASK) modulation. What is its Fourier transform?

The answer is that it is the transform of the d.c. pulse occurring twice centred on $\pm\omega_c$. That is, if $v(t) \leftrightarrow V(\omega)$, then

$$v(t)\cos\omega_c t \leftrightarrow V'(\omega) = V(\omega - \omega_c) + V(\omega - \omega_c)$$

as shown in Figure 4.9.

Here

$$V'(\omega) = \frac{V_0\tau}{2}[\operatorname{sinc}(\omega - \omega_c)\tau/2 + \operatorname{sinc}(\omega + \omega_c)\tau/2]$$

Prove it.

4.7 • The unit impulse, $\delta(t)$

This function often occurs in communications as an idealized voltage impulse because its spectrum contains all frequencies at unity amplitude as shown in Figure 4.10. It is sometimes called the Dirac delta function and is drawn as shown.

The unit impulse is defined as a very narrow pulse of infinitesimal width ($\tau \to 0$), indeterminate height at $t = 0$ but of unit area given by

$$\int_{-\infty}^{\infty} \delta(t)\,dt = 1$$

Its Fourier transform is straightforward to find. First the general expression

$$V(\omega) = \int_{-\infty}^{\infty} v(t)\exp(-j\omega t)\,dt$$

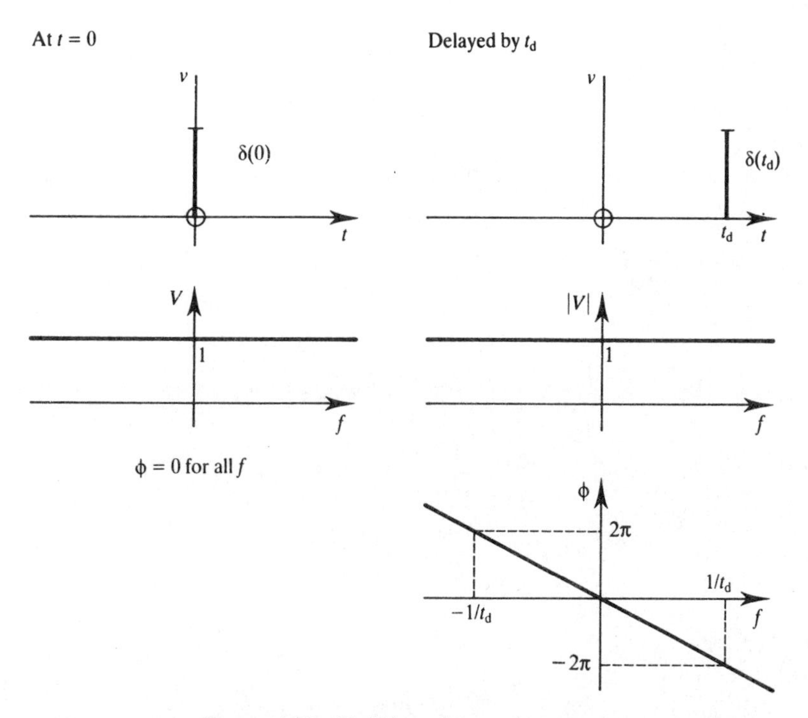

Figure 4.10 Unit impulses and spectra

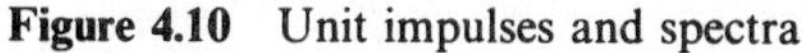

and now this particular one:

$$V(\delta) = \int_{-\infty}^{\infty} \delta(t)\exp(-j\omega t)\,dt$$

$$= \int_{t=0} \delta(t)\,dt$$

(because δ exists at $t = 0$ only, so $\exp(-j\omega t) = 1$). Therefore

$$\boxed{V(\delta) = 1}$$

The same impulse at any other time (e.g. delayed by t_d) has a spectrum given by the time shift theorem so that $V(\delta) = \exp(-j\omega t_d)$. This can also be shown by direct integration of the transform equation.

The usefulness of this function in communications is that it is an idealized theoretical source containing all frequencies at equal amplitude so that the output from a circuit having an impulse as input is the filter characteristic of the circuit. This is always an important requirement and is known as the **impulse response**.

4.8 • Simplification using the differential

Often the differential of a pulse produces a simpler shape whose transform is easier to work out (e.g. no integration by parts) even getting back to impulses. The required transform can then be obtained as follows. If

$$v(t) \leftrightarrow V(\omega)$$

then

$$\frac{dv(t)}{dt} = v'(t) \leftrightarrow V'(\omega) = j\omega V(\omega)$$

Therefore

$$V(\omega) = \frac{V'(\omega)}{j\omega} = -\frac{V''(\omega)}{\omega^2}$$

or

$$V(\omega) = \frac{F[v'(t)]}{j\omega} = -\frac{F[v''(t)]}{\omega^2}$$

This can be proven as follows:

$$\frac{dV(\omega)}{dt} = \frac{d}{dt}\int_{-\infty}^{\infty} v(t)\exp(-j\omega t)\,dt = 0$$

(since $V(\omega)$ is independent of t); therefore

$$\int_{-\infty}^{\infty} \frac{dv(t)}{dt}\exp(-j\omega t)\,dt + \int_{-\infty}^{\infty} v(t)\frac{d(\exp(-j\omega t))}{dt}dt = 0$$

$$\int_{-\infty}^{\infty} \frac{dv(t)}{dt}\exp(-j\omega t)\,dt - j\omega\int_{-\infty}^{\infty} v(t)\exp(-j\omega t)\,dt = 0$$

$$V'(\omega) - j\omega V(\omega) = 0$$

$$V(\omega) = \frac{V'(\omega)}{j\omega}$$

Take as an example the rectangular pulse ($\pi(t)$) used already several times. It can be regarded as the integral of two impulses of area $\pm V_0$ at $\pm\tau/2$ as shown in

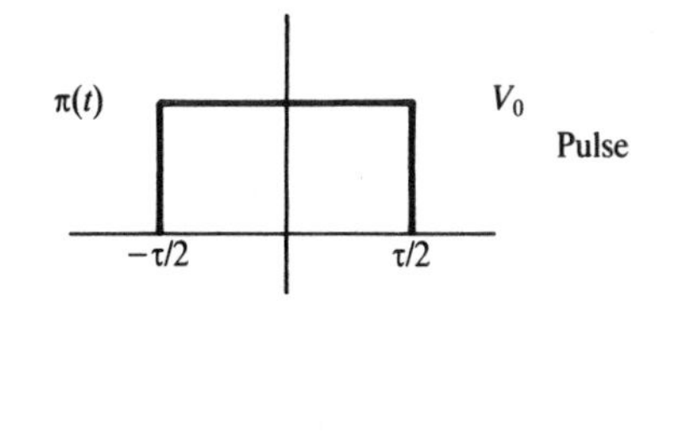

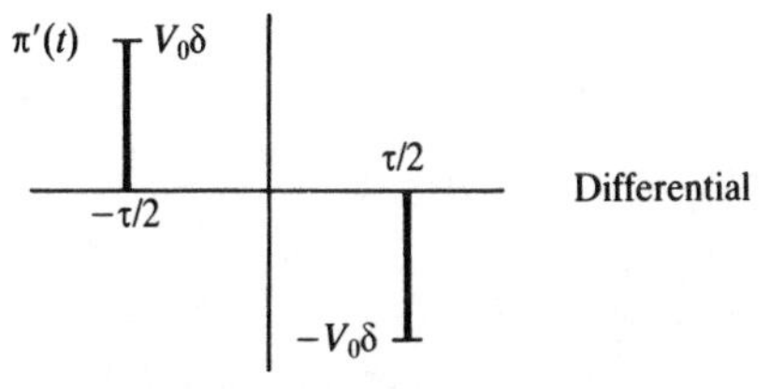

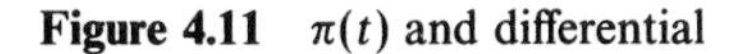

Figure 4.11 $\pi(t)$ and differential

Figure 4.11:

$$v(t) = \int v'(t)\,\mathrm{d}t$$

Therefore

$$V(\omega) = \frac{F[v'(t)]}{\mathrm{j}\omega}$$

$$= \frac{V_0}{\mathrm{j}\omega}[\exp(\mathrm{j}\omega\tau/2) - \exp(-\mathrm{j}\omega\tau/2)]$$

that is

$$V(\omega) = V_0\tau\,\mathrm{sinc}(\omega\tau/2)$$

as before.

This technique can be extended further. For instance, the triangular pulse in Figure 4.12 is the integral of the square pulse and impulse below it. Both directly and by the method above you can show that

$$V(\omega) = \frac{\mathrm{j}V_0}{\omega}[\exp(-\mathrm{j}\omega\tau/2) - \mathrm{sinc}\,\omega\tau/2)]$$

However, there is one difficulty we have not mentioned so far and that is the magnitudes of the differentiated pulses. These can be rather tricky and require care,

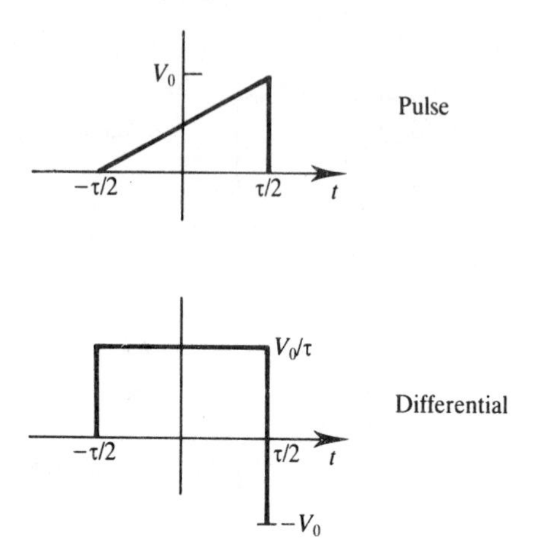

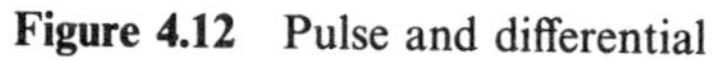
Figure 4.12 Pulse and differential

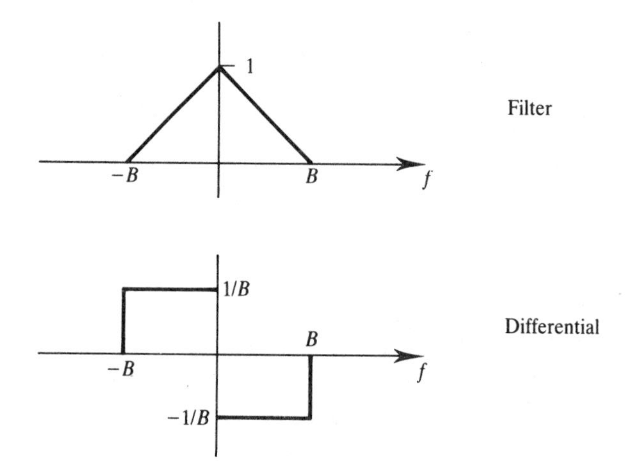

Figure 4.13 Triangular filter and its differential

and it is best to check by integrating mentally afterwards (e.g. here the first impulse of $v'(t)$ takes $v(t)$ to V_0 as it should). Check the problem in the next section in the same way.

This same technique can also be used for inverse transforms using

$$v(t) = -\frac{1}{jt}F^{-1}\left[\frac{dV(\omega)}{d\omega}\right] = -\frac{1}{t^2}F^{-1}\left[\frac{d^2V(\omega)}{d\omega^2}\right]$$

Note the minus sign. Try it for the triangular filter in Figure 4.13.

4.9 • Use of superposition to simplify

This general principle means that if a pulse can be split into the sum of other pulses, these can be transformed separately and the results added to give the transform of the original pulse. The point is that the separate pulses will be simpler to transform.

This has been done already in the last section where the integral pulse was separated into a rectangular pulse and an impulse. But that same pulse can be tackled using superposition by splitting it into the sum of two pulses as shown in Figure 4.14. That is

$$v(t) = v_1(t) + v_2(t)$$

therefore

$$V(\omega) = V_1(\omega) + V_2(\omega)$$

$V_1(\omega)$ is a standard result but $V_2(\omega)$ needs more work. It can be done by parts but the easiest way is to take its differential as in Figure 4.15. Now

$$V_2(\omega) = \frac{F[v_2'(\omega)]}{\mathrm{j}\omega} = \frac{V_0}{2\mathrm{j}\omega}\operatorname{sinc}(\omega\tau/2) - \frac{V_0}{2\mathrm{j}\omega}[\exp(\mathrm{j}\omega\tau/2) + \exp(-\mathrm{j}\omega\tau/2)]$$

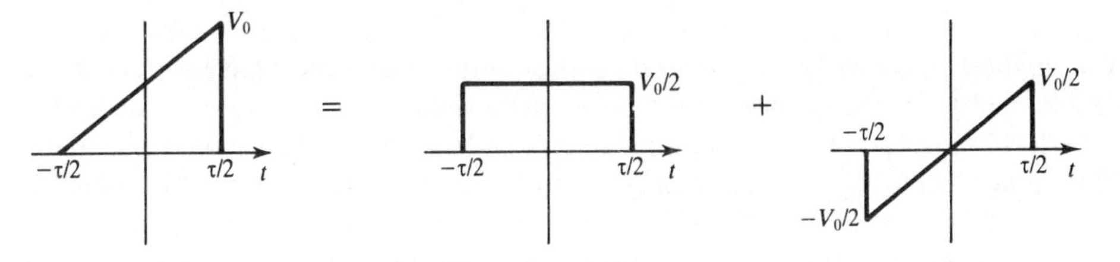

Figure 4.14 Pulse as sum of pulses

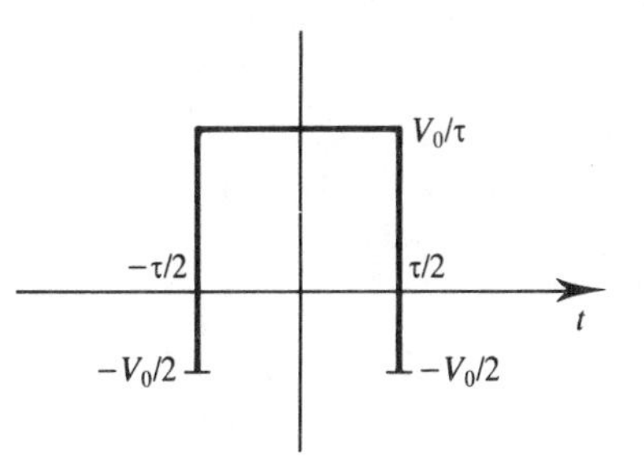

Figure 4.15 Differential of pulse in Figure 4.14

Also

$$V_1(\omega) = \frac{V_0\tau}{2}\operatorname{sinc}(\omega\tau/2)$$

therefore

$$V(\omega) = \frac{\mathrm{j}V_0}{\omega}[\operatorname{sinc}(\omega\tau/2) - \exp(-\mathrm{j}\omega\tau/2)]$$

that is, the same result (go back to exponential forms to prove). It is a little longer than the first method but it illustrates a general principle that you can usually reduce the problem to the transforms of rectangular pulses and impulses that you know, so there is no need to do any involved integration at all. There are some problems at the end of the chapter.

4.10 • Inverse transforms: $V(\omega) \rightarrow v(t)$

Reverse transforms convert a spectrum to the pulse which contains it. They are most often required for the impulse response – the pulse which comes out of a network when all frequencies go into it (see Section 4.7). The defining equation is

$$v(t) = \frac{1}{2\pi}\int_{-\infty}^{\infty} V(\omega)\exp \mathrm{j}\omega t \, \mathrm{d}\omega$$

The methods are clearly very similar to those for the forward transforms, but the details tend to be more tricky. One easier point is that there is no need to bother with the phase terms (except ± 1) because only the time variation of voltages is useful (but "phase" does have meaning in time; see the end of this section). Also, there is a new useful technique – that of **duality** (you get the same result both ways round).

First let us do a simple example, namely the frequency equivalent of a rectangular pulse, the "ideal" low-pass filter as shown in Figure 4.16. From duality you would expect its reverse transform also to be a sinc shape, so let us see.

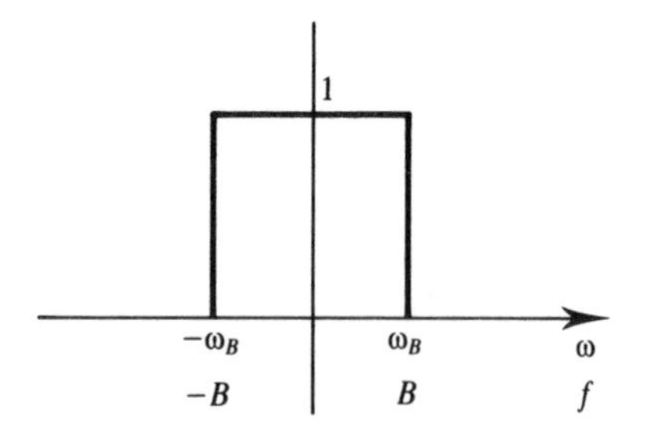

Figure 4.16 Ideal low-pass filter

In general

$$v(t) = \frac{1}{2\pi}\int_{-\infty}^{\infty} V(\omega)\exp \mathrm{j}\omega t \, \mathrm{d}\omega$$

Here

$$v(t) = \frac{1}{2\pi}\int_{-\omega_B}^{\omega_B} 1 . \exp \mathrm{j}\omega t \, \mathrm{d}\omega$$

$$= \frac{1}{2\pi}\left[\frac{\exp \mathrm{j}\omega t}{\mathrm{j}t}\right]_{\omega_B}^{\omega_B}$$

$$= (1/2\pi \mathrm{j}t)[\exp(\mathrm{j}\omega_B t) - \exp(-\mathrm{j}\omega_B t)]$$

$$= (1/\pi t)\sin \omega_B t$$

or

$$v(t) = 2B \operatorname{sinc} \omega_B t$$

Note the exact analogy with the result for the corresponding forward transform if you replace the corresponding constants:

$$V_0 \equiv 1 \qquad \tau \equiv 2B \qquad \text{and} \qquad \omega \equiv t$$

so $v_0\tau \to 2B$ and $\omega\tau/2 \to \omega_B t$.

All the simplifying methods can be used although some care may be needed in applying them because of phase (e.g. when using symmetry). The example above just used the amplitude spectrum assuming phase was zero everywhere within the bandwidth.

The result we have just obtained illustrates the principle of duality (here that "pulses" give sinc transforms either way) and this can be useful to give a result where the integral looks impossible.

For example, what spectrum does a sinc-shaped pulse have? To do it formally you need to integrate a sinc product function itself (it is not in my book of integrals). It is much easier merely to say that if a square pulse transforms to a sinc, then a sinc pulse transforms to a "square" spectrum. These two are shown in Figure 4.17. The only problem is to get the pulse width correctly related to the constants of the sinc.

A similar situation occurs with impulses and spectral lines. If an impulse at $t = 0$ has a constant horizontal spectrum, then a spectral line at $\omega = 0$ must be a d.c. level constant in time. Obvious isn't it? But what if the spectral line is at frequency ω? The pulse now has a meaningful (linear) "phase" spectrum too. This merely means

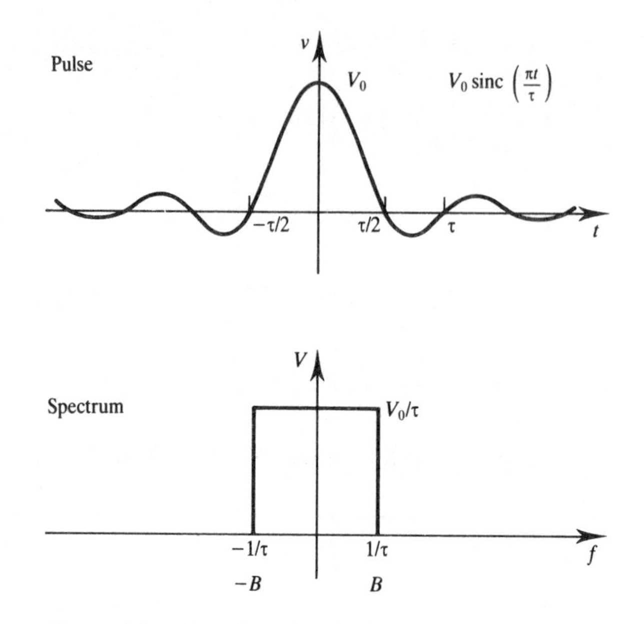

Figure 4.17 Sinc-shaped pulse having square spectrum

that the constant d.c. level is rotating at angular frequency ω rad s^{-1} – the required phasor.

4.11 • Power and energy spectra

A periodic signal carries a fixed finite average power level: it is said to be a **power signal**. A non-periodic pulse signal carries a fixed total energy: it is said to be an **energy signal.**

For both of them power/energy is normalized to a 1 Ω load and can be expressed in either the time or frequency domains, that is either from the expression for the waveform ($v(t)$) or from the spectrum ($V(\omega)$).

This section is concerned with deriving the equations expressing this equality for both types of signal. The results are often referred to as **Parseval's theorem** or **equation.** They are specific cases of his general theorem on the orthogonality of signals.

The results we shall get are as follows.

For **series** or **power signals**:

$$P = a_0^2 + \frac{1}{2}\sum_1^\infty (a_n^2 + b_n^2)$$

$$= a_0^2 + 2\sum_1^\infty |c_n|^2$$

For **pulses** or **energy signals**:

$$E = \int_{-\infty}^{\infty} |v(t)|^2 \, dt$$

$$= \frac{1}{2\pi} \int_{-\infty}^{\infty} |V(\omega)|^2 \, d\omega$$

$$= \int_{-\infty}^{\infty} |V(f)|^2 \, df$$

Proofs are as follows.

Periodic–single sided

$$P = \text{energy flow per unit time}$$

$$= \frac{\int_0^T v^2(t)\, dt}{\int_0^T dt}$$

$$= \frac{1}{T} \int_0^T \left(a_0 + \sum_1^\infty a_n \cos(n\omega_0 t) + \sum_1^\infty b_n \sin(n\omega_0 t) \right)^2 dt$$

$$= \frac{1}{T} \left(\int_0^T a_0^2 \, dt + \sum_1^\infty \int_0^T a_n^2 \cos^2(n\omega_0 t)\, dt + \sum_1^\infty b_n^2 \sin^2(n\omega_0 t)\, dt \right)$$

$$+ \text{integrals in cos cos, cos sin, etc. (all = 0)}$$

$$= \frac{1}{T} \left(a_0^2 T + \frac{T}{2} \sum_1^\infty a_n^2 + \frac{T}{2} \sum_1^\infty b_n^2 \right)$$

Therefore

$$\boxed{P = a_0^2 + \frac{1}{2} \sum_1^\infty (a_n^2 + b_n^2)}$$

But $c_n = \frac{1}{2}(a_n - jb_n)$, so $a_n^2 + b_n^2 = 4|c_n|^2$. Therefore

$$\boxed{P = a_0^2 + 2 \sum_1^\infty |c_n|^2}$$

This last result, for the double-sided spectrum, can be shown directly:

$$P = \frac{1}{T}\int_0^T v^2(t)\,\mathrm{d}t$$

$$= \frac{1}{T}\int_0^T v(t)[v(t)]\,\mathrm{d}t$$

$$= \frac{1}{T}\int_0^T v(t)\left(\sum_{-\infty}^{\infty}[c_n \exp(\mathrm{j}n\omega_0 t)\right)\mathrm{d}t$$

$$= \frac{1}{T}\int_0^T \sum_{-\infty}^{\infty} v(t)c_n \exp(\mathrm{j}n\omega_0 t)\,\mathrm{d}t \qquad v(t) \text{ independent of } n$$

$$= \frac{1}{T}\sum_{-\infty}^{\infty}\int_0^T c_n v(t)\exp(\mathrm{j}n\omega_0 t)\,\mathrm{d}t \qquad \text{rearranged}$$

$$= \frac{1}{T}\sum_{-\infty}^{\infty}\int_0^T v(t)\exp(\mathrm{j}n\omega_0 t)\,\mathrm{d}t \qquad c_n \text{ independent of } t$$

$$= \sum_{-\infty}^{\infty} c_n c_{-n} \qquad \text{then } c_{-n} = c_n^*$$

$$= \sum_{-\infty}^{\infty} c_n c_n^* = \sum_{-\infty}^{\infty} |c_n|^2 = c_0^2 + 2\sum_{n=1}^{\infty}|c_n|^2 \qquad |c_n| = |c_{-n}|$$

$$= a_0^2 + 2\sum_{n=1}^{\infty}|c_n|^2$$

For a pulsed signal the energy is given by

$$E = \int_{-\infty}^{\infty} |v(t)|^2\,\mathrm{d}t$$

$$= \int_{-\infty}^{\infty} v(t)v^*(t)\,\mathrm{d}t$$

$$= \int_{-\infty}^{\infty} v(t)\left(\frac{1}{2\pi}\int_{-\infty}^{\infty} V^*(\omega)\exp(-\mathrm{j}\omega t)\,\mathrm{d}\omega\right)\mathrm{d}t$$

$$= \frac{1}{2\pi}\int_{-\infty}^{\infty} V^*(\omega)\left(\int_{-\infty}^{\infty} v(t)\exp(-\mathrm{j}\omega t)\,\mathrm{d}t\right)\mathrm{d}\omega$$

$$= \frac{1}{2\pi}\int_{-\infty}^{\infty} V^*(\omega).V(\omega)\,\mathrm{d}\omega$$

$$= \frac{1}{2\pi}\int_{-\infty}^{\infty} |V(\omega)|^2\,\mathrm{d}\omega$$

that is

$$\int_{-\infty}^{\infty} |v(t)|^2 \, dt = \frac{1}{2\pi} \int_{-\infty}^{\infty} |V(\omega)|^2 \, d\omega$$

This goes to the df form by writing $d\omega = 2\pi \, df$.

Here are two examples.

1. Square pulse train

$$a_0 = V_0/2 \qquad a_n = 0$$

$$b_n = \frac{2}{T} \int_0^{T/2} V_0 \sin(n\omega_0 t) \, dt \qquad \text{Figure 4.18}$$

$$= \begin{cases} 2V_0/n\pi & \text{for } n \text{ odd} \\ 0 & \text{for } n \text{ even} \end{cases}$$

(Prove these results.) Therefore

$$P = a_0^2 + \frac{1}{2} \sum b_n^2$$

$$= V_0^2[0.25 + (2/\pi^2)(1 + 1/9 + 1/25 + 1/49 + 1/81 + 1/21 + \ldots)]$$

$$= V_0^2(0.250 + 0.203 + 0.022 + 0.008 + 0.004 + 0.002 + 0.002 + \ldots)$$

$$= (0.491 +) V_0^2$$

$$\simeq 0.50 V_0^2$$

which is the actual value. (Check: $P = 1/T(V_0^2 . T/2) = 0.5V_0^2$.)

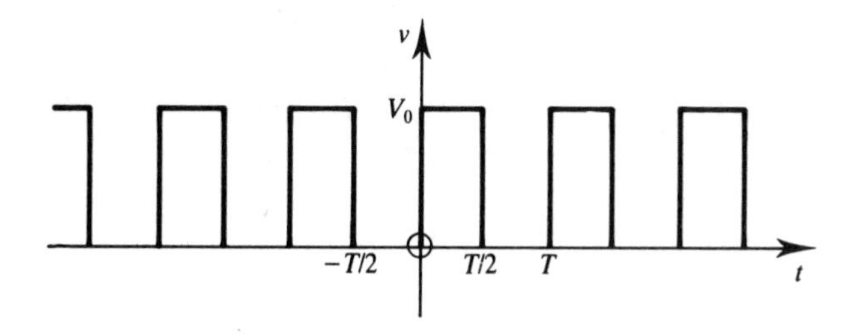

Figure 4.18 Square pulse train

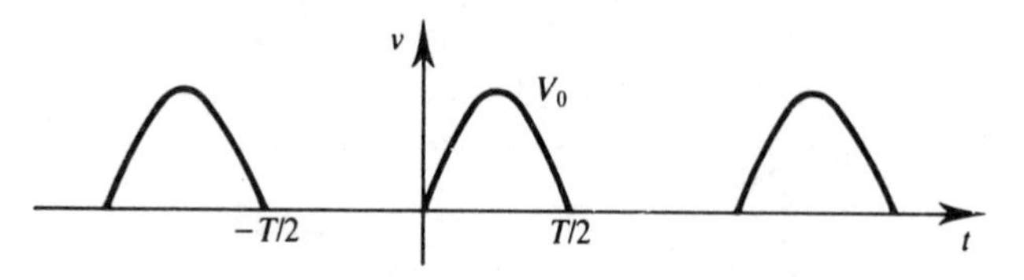

Figure 4.19 Half sine pulse train

2. Half sine pulse train

$$P = \frac{1}{T}\int_0^{T/2} V_0^2 \sin^2(2\pi t/T)\,dt$$

$$= \frac{V_0^2}{2T}\int_0^{T/2} [1 - \cos(4\pi t/T)]\,dt = V_0^2/4$$

Also

$$a_0 = V_0/\pi \qquad a_n(\text{odd}) = 0 \qquad b_1 = V_0/2$$

$$a_n(\text{even}) = -2V_0/\pi(n^2 - 1) \qquad \text{other } b_n = 0 \qquad \text{Figure 4.19}$$

(Prove all these for yourself; $a_1 = 0$ needs care.) Therefore

$$P = a_0^2 + \frac{1}{2}\sum_0^{\infty}(a_n^2 + b_n^2)$$

$$= V_0^2[1/\pi^2 + 1/8 + (2/\pi^2)(1/9 + 1/225 + 1/1225 + \ldots)] = 0.2499V_0^2!!$$

4.12 • Summary of useful expressions

The forward transform: $V(\omega) = \int_{-\infty}^{\infty} v(t)\exp(-j\omega t)\,dt$

The inverse transform: $v(t) = \frac{1}{2\pi}\int_{-\infty}^{\infty} V(\omega)\exp(j\omega t)\,d\omega$

or $v(t) = \int_{-\infty}^{\infty} V(f)\exp(j\omega t)\,df$

Even symmetry: $V(\omega) = 2\int_0^{\infty} v(t)\cos\omega t\,dt$

Odd symmetry: $V(\omega) = -2j\int_0^{\infty} v(t)\sin\omega t\,dt$

Time-shifted pulse: $F[v(t-t_d)] = F[v(t)]\exp(-j\omega t_d)$

Modulated pulse: $F[v(t)\cos\omega_c t] = V(\omega-\omega_c) + V(\omega+\omega_c)$
where $F[v(t)] = V(\omega)$

Unit impulse: $\delta(t)$ where $\int_{-\infty}^{\infty} \delta(t)\,dt = 1$

Differential: $V(\omega) = (1/j\omega)V'(\omega) = (-1/\omega^2)V''(\omega)$

$$v(t) = (-1/jt)F^{-1}\{d/d\omega[V(\omega)]\}$$

Energy:

$$E = \frac{1}{2\pi}\int_{-\infty}^{\infty} |V(\omega)|^2\,d\omega = \int_{-\infty}^{\infty} |V(f)|^2\,df$$

$$= \int_{-\infty}^{\infty} |v(t)|^2\,dt$$

Power:

$$P = a_0^2 + \frac{1}{2}\sum_1^{\infty}(a_n^2 + b_n^2) = a_0^2 + 2\sum_1^{\infty}|C_n|^2$$

4.13 • Conclusion

We have now seen how to deal with a whole variety of pulsed signals and filter characteristics to obtain, respectively, their spectral distributions and their transfer functions – the Fourier transforms. As with Fourier series the basic ideas are quite straightforward but the skill comes in learning to use a variety of simplifying techniques (e.g. differentiation) to obtain results as quickly and easily as possible. These require knowledge and care in application.

But what are all these signals – periodic and aperiodic – we keep referring to? We now need to look at the types of signal which actually occur in communication systems – and that we do in the next chapter.

4.14 • Problems

4.1 Carry out the simple forward transformations in Figure 4.20.

4.2 Carry out the simple inverse transformations in Figure 4.21.

4.3 Carry out the forward transforms in Figure 4.22.

4.4 Carry out the inverse transforms for the filters having the characteristics shown in Figure 4.23.

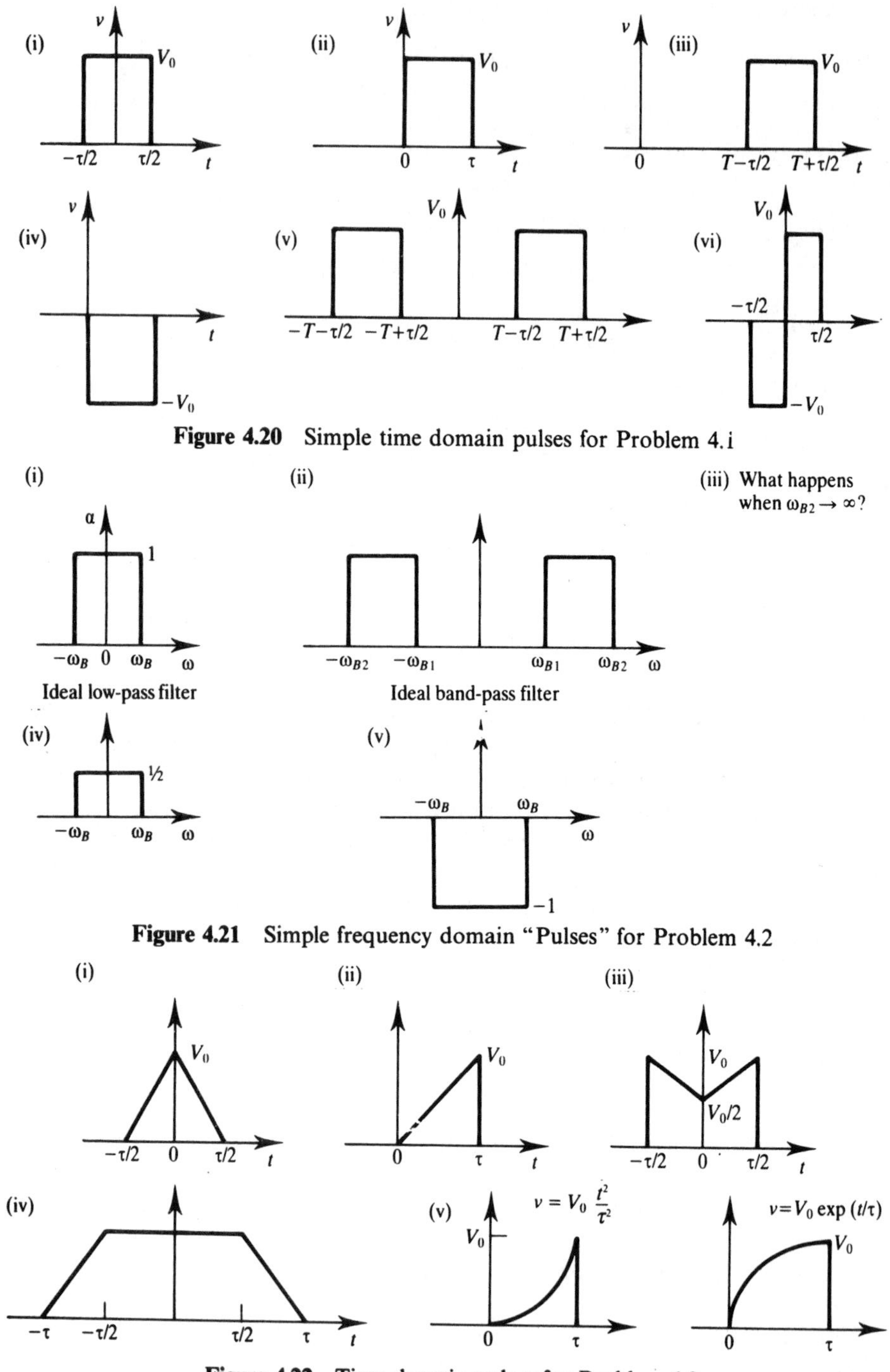

Figure 4.20 Simple time domain pulses for Problem 4.1

Figure 4.21 Simple frequency domain "Pulses" for Problem 4.2

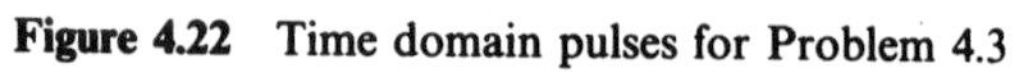

Figure 4.22 Time domain pulses for Problem 4.3

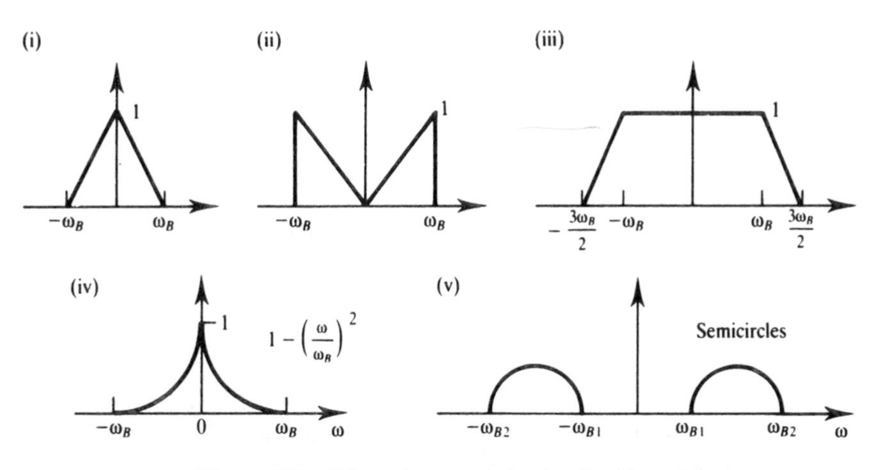

Figure 4.23 Filter characteristics for Problem 4.4

4.5 *Problems on impulse responses*:

(i) If $d(t)$ is defined by $\int_{-\infty}^{\infty} d(t).\mathrm{d}t = 1$ show that $F[d(t)] = 1$.

(ii) Find the Fourier transform of $d(t - T)$.

(iii) Obtain the impulse responses, $(g(t))$, of the networks whose transfer functions are given by the diagrams in Problems 4.2(i), 4.2(ii), 4.4(i), 4.4(ii) and 4.4(iii).

4.6 *Other network responses*:

(i) For the networks in Problem 4.5(iii), what would $g(t)$ be if $v(t)$ is a single sinusoid at $\frac{1}{2}\omega_B$? That is, $v(t) = A\cos\frac{1}{2}\omega_B t$.

(ii) Discuss the form taken by $g(t)$ for a single rectangular pulse (Figure 4.20(i)) into a low-pass filter (Figure 4.21(i)) when

(a) $\omega_B \gg 1/\tau$

(b) $\omega_B \simeq 1/\tau$

(c) $\omega_B \ll 1/\tau$.

4.7 A sinusoidal voltage is given by the expression

$$v_1 = V_0 \cos(\pi t/T)$$

It is a full-wave rectified by a perfect system. Obtain a general expression for the double-sided Fourier spectrum of this rectified waveform and sketch its spectrum with suitable values. If the voltage is now restricted to a single pulse (v_2) in the range $-\frac{1}{2}T$ to $\frac{1}{2}T$ show that its spectrum is given by

$$V_2 = \frac{V_0 T}{2}[\mathrm{sinc}(\omega T/2 - \pi/2) + \mathrm{sinc}(\omega T/2 + \pi/2)]$$

4.8 Figure 4.24 shows the frequency characteristic of a cosine-squared filter given by

$$H(\omega) = \begin{cases} \cos^2\left[\dfrac{\pi\omega}{2\omega_B}\right] = \left\{\dfrac{1}{2} + \dfrac{1}{2}\cos\left[\dfrac{\pi\omega}{\omega_B}\right]\right. & \text{for } -\omega_B \leqslant \omega \leqslant \omega_B \\ 0 & \text{for all other } \omega \end{cases}$$

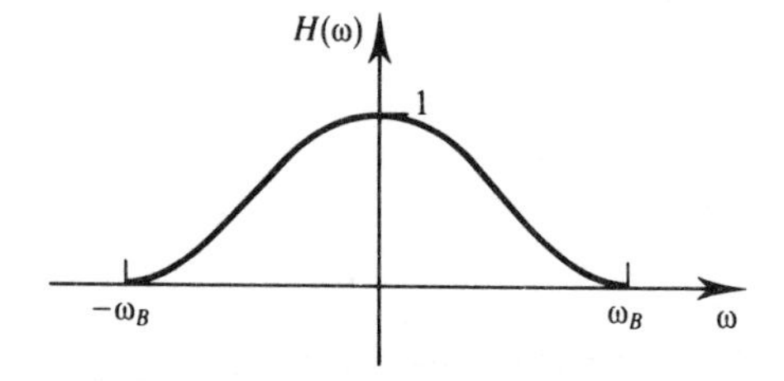

Figure 4.24 Filter characteristics for Problem 4.8

Show that its impulse response is h(t) where

$$h(t) = B\,\mathrm{sinc}(\omega_B t) + \tfrac{1}{2}B[\mathrm{sinc}(\omega_B t + \pi) + \mathrm{sinc}(\omega_B t - \pi)]$$

Sketch the form of this impulse response.

4.9 Show that the impulse response of the filter whose spectrum is shown in Figure 4.25 is given by

$$h(t) = \frac{\omega_0 \pi}{2}\,\mathrm{sinc}(\omega_0 t)\frac{1 + 2\cos(\omega_0 t)}{\pi^2 - \omega_0^2 t^2}$$

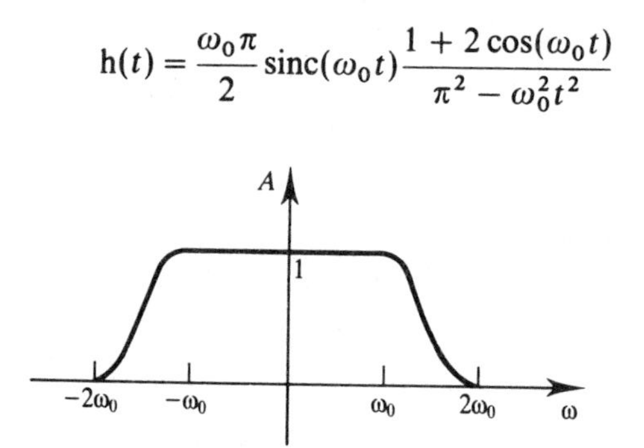

Figure 4.25 Filter characteristics for Problem 4.9

Between $-2\omega_0$ and $-\omega_0$, and between ω_0 and $2\omega_0$, the filter characteristic is of raised cosinusoidal shape given by

$$A = \tfrac{1}{2}[1 - \cos(\pi\omega/\omega_0)]$$

4.10 A low-pass filter has a transfer function as shown in Figure 4.25 where $G(\omega) = 0$ for $\omega < -\omega_B$ and $\omega > \omega_B$.

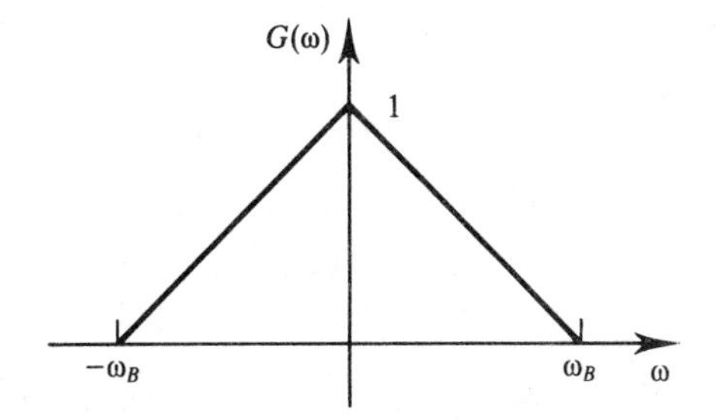

Figure 4.26 Filter characteristics for Problem 4.10

Show that a voltage impulse into this filter produces an output voltage pulse whose shape is $g(t)$ where

$$g(t) = \frac{1 - \cos \omega_B t}{\pi \omega_B t^2} = \frac{B}{2\pi} \operatorname{sinc}^2\left(\frac{\omega_B t}{2}\right)$$

Suggest an advantage of this type of filter over the ideal rectangular filter transfer function.

4.11 An exponentially decaying pulse is given by $v(t) = 0$ for $t < 0$ and $\exp(-at)$ for $t \geqslant 0$. It is then multiplied by a high-frequency cosinusoid, $\cos(\omega_d t)$. Determine the Fourier transform of the resultant waveform. Make $v(\tau)$ symmetrical about $t = 0$.

4.12 A pulse generator, capable of producing rectangular pulses of amplitude V and width τ, either singly or repetitively with period T, is connected to an ideal transmission system of bandwidth 0 to βHz.

(i) Explain the meaning of the term "ideal" in this context. Indicate how a practical system might differ from the ideal.

(ii) Obtain an expression for the spectrum of the input pulse and sketch both it and that of the output pulse deciding the relationship between $1/\tau$ and B for yourself.

(iii) Find the minimum value of τ (τ_m) in terms of B, consistent with an adequate representation of the pulse at the output of the transmission system. Explain the criterion you use for "adequacy" in this context.

(iv) If τ is made much smaller than τ_m, show that the frequency spectrum at the input approaches a constant value, $V\tau$.

(v) Discuss the relationship between τ and B and the repetition period, T, to minimize interference between pulses.

• 5 •

Voltage transfer functions

5.1 • Introduction

A frequent requirement in communication systems is to be able to define the output signal from a particular part of the system when a given input signal goes into it. Such separate parts are given the general name of **networks** with the implication that they are **two-port networks** which are a special class of four-terminal networks having one terminal of each port in common.

Channel filters are a well-known example of such a network in communications as are amplifiers, demodulators and even noisy feeders.

The general representation is given in Figure 5.1.

v_1 and v_2 are often written v_i and v_o and v_{IN} and v_{OUT} respectively. The first will be used here as they are marginally more useful when networks are cascaded.

What we want to know is the relationship of v_1 to v_2 in terms of the properties of the network, N. These properties are expressed as the network's **voltage transfer function** (H) when the output port is open circuited (i.e. $i_2 = 0$). This situation is usually represented in general terms as shown in Figure 5.2.

Lower-case v and h are used for the variables when expressed in the time domain and upper-case V and H for frequency domain use. We deal mostly with frequency-dependent expressions as explained later (Section 5.6). The condition of open-circuit use is appropriate for use in most parts of a communication system because power is low (e.g. in a receiver) and input impedance high.

Now the TF is defined as H where

$$H(\omega) = \frac{V_2(\omega)}{V_1(\omega)}$$

Thus an expression for H can be found for any circuit for which the ratio V_2/V_1 can be obtained. Let us look at a simple example to illustrate this point.

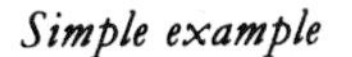

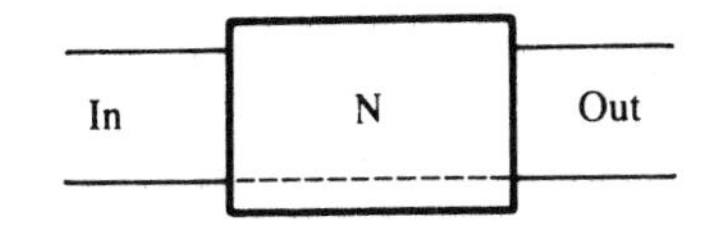

Figure 5.1 General representation of a two-port network

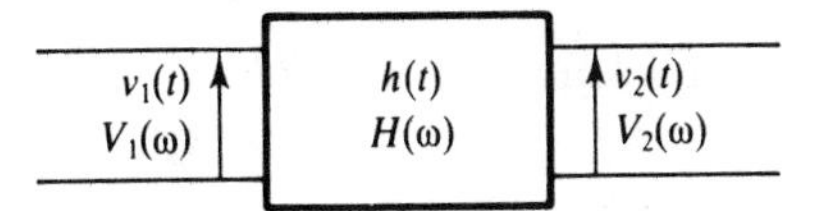

Figure 5.2 Network nomenclature for voltage transfer functions

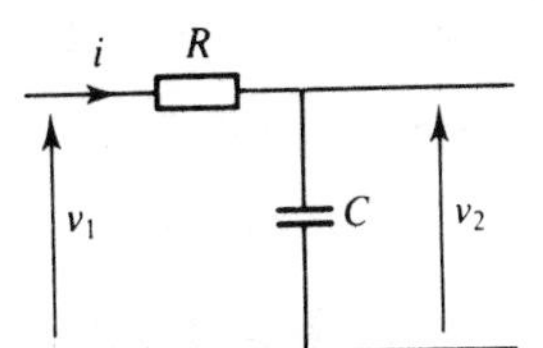

Figure 5.3 Simple *RC* low-pass filter

5.2 • Simple example

Take the simple *RC* circuit in Figure 5.3.

For sinusoidal signals, treating the circuit as a potential divider:

$$V_2 = \frac{1/\mathrm{j}\omega C}{R + 1/\mathrm{j}\omega C} \cdot V_1$$

Therefore

$$\begin{aligned} H = \frac{V_2}{V_1} &= \frac{1}{1 + \mathrm{j}\omega CR} \\ &= \frac{1 - \mathrm{j}\omega CR}{(1 + \mathrm{j}\omega CR)(1 + \mathrm{j}\omega CR)} \\ &= \frac{1}{(1 + \omega^2 C^2 R^2)^{1/2}} \left(\frac{1}{(1 + \omega^2 C^2 R^2)^{1/2}} - \frac{\mathrm{j}\omega CR}{(1 + \omega^2 C^2 R^2)^{1/2}} \right) \\ &= A(\cos\phi + \mathrm{j}\sin\phi) \end{aligned}$$

That is

$$H = A\exp(\mathrm{j}\phi)$$

Transfer function

where

$$A = 1/(1 + \omega^2 C^2 R^2)^{1/2}$$ **Attenuation factor**

and

$$\phi = \tan^{-1}(-\omega CR)$$ **Phase factor**

Now take, as input signal, a single sinusoid with phase. That is

$$v_1 = V_1 \cos(\omega t + \phi_1) = V_1 \exp[\mathrm{j}(\omega t + \phi_1)]$$

Because H is in exponential form we must also use the exponential (rotating phasor) form of V_1. Then

$$\begin{aligned} V_2(\omega) &= V_1(\omega) . H(\omega) \\ &= V_1 \exp[\mathrm{j}(\omega t + \phi_1)] A \exp(\mathrm{j}\phi) \\ &= A V_1 \exp[\mathrm{j}(\omega t + \phi_1 + \phi)] \end{aligned}$$

Therefore

$$V_2(\omega) = A V_1 \cos(\omega t + \phi_1 + \phi)$$ **Output signal**

where

$$V_2 = A V_1$$

and

$$\phi_2 = \phi_1 + \phi$$

[Note that it looks as if we could have gone straight from the cosine version of v_1 by writing

$$v_2 = V_1 \cos(\omega t + \phi_1) A \exp(\mathrm{j}\phi)$$

but this would have led us to the result

$$v_2 = A V_1 \cos(\omega t + \phi_1) \cos \phi$$

which is incorrect. Try it for yourself. The exponential form of the input signal *must* be used.]

Note also that, although we have worked in the frequency domain, time is still there in the form of ωt. We are, in fact, using **rotating phasor notation**. But, as everything is in exponential form, the ωt part can be removed by changing everything to **stationary phasor form** as follows:

$$v_1 = V_1 \exp(j\omega t) \exp(j\phi)$$

Therefore

$$\begin{aligned} v_2 &= AV_1 \exp(j\omega t) \exp[j(\phi_1 + \phi)] \\ &= V_2 \exp(j\omega t) \exp(j\phi_2) \end{aligned}$$

and

$$V_2 \exp(j\omega t) \exp(j\phi_2) = AV_1 \exp(j\omega t) \exp(j\phi_1 + \phi)$$

or

$$V_2 \exp(j\phi_2) = AV_1 \exp[j(\phi_1 + \phi)]$$

so that

$$V_2 = AV_1$$

and

$$\phi_2 = \phi_1 + \phi$$

as before. In practice this is what you usually do – merely treat a signal in terms of its magnitude and stationary phase. Of course this can only be done when frequency is not changing.

5.3 • Transfer function representation

Take the simple example in Figure 5.3. We have

$$\begin{aligned} A &= \frac{1}{(1 + \omega^2 C^2 R^2)^{1/2}} \\ \phi &= \tan^{-1}(-\omega CR) \end{aligned}$$

Obviously these vary with frequency and it is very helpful to your understanding to

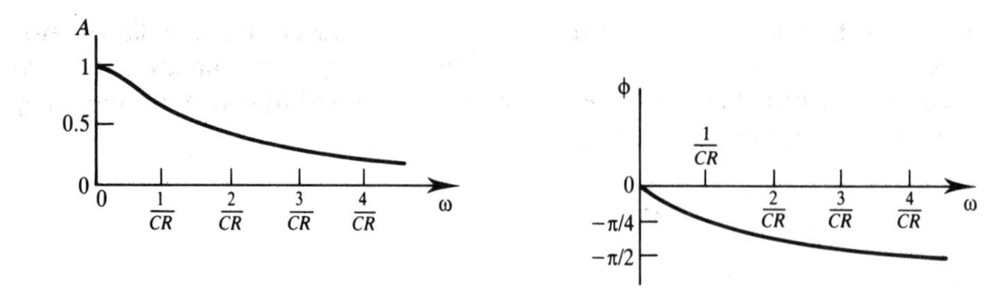

Figure 5.4 Graphical representation of H – linear

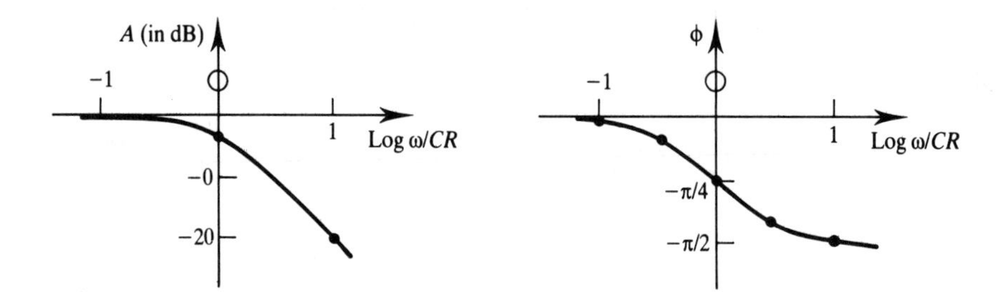

Figure 5.5 Sketches of A (in dB) against $\log \omega$ near $\omega = 1/CR$ (Bode plots)

sketch rough plots of how A and ϕ change over the range $\omega = 0$ to ∞. It is easy to do this by taking a few straightforward guide points. For example

$\omega = 0$	gives $A = 1$ and $\phi = \tan^{-1}(0) = 0$
$\omega \to \infty$	gives $A = 0$ and $\phi = \tan^{-1}(-\infty) = -\pi/2$
$\omega = 1/CR$	gives $A = 1/2$ and $\phi = \tan^{-1}(-1) = -\pi/4$

which are quite enough to give the simple plots in Figure 5.4.

It is often more useful to do these on logarithmic axes, using $\log \omega$, dB, for A but keeping ϕ linear (Figures 5.5 and 5.6). Then, for A, the guide points are:

$\omega \to 0$	$\log \omega \to -\infty$	$A \to 1 \equiv 0\,\text{dB}$
$\omega = 1/10CR$	$\log \omega = -1 - \log CR$	$A = 1/(1.01)^{1/2} = -0.04\,\text{dB} \simeq 0$
$\omega = CR$	$\log \text{w} = -\log CR$	$A = 1/2 \equiv -3\,\text{dB}$
$\omega = 10CR$	$\log \omega = 1 - \log CR$	$A = (1/(101) \equiv -20\,\text{dB}$
$\omega \gg CR$		$A = 1/\omega CR \equiv -(\omega CR)\,\text{dB}$ (i.e. $\log A \alpha \log \omega$)
$\omega \to \infty$	$\log \omega \to \infty$	$A \to 0 = -\infty\,\text{dB}$

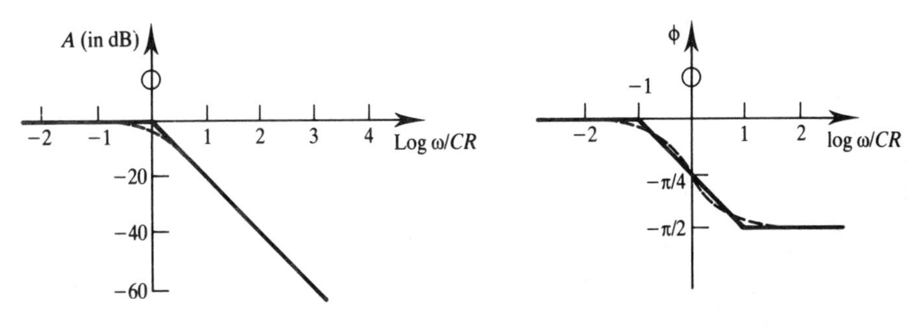

Figure 5.6 Piecewise linear approximations

For ϕ the corresponding points are

$$\begin{array}{ll} \omega = 0 & \phi = 0 \\ \omega = 1/10CR & \phi = \tan^{-1}(-0.1) = 0.1 \text{ rad} \\ \omega = 1CR & \phi = \tan^{-1}(-1) = -\pi/4 \\ \omega = 10/CR & \phi = \tan^{-1}(-10) = -1.47 \text{ rad} \rightarrow -\pi/2 \\ \omega \rightarrow \infty & \phi = \tan^{-1}(\infty) = -\pi/2 = -1.57 \text{ rad} \end{array}$$

Note that these two approximate sketches (Figure 5.6), based on piecewise linear approximations, assume constant values are reached a decade either side of f_0 (i.e. at $1/10CR$ and $10/CR$). These are very useful to give quick rough ideas of a circuit's transfer function form, but note that what we have done is only the basic idea and there is much more to it in detail under various circumstances (e.g. filter roll-off at higher orders, poles and zeros). These details are not included here but are worth finding out about.

Indeed plots of A (linear or dB) against ω (log or linear) are so often used to summarize pictorially the properties of a network that they have come to be called **transfer functions** themselves. When the term is used in a book or paper this is usually what is meant, so do make sure which is being used – graph or algebraic expression (or even just the symbol H itself). Figure 5.7 shows such a "transfer function" for a bandpass filter using linear ω (because ω_1 and ω_2 are usually only fractionally different from ω_0 itself).

The situation is confused even further because a transfer function (especially a pictorial one) will very often be referred to as the impulse response of a network. This is because an impulse ($\delta(t)$) has an amplitude spectrum of constant value so that it acts (mathematically at least) as a source containing all frequencies at the same amplitude (usually 1). This means that, when a circuit has an impulse as an input, its output voltage has the same spectral distribution as that of the circuit's transfer function. That is

$$V_2 = \delta . H = 1 . H$$

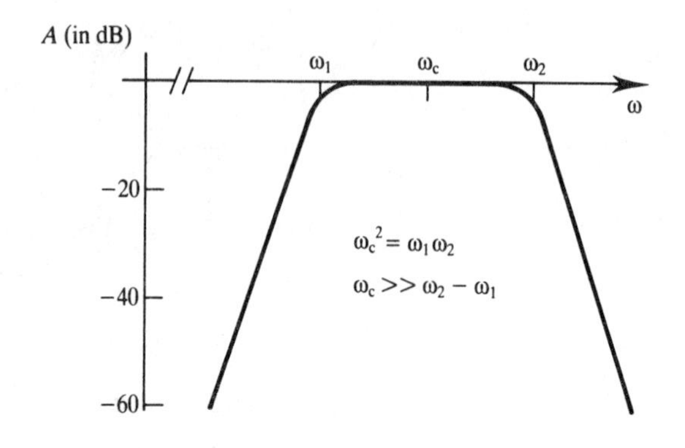

Figure 5.7 "Transfer function" of a bandpass filter

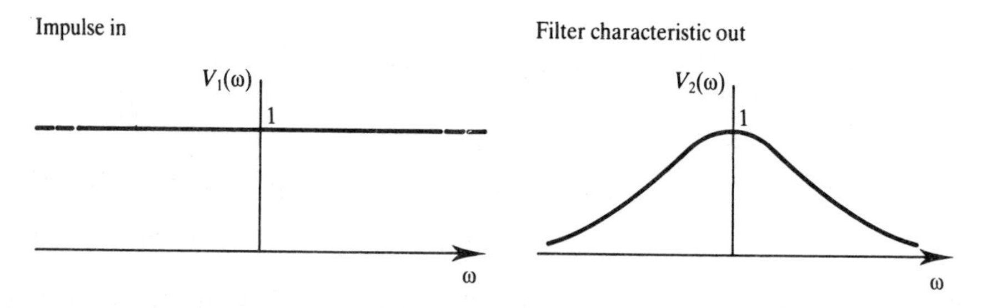

Figure 5.8 An impulse response of a low-pass filter

where δ is defined by

$$1 = \int_{-\infty}^{\infty} \delta . dt$$

so

$$V_2 = H$$

This can be seen more clearly in Figure 5.8.

Some form of pictorial representation is often very illustrative in many situations. For instance, you can show clearly how a bandpass filter can be used to remove some frequencies from a signal as in Figure 5.9.

When an input signal contains more than one frequency (e.g. for a square wave) the full rotating phasor expressions must be used (as in Section 5.2) with each frequency treated independently (the principle of superposition). This is because both A and ϕ are functions of ω and so the full terms in ωt must be kept in. For example, suppose an input signal contains three separate frequencies as below:

$$v_{IN} = V_1 \cos(\omega_1 t + \phi_1) + V_2 \cos(\omega_2 t + \phi_2) + V_3 \cos(\omega_3 t + \phi_3)$$

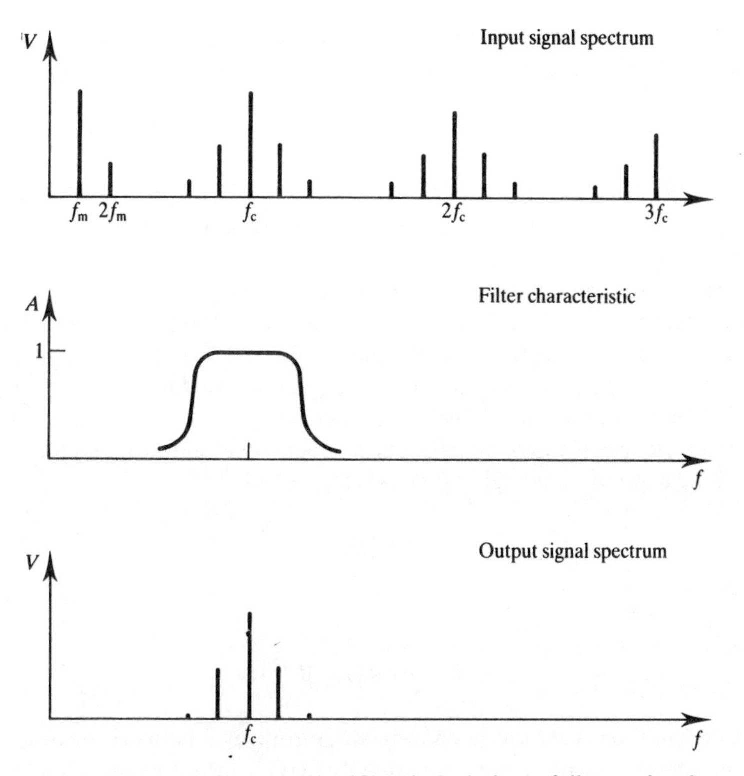

Figure 5.9 Action of bandpass filter in isolating a full a.m. signal

Then

$$v_{\text{OUT}} = A(\omega_1)V_1\cos[\omega_1 t + \phi_1 + \phi(\omega_1)] + A(\omega_2)V_2\cos[\omega_2 t + \phi_2 + \phi(\omega_2)] + A(\omega_3)\cos[\omega_3 t + \phi_3 + \phi(\omega_3)]$$

For example, the simple *RC* filter in Section 5.2 would give

$$v_{\text{OUT}} = \frac{V_1}{(1 + \omega_1^2 C^2 R^2)^{1/2}}\cos[\omega_1 t + \phi_1 - \tan^{-1}(\omega_1 CR)] + \frac{V_2}{(1 + \omega_2^2 C^2 R^2)^{1/2}}\cos[\omega_2 t + \phi_2 - \tan^{-1}(\omega_2 CR)] + \frac{V_3}{(1 + \omega_3^2 C^2 R^2)^{1/2}}\cos[\omega_3 t + \phi_3 - \tan^{-1}(\omega_3 CR)]$$

Note that the three phase changes introduced by the network have been written straight into the cosine terms without going through the full analysis in the exponential form. We can do this because we know what the result is going to be.

5.4 • Time and frequency domain representations of transfer functions

So far we have looked at transfer functions entirely as functions of frequency not of time. Even where time does appear to have been included it comes in merely as an additional multiplying factor as at the end of Section 5.2. This is true also of the multifrequency input just above because, if analysed in full, time would have appeared merely as an exponential multiplier in each term.

It is only because we have been in the frequency domain that we have been able to apply H as a simple multiplying factor. That is, as

$$V_2 = HV_1$$

or, more rigorously,

$$V_2(\omega) = H(\omega) \,.\, V_1(\omega)$$

This is invariably the way you work out the output of a network in practice – and then do an inverse Fourier transform to get it back to the time domain form if you need to. However, you will often see the same relationship quoted in time domain form, although it is hardly ever used directly to find out the effect of a network on a signal because of the difficulty of carrying out the convolution operation. This form is

$$v_2(t) = h(t) * v_1(t)$$

where the asterisk (*) represents the operation of convolution.

To analyse a circuit this way you would first of all have to convert both V_1 and H to time form. This may be fairly easy for V_1 but very hard for H. For example, the RC circuit of Figure 5.2 would need the following inverse Fourier operation:

$$\begin{aligned} h(t) &= \frac{1}{2\pi}\int_{-\infty}^{\infty} H(\omega)\exp(\mathrm{j}\omega t)\,\mathrm{d}\omega \\ &= \frac{1}{2\pi}\int_{-\infty}^{\infty} \frac{\exp(\mathrm{j}\omega t)}{1 + \mathrm{j}\omega CR}\,\mathrm{d}\omega \end{aligned}$$

which is difficult to do. Even if you could, you would then have to perform the convolution – probably even more difficult – and then perhaps the inverse Fourier transform of $v_2(t)$. I am sure you will agree it is better to stick to the frequency domain with its simple multiplication operations. However, it is important to know that the time domain equivalent exists and to be able to recognize it.

5.5 • Summary

Open-circuit **voltage transfer function** (TF or H) defined by

$$\begin{aligned} H(\omega) &= \frac{V_{\text{OUT}}(\omega)}{V_{\text{IN}}(\omega)} \\ &= \frac{V_2 \exp(\text{j}\phi_2)}{V_1 \exp(\text{j}\phi_1)} \\ &= A \exp(\text{j}\phi) \end{aligned}$$

where A (the attenuation factor) equals V_2/V_1 (usually in dB) and ϕ (phase factor) equals $\phi_2 - \phi_1$ (usually in fractions of π). For linear circuits, expressions for how A and ϕ vary with frequency in terms of component values can be worked out by circuit analysis methods.

Plots of A (in dB) and ϕ against ω (or f) form **Bode plots** giving much useful information about the behaviour of a network. **Piecewise linear approximations** to these are very easy to sketch with straight portions which meet at corner frequencies (f_c) with change of attenuation ($-3n$ dB at f_c), slope ($20n$ dB/decade) and phase ($n\pi/2$ total; $n\pi/4$ at f_c) occurring over the range $f_c/10$ and $10f_c$ (i.e. ± 1 in logs). (See Figure 5.6.) The **impulse response** of a network is the same as its Bode plot. This is because an impulse (δ) contains all frequencies at equal amplitudes.

$$1 = \int_{-\infty}^{\infty} \delta \, . \, \mathrm{d}t$$

When the input is δ, the output is H.

The term **transfer function** can be used to mean several ways of looking at the same thing:

1. Just the symbol H itself.
2. The expression for V_2/V_1 in terms of ω.
3. Bode plot or sketch – usually A only and often idealized.
4. The impulse response.

In the time domain a TF is given the symbol $h(t)$ and related to $H(\omega)$ by

$$H(\omega) \leftrightarrow h(t)$$

where

$$V_2(\omega) = H(\omega) . V_1(\omega)$$

and

$$v_2(t) = h(t) * v_1(t)$$

with the asterisk representing the operation of convolution which is not dealt with in this book.

5.6 • Conclusions

This chapter is the last of those looking at the mathematical background needed to understand modulation techniques. Now is the time to start considering real signals and what we are going to do with them and why. So turn to the next chapter.

5.7 • Problems

5.1 Find $H(\omega)$ for the simple CR high-pass filter in Figure 5.10. Sketch changes of A and ϕ against ω.

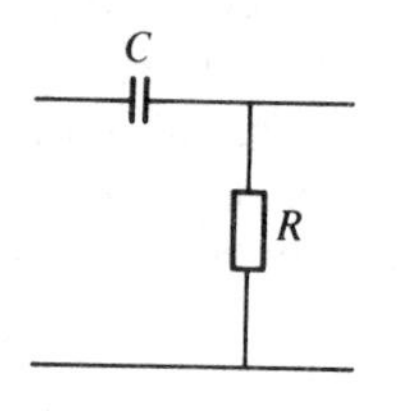

Figure 5.10 For Problem 5.1

5.2 Find $H(\omega)$ for the simple transistor equivalent circuit in Figure 5.11. Sketch forms.

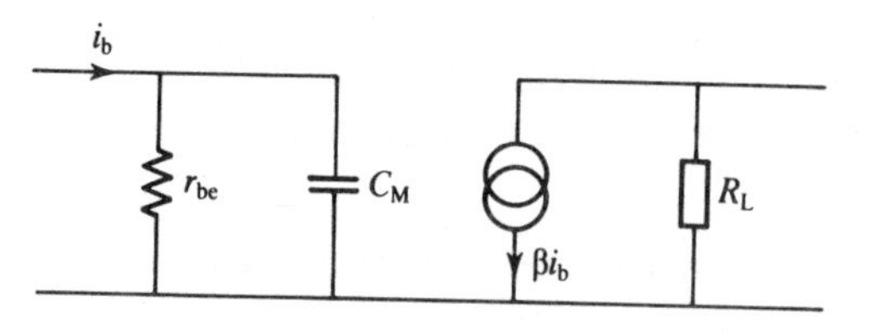

Figure 5.11 For Problem 5.2

5.3 Find $H(\omega)$ for the op-amp differentiator in Figure 5.12. Sketch A and ϕ variation with ω.

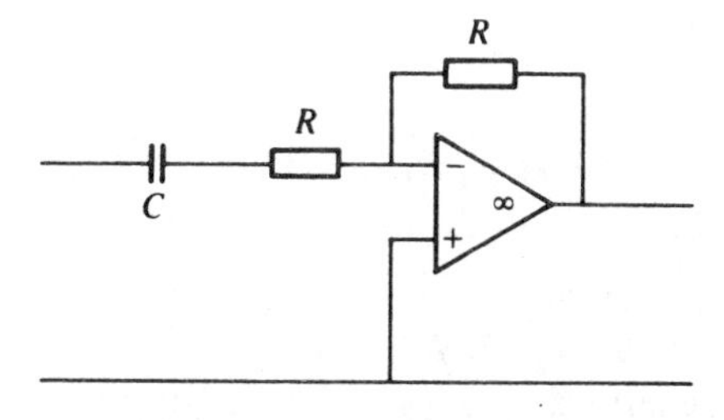

Figure 5.12 For Problem 5.3

5.4 Find $H(\omega)$ for the *RC* circuit in Figure 5.13. Sketch A and ϕ forms. What kind of circuit is it?

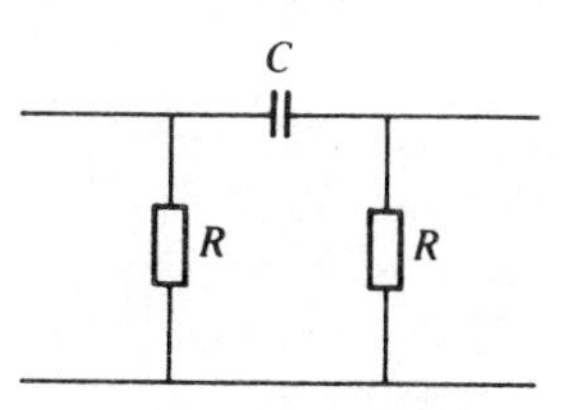

Figure 5.13 For Problem 5.4

5.5 Find $H(\omega)$ for the *LC* circuit in Figure 5.14. Sketch. What does it do?

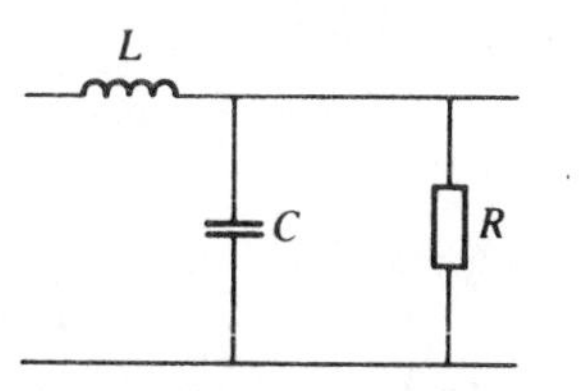

Figure 5.14 For Problem 5.5

5.6 For the *RC* low-pass circuit of Figure 5.3 find $v_2(t)$ for

(i) $v_1(t) = V_1 \sin \omega t$
(ii) $v_1(t) = \cos^2 \omega t$
(iii) $v_1(t) = \exp(-t/2CR)$ (find $V_2(\omega)$ only)
(iv) $V_1(\omega) = \omega CR$.

Sketch the time variation of $v_2(t)$ for each.

5.7 Find $v_2(t)$ for the circuit of Problem 5.3 for the following inputs:

(i) $v_1(t) = V_0$ (i.e. d.c.)
(ii) $v_1(t) = \delta(t)$
(iii) $V_1(\omega) = \omega CR$
(iv) $v_1(t) = V_1 \cos \omega t$.

5.8 Sketch the following ideal filter characteristics:

(i) LP – $f_c = 5$ kHz
(ii) HP – $f_c = 15$ kHz
(iii) BP – $f_0 = 10$ kHz and B/W = 6 kHz
(iv) Notch – at general f_c.

Then sketch the effect of each of these on a square wave input of repetition frequency 2.5, 7.5, 12.5 and 17.5 kHz each. Sketch both $V_2(\omega)$ and $v_2(t)$ where possible.

5.9 Give the impulse responses of the circuits in Problems 5.1–5.5.

• 6 •
Signals and modulation

6.1 • Introduction

There are four aims of this chapter. The first is to give an overall view of the various types of baseband signal which can form the information to be sent down a communication system. The second is allied to this and looks at the baseband bandwidths involved and how they come about. The third discusses the need for modulation to enable these basebands to be sent, giving the various types of modulation used and showing how they relate to each other. The fourth looks at the general basis of the two classes of modulation technique.

6.2 • Baseband signal types

Here we are considering the signals only after they have become electrical. Many will start off in non-electrical form (e.g. voice, temperature) but will have been converted to voltage using some form of **transducer** (e.g. microphone, thermocouple).

These electrical signals form the original information which is to be sent, the **baseband**. Once obtained they will always fall into one of two broad **classes**.

1. Analogue signals

In these the electrical signal shows a continuous variation in time over a wide range of magnitudes. Often this time variation is the same as that of an original non-electrical signal (e.g. air pressure and voltage for a microphone) so that the two are said to be **analogous**; hence the name of analogue signal.

2. Digital signals

In these the electrical signals consist of discontinuous pulses (or digits) each constant in value but changing abruptly from one digit to the next. They are usually coded signals as in a teleprinter. The most common type is, of course, binary coding where only one type of pulse occurs but they can, in general, be multilevel (M-level) digits. Often a signal will start out analogue and be converted to digital (Figure 6.1).

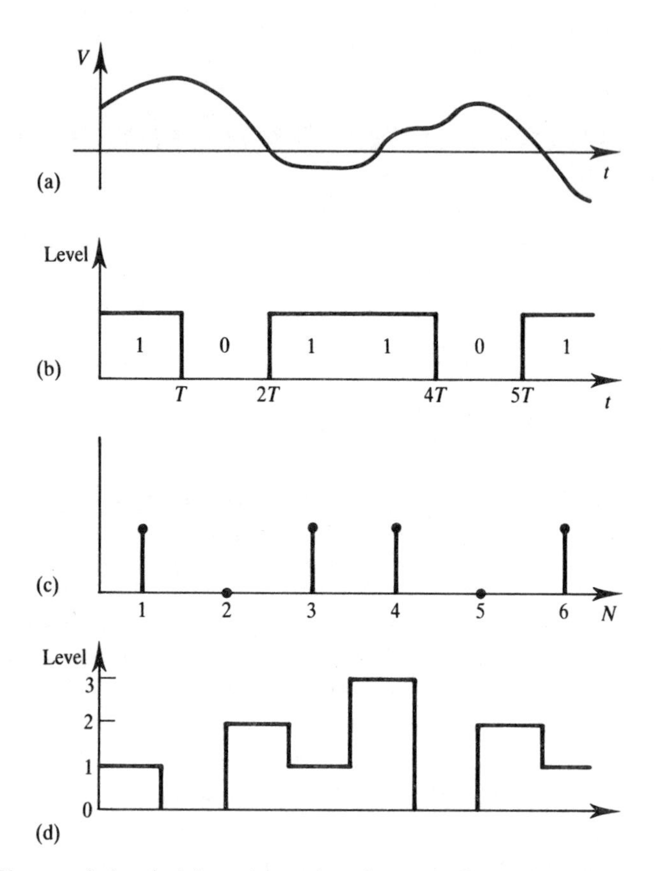

Figure 6.1 Classes of signal: (a) analogue (continuous); (b) analogue (discrete) or digital (binary); (c) functional representation of binary; (d) M-level digital (4-PAM: $M = 4 = 2^2$)

Within these two broad classes there are only a few distinct signal types and the main ones are listed in Table 6.1 which follows, although the boundaries are somewhat blurred in places (e.g. in telemetry).

For some types the baseband bandwidths required depend greatly on the application (e.g. telemetry), but for others it is much more closely defined (e.g. television). The next section shows how some of these standard bandwidths are obtained.

6.3 • Baseband and bandwidth terminology

The term **baseband** has both a general and a specific meaning. Generally it is used, as above, to refer to the original signal. Specifically it means the band (or range) of frequencies occupied by the baseband signal. The actual limits of this range need to be specified separately.

Table 6.1 Types of baseband signal

Signal type	Nature	Class	Baseband bandwidth	Usual mod. type
Morse	pulsed c.w.	digital	0–50 Hz	ASK(OOK)
Teletype	pulses from keyboard	digital	0–120 Hz	FSK/PSK
Facsimile	still copies	digital	0–9.6 kHz	FSK/PSK
Telephone	voice frequencies	analogue	0–4 kHz	SSB/FDM
Audio	music	analogue	0–15 kHz	FM
Radio	AM (LF–HF)	analogue	0–4.5 kHz	AM
broadcast	FM (VHF)	analogue	0–15 kHz	FM
Radio	amateur	analogue	0–3 kHz	SSB/NBFM
Radio	CB	analogue	0–4 kHz	NBFM
Radio comms	mobile	analogue	0–3 kHz	AM/FM
	HF	analogue	0–3 kHz	AM
PCM	digitized audio	digital	0–64 kHz	PSK
Telemetry	data	digital	to MHz	ASK/PSK
Television	moving pictures	analogue	0–6.5 MHz (UK)	VSB
			0–5.5 MHz (US)	VSB
Radar	pulsed c.w.	digital	to GHz	ASK

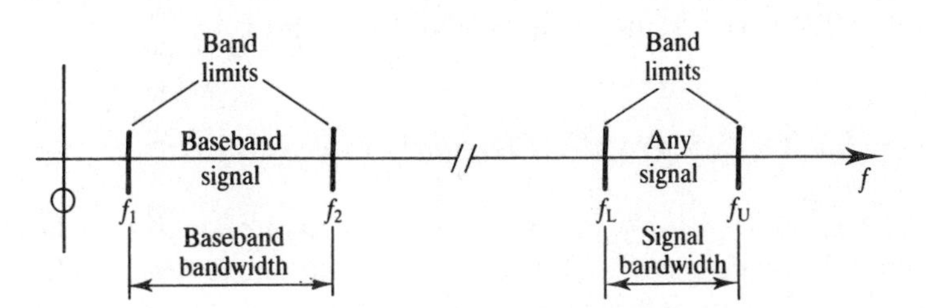

Figure 6.2 Bandwidths, baseband, and band limits

The term **bandwidth** refers merely to the frequency range within any band, without specifying the limits. It is a term which can be used at any frequency and not just at baseband (e.g. 9 kHz bandwidth for a.m. radio) and therefore, strictly speaking, we should use the term **baseband bandwidth** when specifically referring to baseband signals. Figure 6.2 illustrates these terms.

The two terms are often used loosely as if they were synonymous, but there is an important distinction between them. For instance, a voice frequency (telephone) channel always has a nominal bandwidth of 4 kHz but the actual band it occupies at various stages of a telephone network may be very much higher, even megahertz. It is only at the beginning that it is a baseband.

6.4 • Baseband bandwidth evaluation

Most baseband signals used in commercial communication systems are assigned to standard bandwidths fixed by international agreement. This is done by URSI (Union Radio Scientific International) through its two consultative bodies, CCITT (Comité Consultatif International de Telegraphie et Telephonie) and CCIR (Comité Consultatif International des Radiocommunications).

The reasons for the choices of bandwidth are partly subjective (i.e. how much needs to be sent to get the information across in acceptable form) and partly technical based on the way in which the information is sent. In this section the reasons for some of the choices are discussed and values obtained for the relevant baseband bandwidths.

6.4.1 • *Telephony (Analogue)*

Here the reasons for choice of bandwidth are mostly subjective. Human hearing can extend from about 20 to 20 000 Hz, but speech is quite clear and individual voices quite recognizable, whether pitched high or low, if only the middle frequencies are transmitted. The exact limits vary a little from country to country (300–3400 Hz in UK) but the addition of guard bands to allow for filtering and multiplexing means that we can have international agreement on a standard **voice frequency channel** (Figure 6.3) of **0–4 kHz** – standard throughout the world.

6.4.2 • *Telegraphy (Digital) (i.e. teleprinters)*

Here the criteria used are only partly subjective. Operators can type fast but only with continuously varying symbol rates, so, to standardize matters, buffer amplifiers are used to allow data to be transmitted at one of several standard baud rates (e.g. 50 or 100 symbols per second) into the teleprinter. Below we shall look at 50 baud transmission but expressed as an information rate of 66 words per minute. Then

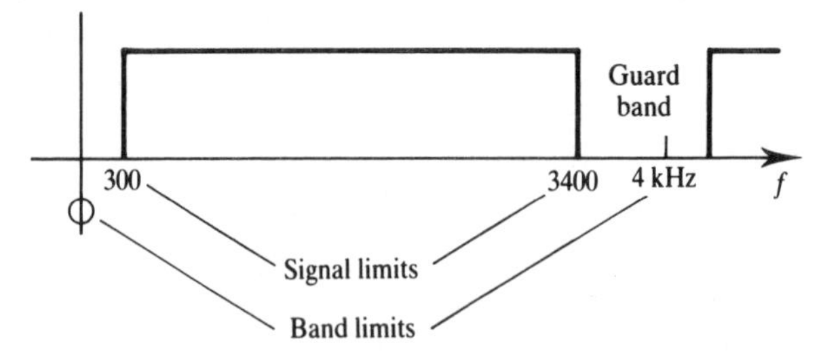

Figure 6.3 Standard voice frequency (telephone) channel

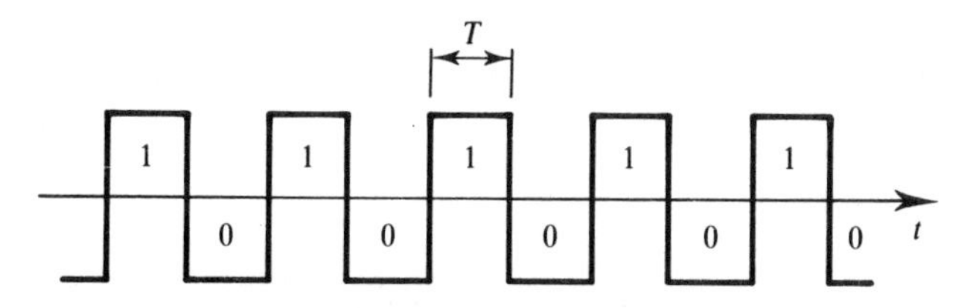

Figure 6.4 "Worst case" bit stream

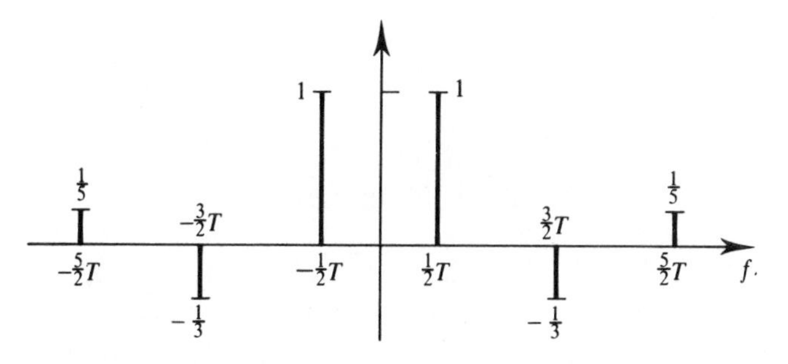

Figure 6.5 Spectrum of "worst case" square wave

some assumptions have to be made about the average number of characters per word (5 here) and the number of symbols per character used by the code chosen (7.5 here). Hence we take the particular case of

$$
\begin{aligned}
66 \quad & \text{words per minute} \\
5+1=6 \quad & \text{characters per word (1 for space)} \\
5+1+1.5=7.5 \quad & \text{symbols per word (1.5 for ``start'' and 1 for ``stop'')}
\end{aligned}
$$

$$
\begin{aligned}
\text{no. words per minute fixed at} &= 66 \\
\text{no. characters per minute} &= 66 \times 6 = 396 \\
\text{no. symbols per minute} &= 66 \times 6 \times 7.5 = 2970 \\
\text{no. symbols per second} &= \frac{66 \times 6 \times 7.5}{60} = 49.5 \text{ (50 baud)}
\end{aligned}
$$

$$\text{Therefore, symbol length} = 1/49.5 \simeq 20 \text{ ms} = T$$

To translate this into bandwidth it is necessary to consider the "worst case" situation. This is the one for which the signal is changing the most rapidly, which is when the symbol stream consists of alternate zeros and ones as shown in Figure 6.4 – drawn bipolar.

Figure 6.4 is equivalent to a square pulse voltage stream of period $2T$ which will have standard spectrum as shown in Figure 6.5 and a fundamental frequency (f_0) of $1/2T$ or **half** the symbol rate.

The question then is how much of this spectrum needs to be retained as it comes out of the teleprinter still in baseband form (e.g. to operate a printer). Here there is another somewhat subjective criterion that the symbols may need to retain a reasonably "square" shape. To do this it is usually regarded as adequate to send up to the third harmonic only. Thus

$$\text{baseband bandwidth} = 3f_0 = 3/2T = 3/(2 \times 0.2) = 75\,\text{Hz}$$

But for an actual teleprinter communication channel some allowance must be made for guard bands and multiplexing so that this bandwidth is increased in practice to a standard value of 120 Hz. That is, **teleprinter channel = 0–120 Hz**. Note that the d.c. term has to be retained in the spectrum.

6.4.3 • *Television (Analogue)*

What baseband bandwidth is needed to send the luminance (light intensity) information in the standard UK TV signal? Very similar reasoning is used to that above for a teleprinter. Subjective criteria are used to decide values for the factors controlling the signal. Then the symbol length is calculated and the bandwidth obtained from this. Here the subjective factors are more involved. There are four of them:

1. Sufficient picture detail obtained by using 625 line scans per frame (525 in USA).
2. Vertical and horizontal resolutions kept the same by assuming smallest areas of constant intensity (picture elements or **pixels**) are square.
3. Most scenes framed adequately by using a rectangular picture shape of aspect ratio 4:3.
4. Unacceptable flickering avoided by making use of the eye's persistence of vision at the slowest convenient picture refresh rate of 25 frames per second (30 in Japan and North America).

Three of these criteria are illustrated in Figure 6.6.

The calculation of bandwidth now proceeds as follows:

$$\begin{aligned}
&\text{Each line has } 625 \times 4/3 \text{ pixels in it} && = 833 \text{ pixels}\\
&\text{Each scan has } 625 \times 625 \times 4/3 \text{ pixels in it} && = 520\,\text{kpixels}\\
&\text{Each scan takes } 1/25\,\text{s to complete} && = 40\,\text{ms}\\
&\text{Each pixel takes } 40\,\text{ms}/520\,\text{ks to scan} && = 76.8\,\text{ns}\\
&\qquad \text{That is, pixel length } (t_p) = 76.8\,\text{ns}
\end{aligned}$$

Again we take the "worst case" situation in which the pixel intensity is changing most often and by the largest amount each time. This obviously occurs when alternate black and white elements occur as in Figure 6.6. The result is a sort of "square wave" of light intensity which must be supplied by a square wave voltage signal of period

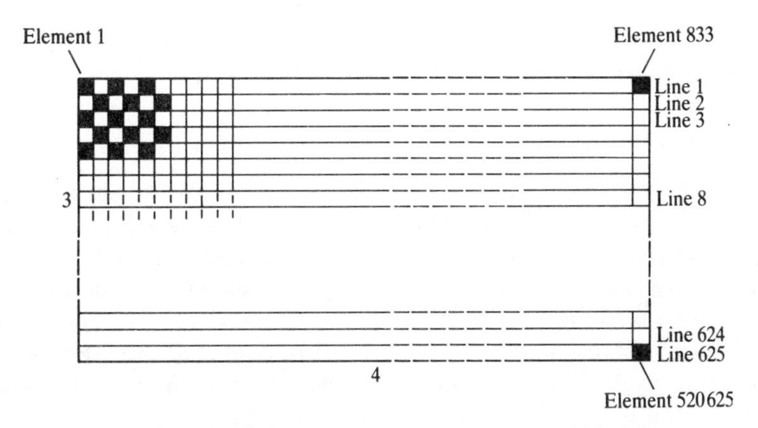

Figure 6.6 Scanning lines and pixels on a UK TV screen

$2t_p$ or 153 ns. This square wave therefore has a fundamental frequency of

$$f_0 = 1/153\ \text{ns} = 6.54\ \text{MHz}$$

but because it is visually quite acceptable for pixel intensity to fall off at the edges it is now only necessary to send f_0 itself without any of its harmonics. There is a need, however, to use more bandwidth to allow for channel separation, VSB modulation and sound signals. Thus the actual r.f. frequency range allowed is

$$\text{UK TV bandwidth} = 8.0\ \text{MHz (6.5 in USA)}$$

The baseband bandwidths of other signals in Table 6.1 can be calculated in similar ways – partly subjective and partly technical.

6.5 • The need for modulation

In general a baseband signal cannot be transmitted usefully without modification. The exceptions are some simple dedicated systems such as an internal telephone system or the line from a teletype machine to a printer. But for the vast majority of systems communication would be either prohibitively expensive or actually impossible without the kind of changes which come under the heading of modulation. For example, a public telephone system would need a separate wire connection for each conversation and a radio link would require huge aerials and enormous power – and even then only one station could operate at a time to avoid interference.

The answer is to change the baseband signal in some way to enable efficient economic communication methods to be used. Two distinct classes of method come under this heading:

1. **Frequency translation**: moving the whole baseband up to a much higher frequency range.
2. **Digitizing**: changing the baseband to digital form, usually binary, by sampling.

Both are given the name of modulation although the second keeps the signal in a baseband frequency region of different bandwidth. The baseband is changed in nature by a process which is strictly a form of source coding and not really covered by the *Oxford English Dictionary* definition which is: "**modulation**, alteration in amplitude or frequency of a wave by a frequency of different order." This is a good description of most frequency translation methods but not of digitizing ones. Perhaps we can forgive such a non-technical publication for not being entirely up to date in our speciality.

6.6 • Classification of modulation types

All the methods of modulation described later in this book fall into one of the two categories mentioned in the last section. First we need a slightly fuller definition of each category:

1. **Frequency translation**: The baseband is moved to a higher-frequency range by arranging for it to alter some property of a higher-frequency carrier.
2. **Sampling**: The baseband waveform voltage is allowed through for short periods of time at regular intervals and these values only are sent, either coded or uncoded. The signal, although much changed in form, still remains essentially baseband in nature.

These categories now subdivide into the specific modulation types related by the techniques used, as shown in Figure 6.7.

In addition, all the methods resulting from sampling can themselves be further modulated by frequency translation methods related to the keying methods already included for binary signals.

The full meaning of the type abbreviations in Figure 6.7 are as follows:

AM Amplitude modulation
PM Phase modulation
FM Frequency modulation
ASK Amplitude shift keying
PSK Phase shift keying
FSK Frequency shift keying
PAM Pulse amplitude modulation
PDM Pulse duration modulation
PPM Pulse position modulation

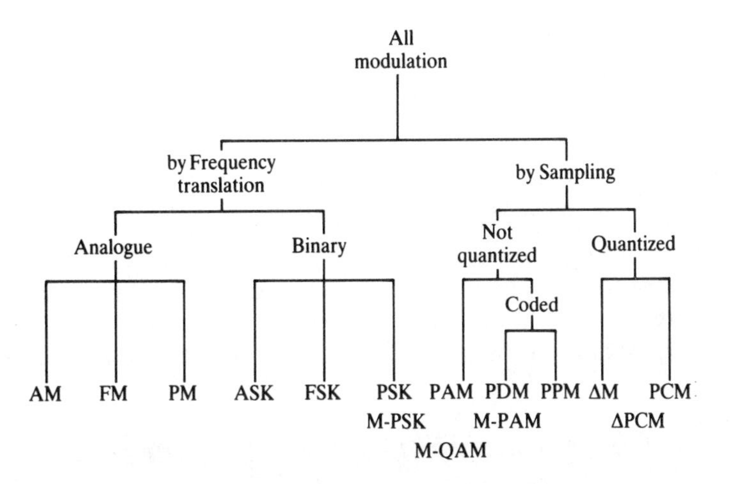

Figure 6.7 Family tree of modulation methods

PWM	Pulse width modulation, another name for PDM
ΔM	Delta modulation
PCM	Pulse code modulation
AΔPCM	Adaptive delta PCM
QAM	Quadrature amplitude modulation
M-...	Multilevel signal (e.g. M-QAM).

6.7 • Advantages of using modulation

There are several general reasons for modulating a baseband signal. Some have been mentioned already but are included again in this summary.

1. Advantages produced by frequency translation

(i) Use of frequency division multiplexing (FDM): This allows many signals to be sent simultaneously down the same communication channel. It gives economic use of equipment and enables complex systems to be designed.

(ii) Use of correct transmission frequency to give best transmission conditions: This is especially important in radio links where aerial efficiency increases with frequency and the best frequencies may need to be selected for propagation through the troposphere or via the ionosphere.

2. Advantages produced by sampling

(iii) Use of time division multiplexing (TDM): This allows many signals to be sent simultaneously along the same communication link by interleaving them in time. Gives similar economic and design advantages as (i).

3. Advantages produced by coding

(iv) Reliability of transmission greatly increased. Noise corruption very much less likely. Received baseband reproduces original signal very accurately.
(v) Signal processing much easier using standard logic and computing techniques. Facilitates design and production of complex systems at their most economic and reliable as in modern telephone systems.

In general it can be said that electrical and electronic communication systems would be virtually impossible without modulation. It is a vital part of most systems. The exact method used will depend on the application, and the next few chapters will look in detail at these various methods. But first we must consider the following.

6.8 • Analogue modulation – general

In analogue modulation a continuously varying analogue baseband signal changes some aspect of a carrier signal so that, when sent through a transmission system, the baseband can be recovered intact from the carrier. The carrier is a single frequency which can be represented generally as

$$v_c = E_c \cos(\omega_c t + \phi_c)$$

AMPLITUDE FREQUENCY PHASE

To modulate it something must be changed by the baseband signal. There are only three things which can be changed as shown above: amplitude, frequency or phase reference. Each of these leads to one class of **analogue modulation**:

Changing E_c	→	Amplitude modulation
Changing ω_c (i.e. f_c)	→	Frequency modulation
Changing ϕ_c	→	Phase modulation

A further general classification occurs if the carrier is rewritten as

$$v_c = E_c \cos(\theta_c)$$

From this you can see that both FM and PM alter the total phase angle (θ_c) in different ways. Thus they are closely related and are often classified together as **angle modulation**. Thus we have the three A's of continuous modulation:

Analogue = Amplitude + Angle

Finally, note that many books write the equation for the carrier in sine form as

$$v_c = E_c \sin(\omega_c t + \phi_c)$$

Do not be confused by this. It is merely the same form as used here but with a phase difference of $\pi/2$ at time zero as $\cos\theta = \sin(\theta + \pi/2)$. The cosine form seems to make the algebra marginally easier.

6.9 • Digital modulation – general

For these techniques the single common feature is the process of sampling – sending short bits of the analogue baseband at regular intervals. In the analogue sense above there is no carrier and the baseband not only changes greatly in form but also remains baseband in nature – then it effectively **becomes** the baseband.

In another sense the sampling signal (the one which chops the bits out of the analogue baseband) acts as the carrier, as we shall see. It is this act of sampling which is analysed first in the digital part of the book in Chapter 15.

6.10 • Summary

Signals are of two broad classes – analogue and digital. They vary greatly in waveform, bandwidth and modulation used depending on purpose (see Table 6.1).

A **baseband** is the original information signal in a communication system. The band of frequencies it occupies is the **baseband bandwidth** which usually starts from 0 Hz and is often quoted as a single figure, B Hz. Some very common ones are:

Telephone channel	4 kHz
Hi-fi music on VHF	15 kHz
TV in USA	6.5 MHz
TV in UK	8 MHz

Modulation refers to changes produced by or in the baseband to facilitate its transmission. There are two broad classes:

Frequency translation	AM, FM, PM, etc.
Sampling and coding	PAM, PCM, etc.

with a third group combining the two in

Binary (or keying) modulation	ASK, FSK, PSK, etc.

(see the end of Section 6.6 for the full meanings).

Analogue modulations are frequency translation methods caused by changing the appropriate quantity in a carrier signal as

$$v_c = E_c \cos(\omega_c t + \phi_c)$$

$$\begin{array}{ccc} \uparrow & \uparrow & \uparrow \\ \text{AM} & \text{FM} & \text{PM} \end{array}$$

Digital modulation is the result of changing analogue signals into binary ones by sampling and coding.

Keying modulations are digital signals subsequently modulated by frequency translation using one or other analogue method above.

6.11 • Conclusions

This chapter gives an overall picture of signals used and the purpose and types of modulation they can undergo. Now we start the long process of looking at these modulation methods in detail, starting with amplitude modulation in the next chapter.

6.12 • Problems

6.1 For standard telephone channels answer the following questions:

(i) Give the baseband frequency limits for a single telephone channel, then give its bandwidth and band limits if frequency translated by 60 kHz.
(ii) A standard telephone group consists of 12 adjacent channels. Give its bandwidth and its band limits if its lowest frequency channel is that in (i) above.
(iii) What is the width of each guard band in (ii)?
(iv) Give the bandwidth of a telephone supergroup (= five groups).
(v) How many standard teleprinter channels can be sent down one telephone channel?

6.2 Morse code is usually transmitted with a dash equal in length to three dots; a space of one dot length between symbols; three dot lengths between characters and five between words. A skilled operator can sent at 25 words per minute, assuming an average of 5 characters per word. Making reasonable assumptions calculate the overall symbol rate and assign a suitable baseband bandwidth.

6.3 A facsimile transmission system uses a cylinder 15.2 cm in diameter. The picture, on the outer surface of the cylinder, is scanned at 90 rpm with a lateral speed of 38 rev per cm. Each picture element is square. Calculate the minimum baseband bandwidth.

6.4 An image measuring $100 \times 50\ \text{cm}^2$ in area is made up of alternate "light" and "dark" elements 1.0 mm square. Calculate how long it will take to transmit this picture, by facsimile, through a channel band limited to 10 kHz.

6.5 One modern Data Facsimile machine claims to be able to transmit an A4 page up to 33 s using an information rate of 9600 baud. By making reasonable assumptions show that the scan rate is around 100 lines per second. What linewidth does that represent?

6.6 Assign a suitable baseband bandwidth for a 525 line TV signal using alternate line scanning at 60 Hz. Use 4:3 aspect ratio and square pixels.

6.7 The width-to-height ratio of a 405 line TV screen is 1.6:1. The horizontal definition is half that of the vertical definition. If the TV operates at 32 frames per second, determine the highest baseband frequency.

6.8 A facsimile document transmission system will send an A4 size picture in 6.5 s using pixels 1 mm square. Calculate the signal bandwidth required.

• 7 •
Amplitude modulation theory

7.1 • Introduction

In an amplitude-modulated signal the baseband information which is to be conveyed is impressed on to the carrier by varying its instantaneous amplitude. This leads to extra frequency components (the sidebands) which actually contain the information.

To economize in power and bandwidth some frequency components may be removed to give other forms of AM signal which do not show their AM nature as clearly.

7.2 • Types of amplitude modulation

There are three types:

1. "Full" amplitude modulation (AM)
2. Double sideband with suppressed carrier (DSBSC)
3. Single sideband (SSB).

All three types are analysed below and their waveforms and spectra illustrated as well as other aspects.

7.3 • "Full" amplitude modulation

This type of modulation was the first one in use in the very early days of broadcasting in the 1920s and has developed several names in the course of time. It is often called just simply amplitude modulation (AM) but sometimes envelope modulation or even double sideband with carrier (DSBWC). Also the term full AM is often used to mean maximum amplitude modulation (i.e. $m = 1$).

Whatever it is called, it is obtained by taking a single frequency carrier and altering its amplitude instantaneously in proportion to the instantaneous magnitude of a baseband signal. First take the simple situation where the baseband is also a signal frequency sinusoid. Then we can represent them both as

$$\text{Carrier} \quad v_c = E_c \cos \omega_c t$$
$$\text{Baseband} \quad v_m = E_m \cos\omega_m t$$

[Note that you may also find v_m written as $v_m(t)$ or $m(t)$ in different books.] Now we arrange for E_c to vary with v_m but with the condition that it never becomes negative because that would lead to overmodulation. This means that E_c is replaced by $E_c + E_m \cos \omega_m t$ with $E_m \leqslant E_c$, giving a modulated signal which is now written

$$\begin{aligned} v_{AM} &= (E_c + E_m \cos \omega_m t) \cos \omega_c t \\ &= E_c(1 + E_m/E_c \cos \omega_m t) \cos \omega_c t \\ &= E_c(1 + m \cos \omega_m t) \cos \omega_c t \end{aligned}$$

where m is the **modulation factor** and must have a value between 0 and 1 (to avoid E_c becoming negative as stipulated above). m is a very important characteristic of any system which uses full AM. It is often quoted as a percentage (e.g. $m = 0.6$ means 60% modulation) and may be written m_a, β_m or even k_m in other books. Some people use full AM to mean $m = 100\%$ only and not any level of envelope modulation as here. In the equation above

$$m = E_m/E_c$$

[But beware of using this definition as a basis for measurement of m. The two amplitudes are those at the point of modulation and not at the input terminals of the modulator or anywhere else.]

Thus the modulated signal is

$$v_{AM} = E_c(1 + m \cos \omega_m t) \cos \omega_c t$$

This produces **waveforms** as shown in Figure 7.1 for two values of m.

See how clearly this diagram shows the idea of this method as envelope modulation. Note, however, that the modulating waveform is not actually there; it just looks as if it is because the locus of the peaks of the modulated carrier follows it. It forms an envelope containing the carrier and its shape is very clearly defined, especially if $f_c \gg f_m$ so that the carrier peaks are very close together.

m can be **measured** directly from this waveform by measuring the peak-to-peak voltages A and B in Figure 7.2.

This can be seen as follows:

$$m = \frac{E_m}{E_c} = \frac{\frac{1}{2}(E_c + E_m) - \frac{1}{2}(E_c - E_m)}{\frac{1}{2}(E_c + E_m) + \frac{1}{2}(E_c - E_m)} = \frac{\frac{1}{4}A - \frac{1}{4}B}{\frac{1}{4}A + \frac{1}{4}B}$$

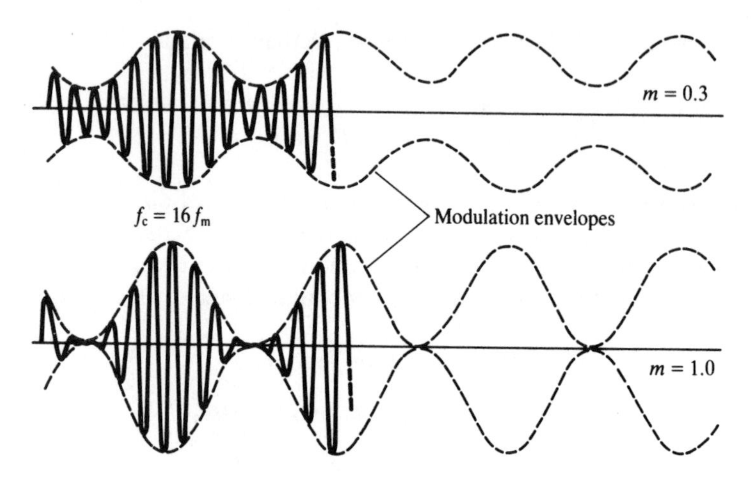

Figure 7.1 Full AM waveforms

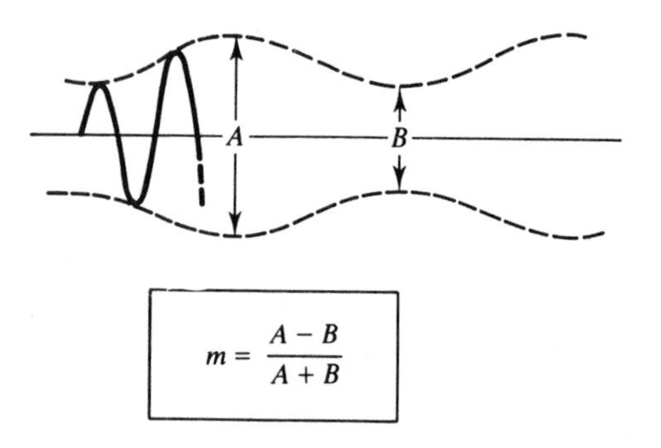

Figure 7.2 Measuring modulation factor

Therefore

$$m = \frac{A - B}{A + B}$$

This modulated carrier is obviously not a single frequency. Its spectrum can be obtained from the following analysis:

$$\begin{aligned} v_{AM} &= E_c(1 + m\cos\omega_m t)\cos\omega_c t \\ &= E_c\cos\omega_c t + mE_c\cos\omega_m t\cos\omega_c t \\ &= E_c\cos\omega_c t + \tfrac{1}{2}mE_c[\cos(\omega_c - \omega m)t + \cos(\omega_c + \omega_m)t] \end{aligned}$$

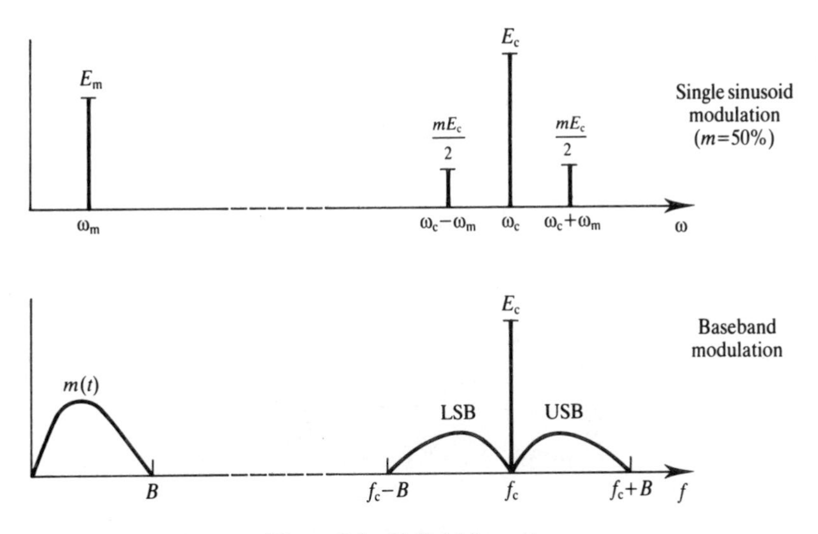

Figure 7.3 Full AM spectra

or

$$v_{AM} = E_c \cos \omega_c t + \frac{mE_c}{2} \cos(\omega_c - \omega_m)t + \frac{mE_c}{2} \cos(\omega_c + \omega_m)t$$

Now, in addition to the original unmodulated carrier, there are two new frequencies present, one greater than f_c by the amount of the modulating frequency, f_m, and one less than it by the same amount. These are the **upper and lower side frequencies**, $(f_c + f_m)$ and $(f_c - f_m)$, each of amplitude $m/2$ less than E_c, and are shown in Figure 7.3 together with the analogous situation for a baseband which is a **band of frequencies** and not just a single sinusoid. For both cases the baseband is shown as well but is not, of course, part of the modulated spectrum.

The lower spectrum is the more usual practical situation because most actual basebands have a spread of frequencies and often contain a continuous band of frequencies from near zero up to some band limit of B Hz (e.g. voices). Then each frequency of the band produces its own side frequency pair above and below f_c. The effect is to produce **sidebands**, the **upper** one of which (USB) has the same spectral shape as v_m whilst the **lower** one (LSB) has this shape reversed in mirror-image form.

But **be warned** that, although widely used, there is some confusion in this second spectrum type because the baseband is now a spectral density distribution with a vertical axis of volts per hertz (V Hz^{-1}). The single frequency carrier is still plotted with a vertical axis of volts only so that the modulated signal has two different vertical scales on the same plot. Thus this spectrum is really a hybrid which must only be used for what it is – a useful diagrammatic way of illustrating what happens. Do not use it quantitatively – for example, the upper spectrum can show clearly that the side frequency cannot have amplitude greater than $E_c/2$, but this cannot be done with the lower one.

So let us safely take the upper single frequency situation and show how little power there is in the sidebands, which carry the baseband information. This is done as follows:

$$\text{carrier power} = \left(\frac{E_c}{\sqrt{2}}\right)^2 = \frac{E_c^2}{2}$$

$$\text{side frequency power} = 2\left(\frac{mE_c}{2\sqrt{2}}\right)^2 = \frac{m^2E_c^2}{4}$$

$$\frac{\text{carrier power}}{\text{s.f. power}} = \frac{E_c^2/2}{m^2E_c^2/4} = \frac{2}{m^2} \geqslant 2 \quad \text{for all } m$$

Thus at least two-thirds of the transmitted power is in the carrier and conveys no information. Therefore it can be dispensed with (see Section 7.4).

m can also be **measured** from the modulated spectrum (as well as from the waveform, Figure 7.2) merely by taking the ratio of the sideband amplitude to the carrier amplitude (for a single sinusoid baseband, of course). That is

$$\frac{\text{sideband amplitude}}{\text{carrier amplitude}} = \frac{mE_c/2}{E_c} = \frac{m}{2}$$

In practice this particular measurement is done on a spectrum analyser using the log amplitude (dB) scale so that the ratio is the difference (in dB) as

$$\text{difference (in dB)} = 20\log_{10}(\text{ratio}) = 20\log_{10}(2/m)$$

$$m = \frac{2}{\text{antilog(diff./20)}}$$

[For example, 6 dB makes $m = 2/\text{antilog}(0.3) = 2/2 = 1$ or 100%.]

The **bandwidth** needed to transmit a full AM signal can also be seen from the spectrum. It is $2B$ Hz which is twice that of the baseband. That is

$$B/W = 2B$$

Again this is using more of one's resources than is necessary because v_m occurs independently in each sideband, so only one really needs to be transmitted to send the information. This would reduce the bandwidth by half (see Section 7.5).

A full AM signal, with single sinusoid baseband, can be well described graphically using a quasi-**stationary phasor** picture as in Figure 7.4.

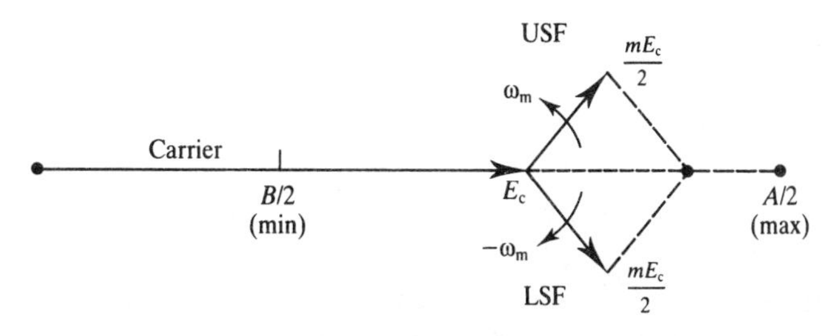

Figure 7.4 Phasor representation of full AM

The carrier itself is represented by a stationary phasor of constant amplitude E_c and at phase zero. Then the side frequencies are smaller phasors rotating symmetrically about the end of E_c at the same speed but in opposite directions, that is at ω_m and $-\omega_m$ as shown. Each will be of length $\frac{1}{2}mE_c$ and of opposite phase relative to E_c. At any instant the actual signal amplitude will be the phasor sum of these three so that the points marked $A/2$ and $B/2$ correspond to the maximum and minimum amplitudes already shown in Figure 7.2. Compare this with the phasor diagram for NBFM in Chapter 12.

7.4 • Double sideband suppressed carrier (DSBSC) modulation

On the whole full AM is fairly easy to visualize and, as it turns out later, very easy to demodulate, but it does have the two disadvantages mentioned earlier: it wastes both power and bandwidth. Power sent as carrier contains no information. Each sideband carries the information independently giving unnecessary duplication.

Part of this problem can be overcome by transmitting only the sidebands. One way to do this would be to remove the carrier from a full a.m. signal, leading to the usual name for this type of modulation – **double sideband suppressed carrier** or DSBSC for short. In practice it is usually obtained more directly by multiplying the carrier and baseband signal together in a balanced modulator. That is

$$v_{\mathrm{DSBSC}} = v_{\mathrm{AM}} - v_c$$

Also

$$\begin{aligned} v_{\mathrm{DSBSC}} &= v_m \times v_c \\ &= E_m \cos \omega_m t \times E_c \cos \omega_c t \\ &= \tfrac{1}{2} E_m E_c [\cos(\omega_c - \omega_m)t + \cos(\omega_c + \omega_m)t] \end{aligned}$$

which is most simply written as

$$v_{\mathrm{DSBSC}} = E_{\mathrm{D}} \cos(\omega_c - \omega_m)t + E_{\mathrm{D}} \cos(\omega_c + \omega_m)t$$

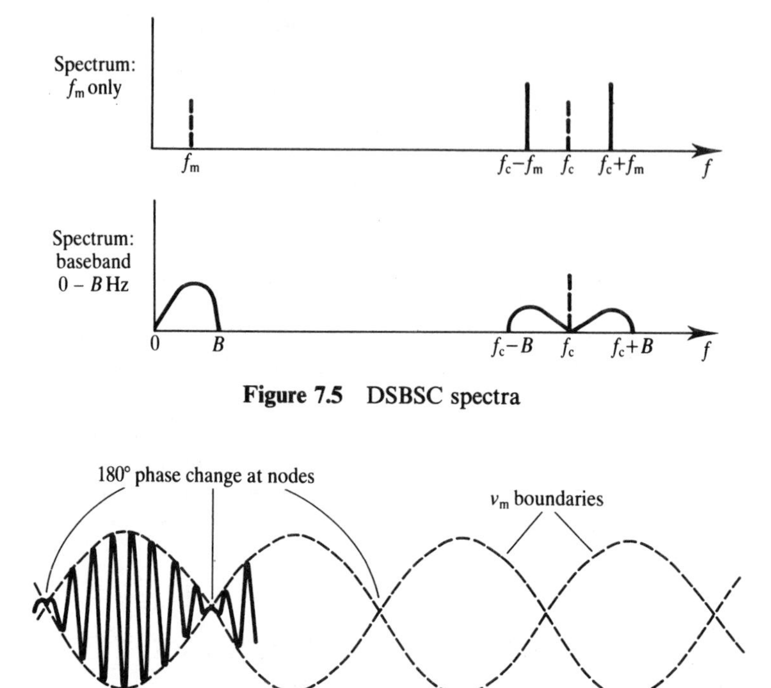

Figure 7.5 DSBSC spectra

Figure 7.6 DSBSC waveform for single sinusoid baseband

E_D is merely an amplitude factor proportional to both E_c and E_m with the dimensions of voltage not voltage squared (a popular misconception). This gives the characteristic double sideband spectrum as in the upper part of Figure 7.5.

The waveform itself is even more characteristic and is shown in Figure 7.6 for a single sinusoid baseband.

This waveform is merely the carrier signal with its instantaneous amplitude fixed by the baseband voltage. The zero crossing times stay fixed but the peaks vary in height and position because the instantaneous signal amplitude is the baseband voltage. Thus the modulated signal fits within an envelope of the baseband waveform, positive and negative. One effect of this is that the signal has an abrupt 180° phase change when the baseband passes through zero and changes sign. This occurs at the nodes of the waveform and is illustrated in Figures 7.6 and 7.7. Another way to look at the characteristic "blobby" waveform in Figure 7.6 is that it is the beat signal between two signals close together in frequency – the sidebands.

Also included in Figure 7.7 is the zero signal part of a waveform of a full AM signal for which $m = 1$. Both that and the DSBSC nodes are very similar in appearance but with two **important differences**:

1. Envelope zeros occur every $\lambda_m/2$ for DSBSC but only every λ_m for full AM.
2. Envelope lines actually cross for DSBSC but touch asymptotically for full AM.

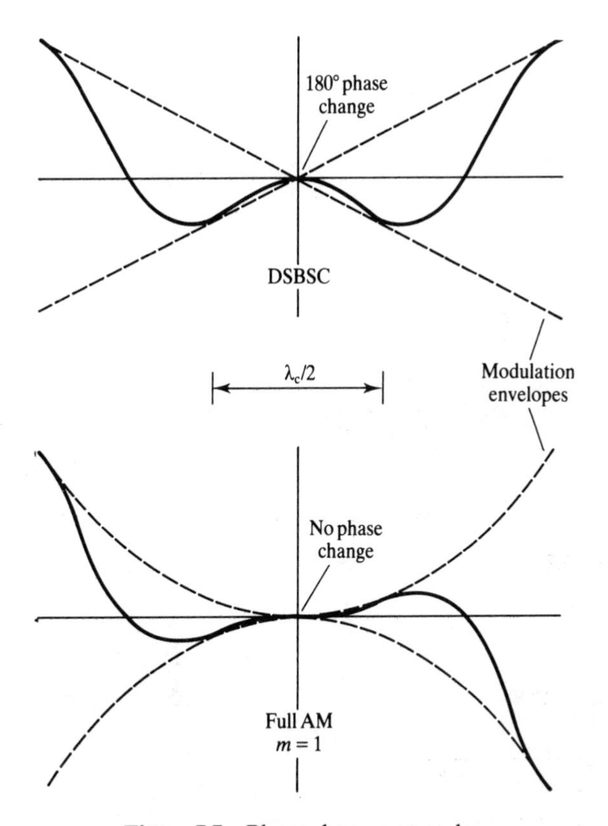

Figure 7.7 Phase changes at nodes

Of course, for a **non-sinusoidal baseband** the spectrum and waveform are more complicated in a very similar way to that for full AM. Figure 7.5 has already shown a spectrum and now Figure 7.8 shows a waveform.

As can be seen from the spectra the **bandwidth** of a DSBSC signal is exactly the same as that for a full AM one, that is

$$B/W = 2f_{max}$$

One familiar use of this method of modulation is in subcarrier modulation of the L–R signal in stereo VHF FM radio.

7.5 • Single sideband modulation

The final simplification in amplitude modulation is to send one sideband only. This is possible because it contains all the signal information (E_m and f_m) with less signal power and half the bandwidth.

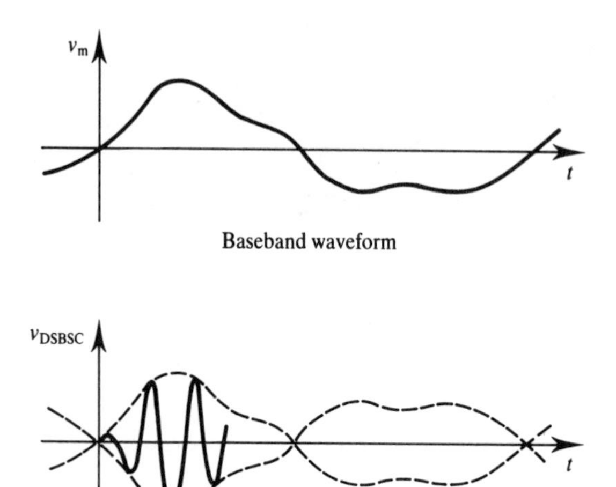

Baseband waveform

DSBSC waveform and envelope

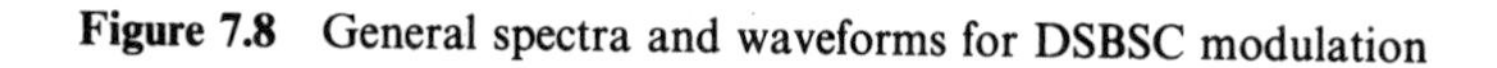

Figure 7.8 General spectra and waveforms for DSBSC modulation

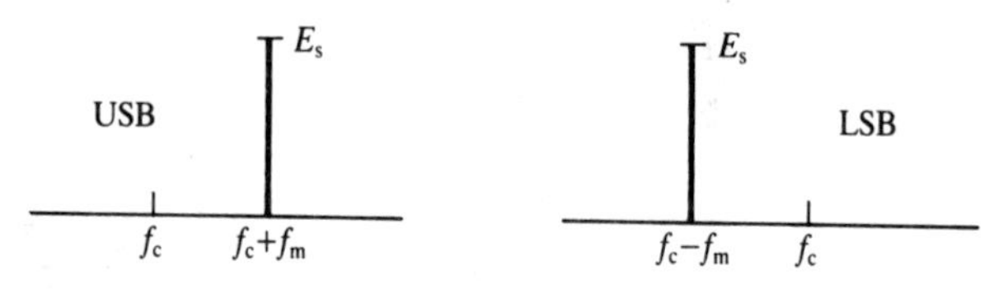

Figure 7.9 SSB spectra (after Holdsworth and Martin, 1991)

The simplest way to obtain an SSB signal is, in principle at least, to take a DSBSC signal and remove one sideband by filtering. This leaves the other sideband only – either the upper or lower – producing the two possible SSB signals. For a single sinusoid baseband these can be written simply as

$$E_S \cos(\omega_c - \omega_m)t \quad \text{the } \textbf{lower sideband} \text{ (LSB)}$$
$$E_S \cos(\omega_c + \omega_m)t \quad \text{the } \textbf{upper sideband} \text{ (USB)}$$

where E_S is proportional to E_c and E_m as for DSBSC. These produce spectra as in Figure 7.9.

Relevant waveforms are given in Figure 7.10 and, at first sight, seem to have little meaning, looking just like the unmodulated carrier. On close inspection, in a suitable experimental set-up, it is possible to see that the frequency varies by very small amounts (i.e. f_m changes; hence so does $f_c + f_m$ or $f_c - f_m$, but not much because $f_c \geqslant f_m$). It is easier to see that the signal amplitude changes proportional to changes in baseband voltage (v_m).

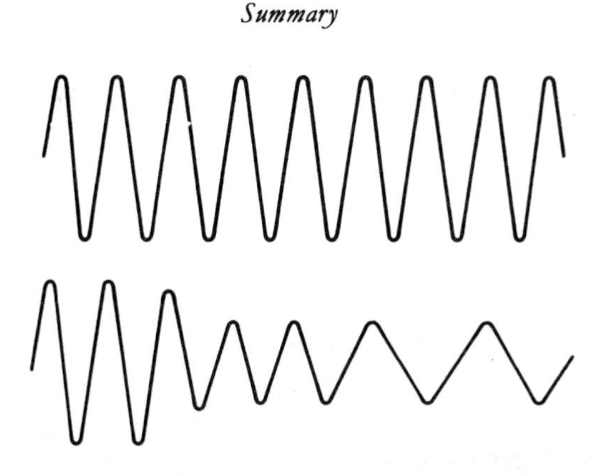

Figure 7.10 SSB waveforms: upper, for an unchanging baseband; lower, with variations in amplitude and frequency of the baseband (after Holdsworth and Martin, 1991)

The bandwidth is halved to become

$$B = f_{\max} = \text{baseband bandwidth}$$

SSB is used in many applications. One we use nearly every day occurs in our telephone system where signals are LSB modulated on to suppressed subcarriers to enable many to be sent down the same communication channel in FDM. Another well-known application is in radio communication where SSB is used to conserve bandwidth – a precious commodity.

7.6 • Summary

There are three types of amplitude modulation:

1. Full AM
2. Double sideband suppressed carrier (DSBSC)
3. Single sideband (SSB).

For a single sinusoid baseband a carrier modulated by them is given by the following.

Full AM

$$\begin{aligned} v_{\mathrm{AM}} &= E_c(1 + m\cos\omega_m t)\cos\omega_c t \\ &= E_c\cos\omega_c t + \tfrac{1}{2}mE_c\cos(\omega_c - \omega_m)t + \tfrac{1}{2}mE_c\cos(\omega_c + \omega_m)t \end{aligned}$$

m is the modulation factor:

$$m = \frac{E_m}{E_c}$$

at the point of modulation ($m \leqslant 1$)

$$m = \frac{A - B}{A + B}$$

on the waveform (see Figure 7.2)

$$m = \frac{2 \times \text{s.f. amplitude}}{\text{carrier amplitude}}$$

on the spectrum (see Figure 7.3), and

$$m = \frac{2}{\text{antilog (amp. ratio in dB/20)}}$$

$$B/W = 2f_m = 2B$$

where B is the baseband bandwidth.

DSBSC

$$v_{DSBSC} = E_D \cos(\omega_c - \omega_m)t + E_D \cos(\omega_c + \omega_m)t$$

The waveform has a characteristic 180° carrier phase reversal at zero crossings of the modulation envelope (Figures 7.6 and 7.7):

$$B/W = 2f_m = 2B$$

SSB

Either

$$v_{SSB} = v_{LSB} = E_S \cos(\omega_c - \omega_m)t$$

or

$$v_{SSB} = v_{USB} = E_S \cos(\omega_c + \omega_m)t$$

$$B/W = f_m = B$$

For stationary phasor representation see Figure 7.4. For the effect of actual basebands see Figure 7.3 etc.

7.7 • Conclusion

Two of these amplitude modulation methods are in very common use: full AM for broadcasting, and SSB for telephone and radio communication. But they all have serious deficiencies in dynamic range and in noise immunity. To improve matters we need to go to an entirely different method of modulation – FM – as described later, but first we will discuss amplitude modulators and demodulators.

7.8 • Problems

7.1 Full AM is produced by a signal, $v_m = 3.0\cos(2\pi \times 10^3)t$ volts, modulating a carrier, $v_c = 10.0\cos(2\pi \times 10^6)t$ volts. Calculate:

(i) m.
(ii) Side frequencies and bandwidth.
(iii) Ratio of sideband amplitude to carrier amplitude.
(iv) Maximum and minimum peak-to-peak amplitude of the modulated waveform.
(v) The percentage of the total power in the side frequencies.

7.2 An AM signal voltage is given by $v = 100\sin(2\pi \times 10^6)t + [20\sin(6250t) + 50\sin(12\,560t)]\sin(2\pi \times 10^6)t$:

(i) What type of amplitude modulation is being used?
(ii) Draw its amplitude spectrum.
(iii) Sketch its waveform over one modulation cycle showing quantitative values for important times and amplitudes.
(iv) Work out the peak and average powers into a $100\,\Omega$ load.
(v) Does m have values? If so, what are they?
(vi) What is the significance (if any) of using sines not cosines?

7.3 Repeat the last problem but with the carrier suppressed (questions (i) to (v) only).

7.4 Discuss the relative merits of the three main types of amplitude modulation. For each give at least one unique advantage relative to the other two, and one disadvantage.

7.5 A carrier, f_c, is modulated by a baseband extending continuously in frequency from f_1 to f_2, with amplitude proportional to frequency:

(i) Draw the baseband amplitude spectrum.
(ii) Draw the full AM spectrum.

(iii) Give the bandwidths for full AM and DSBSC.

7.6 A radar signal consists of 1 μs pulses of a 10 GHz carrier. The pulses are of constant amplitude and are spaced by 12 μs gaps during which nothing is transmitted.

(i) Sketch the modulation envelope.
(ii) Draw the modulation spectrum showing frequencies and relative amplitudes.
(iii) Assign a bandwidth giving reasons.
(iv) Repeat for an interval of 99 μs between pulses.

7.7 A signal is band limited to the frequency range 0–5 kHz. It is frequency translated by multiplying it by the signal $v_c = \cos 2\pi f_c t$. Find f_c so that the bandwidth of the translated signal is 1% of f_c.

7.8 A DSBSC transmitter delivers 10 W when the modulating signal is a d.c. voltage of 1.0 V. Determine the power delivered when the modulating signal is a tone of RMS value 1.6 V.

7.9 Show that the signal

$$v = \sum[\cos \omega_c t \,.\, \cos(\omega_i t + \theta_i) - \sin \omega_c t \,.\, \sin(\omega_i t + \theta_i)]$$

is an SSBSC signal. Is it upper or lower sideband? Write an expression for the other sideband. Obtain an expression for the total DSBSC signal.

7.10 Two transmitters, one generating a full AM signal and the other an SSB one, have equal mean output power ratings. A single sine wave modulating signal causes the AM signal to have a modulation factor of 0.8. If both transmitters are operating at maximum output, compare (in dB) the power contained in the sidebands of the full AM signal with that contained in the SSB one.

•8• Amplitude modulators

8.1 • Introduction

To achieve modulation, it is necessary to have a circuit in which two signals (carrier and baseband) can interact with each other (and not merely add). This requires some form of non-linear circuit in which the level of one signal provides a variable bias which decides the level of the other signal. Any of the standard active electronic devices can be used for this purpose, operating either non-linearly or as switches.

Until about 1960 these devices were always thermionic valves and they are still the best for some applications (e.g. radio transmitters). But nowadays mostly transistors (BJT or FET) are used, either as **discrete devices** or in specially designed **integrated circuits**. Most modulation circuits use these devices in either a form of switching circuit or ones which make use of the variable slope resistance (r_d or $1/g_m$) of the device characteristic.

Whatever basic circuit is used it is usually part of some form of **balanced modulator** reducing the number of modulation products by cancellation. Their primary output is simple product modulation or DSBSC and hence this can be regarded as the basic form of amplitude modulation. Full AM is then regarded as just unbalanced product modulation as we shall see later.

We start by looking at the basic ideas using simplified circuits containing "square-law" devices only.

8.2 • Square-law diode modulation

For small voltages the characteristic of a germanium or silicon diode can be represented by a simple square law as in Figure 8.1. This is, of course, very much an idealized approximation, but it is useful to show how modulation occurs in a non-linear device.

A simplified modulation circuit using one of these devices is shown in Figure 8.2. The output voltage is proportional to the diode current which is, in turn, related

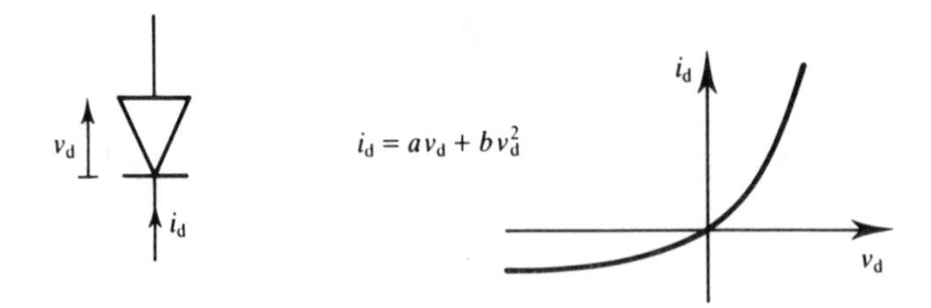

Figure 8.1 Square-law diode

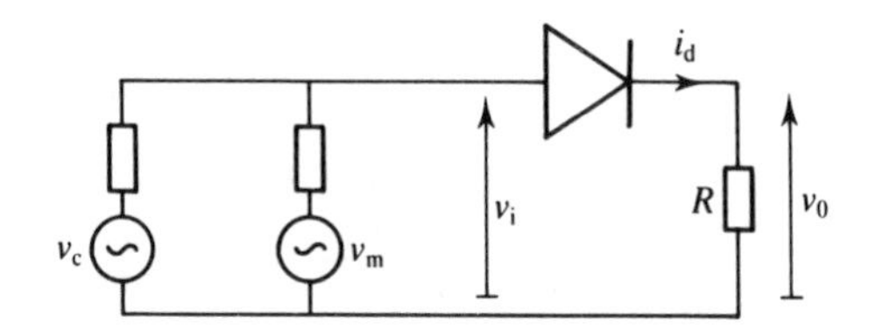

Figure 8.2 Simple diode demodulator

to the diode voltage by a quadratic relationship of the form

$$i_d = av_d + bv_d^2$$

The analysis proceeds as follows:

$$\begin{aligned}
v_O &= Ri_d \\
&= R(av_d + bv_d^2) \\
&\simeq R(av_i + bv_i^2) \qquad (v_i \simeq v_d \text{ as } R \text{ chosen to make } v_o \leqslant v_i) \\
&= Ra(v_c + v_m) + Rb(v_c + v_m)^2 \\
&= Rav_c + Rav_m + Rbv_c^2 + 2Rbv_cv_m + Rbv_m^2 \\
&= \underbrace{Rav_m + 2Rbv_m^2 + 2Rbv_c^2}_{\text{unwanted terms}} + \underbrace{Rav_c + 2Rbv_mv_c}_{\text{full AM}}
\end{aligned}$$

Thus v_o contains full AM plus other unwanted frequencies which can be removed by filtering. As an exercise replace v_c and v_m by single sinusoids and work out all the frequencies in v_o – then draw its spectrum.

A similar analysis can be done for a transistor, representing the output current as a square-law function of the input voltage as shown in Figure 8.3.

These characteristics are now expressed as

$$\begin{aligned}
&\text{for FET:} \quad i_D = av_{GS} + bv_{GS}^2 \\
&\text{for BJT} \quad i_c = av_{BE} + bv_{BE}^2
\end{aligned}$$

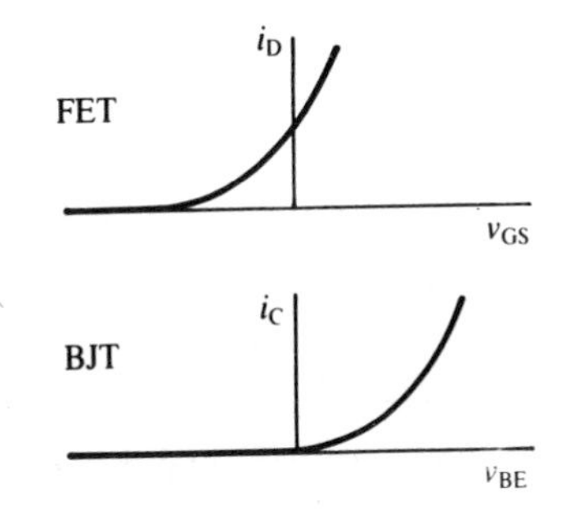

Figure 8.3 Non-linear transistor characteristics

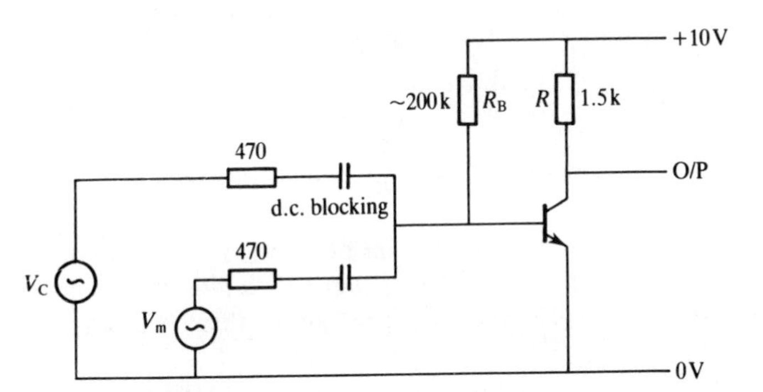

Figure 8.4 Simple transistor modulator

This is merely another way of saying that g_m varies with input voltage level. The advantage of using transistors is that gain occurs as well as modulation and it is easy to rig up a simple circuit which will show this (e.g. as in Figure 8.4).

8.3 • Balanced modulation

For the simple characteristic above it is easily possible to remove the unwanted frequencies by filtering, but "real" characteristics are more complex and filtering is not so easy. Try the same analysis as above but using a cubic law of the form below to illustrate the point:

$$i_D = av_D + bv_D^2 + cv_D^3$$

This problem is overcome by the technique of *balanced modulation* mentioned in Section 8.1. The principle of this method is that unwanted terms are removed by cancellation leaving only those frequencies which form the required modulation. This principle can be illustrated with the "square-law" diode of the last section when

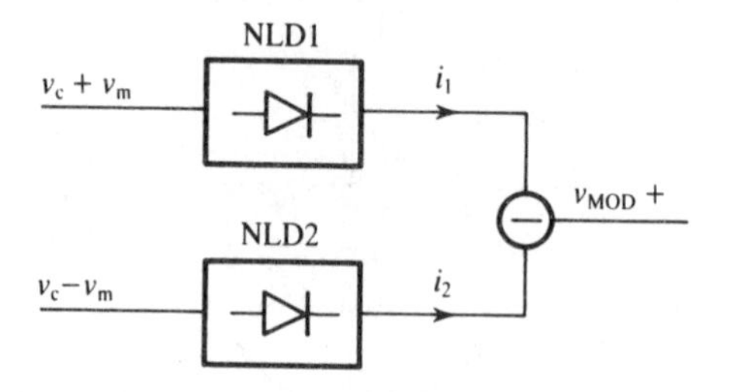

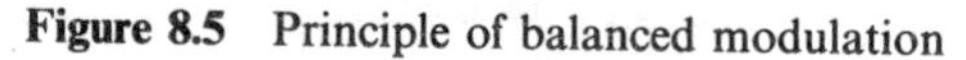

Figure 8.5 Principle of balanced modulation

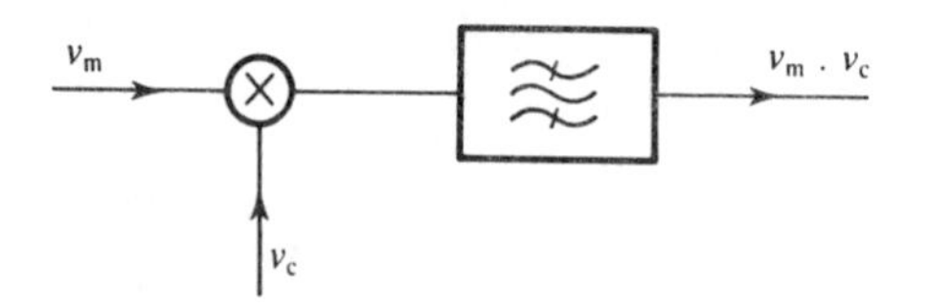

Figure 8.6 DSBSC by multiplier modulator

connected as two identical modulators in parallel as shown diagrammatically in Figure 8.5. One modulator is fed with the **sum** of the two input signals (carrier and baseband) and the other with their difference giving their two output voltages as

$$v_{o1} = Rav_m + Rbv_m^2 + Rbv_c^2 + Rav_c + 2Rbv_c v_m$$

$$v_{o2} = -Rav_m + Rbv_m^2 + Rbv_c^2 + Rav_c - 2Rbv_c v_m$$

Then

$$\begin{aligned} v_o &= v_{o1} - v_{o2} \\ &= 2Rav_m + 4Rbv_c v_m \end{aligned}$$

Now we have a much simpler output containing only DSBSC and the baseband. This is an example of a **single balanced modulator** whereby one of the original signals has been balanced out but not the other. [Try using $-(v_c + v_m)$ as the second input and show that you also get single balanced modulation but now with v_m balanced out. What are you left with?]

If both inputs are balanced out the circuit is said to be a **double balanced modulator** and produces a pure DSBSC output, but this cannot be done with the simple circuit of Figure 8.5.

The general principle of double balanced modulation by product modulator is shown in Figure 8.6. Actual circuits are described later.

Paradoxically full AM can be obtained from a balanced modulator merely by unbalancing it to allow the carrier through. This is easily done with the 1496 described later.

8.4 • Piecewise linear modulator

At larger input voltage levels, a diode characteristic approximates much more accurately to two straight lines, one of zero current along the negative voltage axis and the other of slope equal to the reciprocal of the load resistance in the forward direction. This is a **piecewise linear characteristic**, shown in Figure 8.7 together with a skeleton modulation circuit using it.

Modulation occurs because the effect on an input signal containing the sum of carrier and baseband is to rectify it, producing a pulsed signal, at the carrier frequency, containing many harmonics and modulation products. Among these will be the carrier and sidebands of full AM which can be filtered out. The basic operation is shown in Figure 8.8. However, this is a highly inefficient method producing a whole range of unwanted harmonics and sidebands.

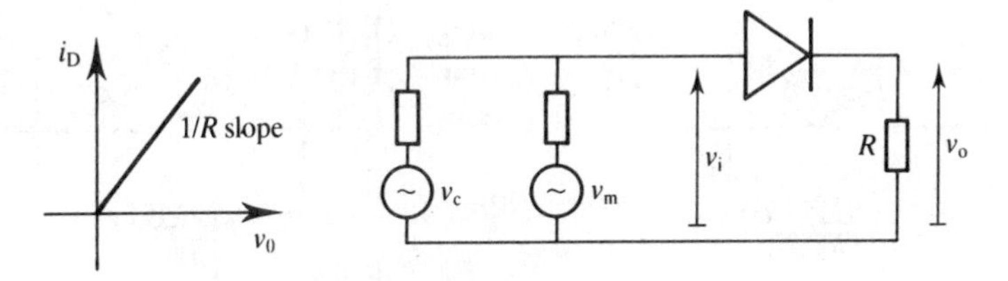

Figure 8.7 Modulation using piecewise linear diode

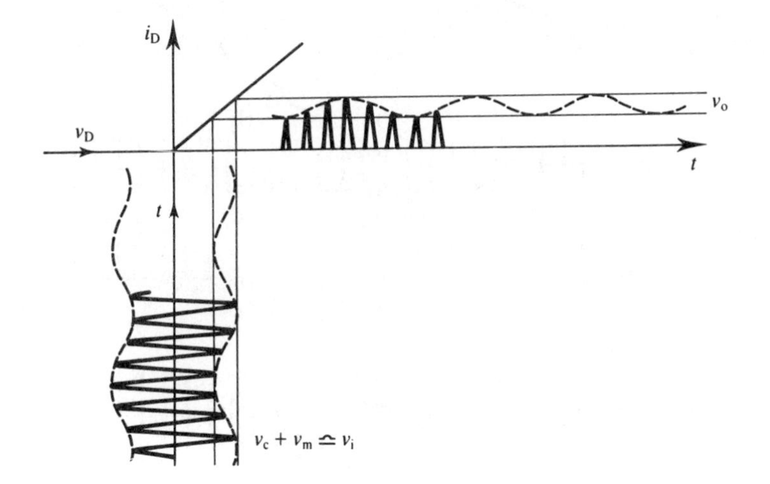

Figure 8.8 Operation of rectifying modulator

8.5 • Chopper modulators

Another way of using diodes to modulate is merely to use them as switches. In this case they are called **chopper modulators** because their action is to chop the baseband at the carrier rate, so introducing many harmonics and modulation products including full AM and/or DSBSC. What happens is that diodes are arranged in a circuit in such a way that they are switched by the carrier allowing the baseband through in square pieces at the carrier frequency. The two most common circuits are shown in Figure 8.9.

The Cowan circuit balances out f_c only to produce DSBSC but with v_m also present (as well as harmonics of f_c). Thus it is a **single balanced** modulator. It requires more filtering at the output than the Ring circuit which balances out both carrier and baseband and so acts as a **double balanced** modulator. Relevant waveforms are shown in Figure 8.10. Work out the action of these circuits for yourself.

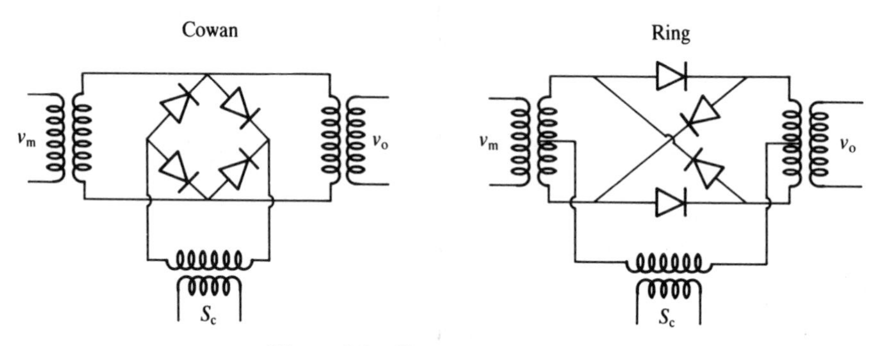

Figure 8.9 Chopper modulators

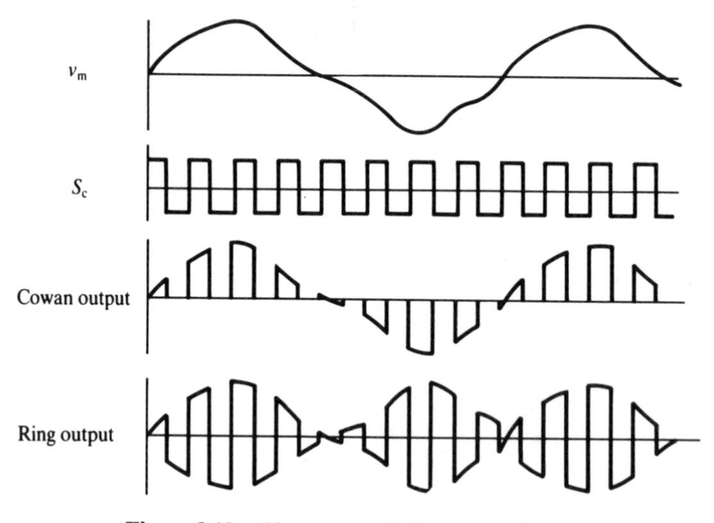

Figure 8.10 Chopper modulator waveforms

These diode rings are easily realized in integrated circuit form enabling them to be made with identical characteristics (e.g. turn-on voltage) to produce cheap reliable modulators.

8.6 • Variable transconductance modulators

Nowadays one of the easiest ways to modulate a carrier with a baseband signal is to use an integrated circuit such as MC1496 or SG1402. These are double balanced modulators using the variable transconductance properties of bipolar transistors, that is g_m varies with i_c. Being in integrated circuit (i.c.) form this technique not only uses standard bipolar technology but enables devices with acceptably identical properties to be produced. This is essential for the successful operation of the modulator. The output is, of course, DSBSC.

Its operation depends upon the fact that, to a good approximation, the value of g_m is proportional to the size of the total collector current, i_c, according to these simple theoretical relationships:

$$g_m = (e/kT)I_c \qquad (I_c \text{ is d.c. value})$$

therefore

$$g_m = I_c/v_T \qquad (v_T \text{ is "voltage equivalent" of temperature})$$

or

$$g_m = 40I_c \qquad (\text{at room temperature})$$

The **basic circuit** used is the long-tailed pair shown in outline form in Figure 8.11.

The transistors have identical characteristics which means that, with input voltage zero,

$$v_1 = 0$$

$$v_{BE1} = v_{BE2}$$

$$I_1 = I_2 = I_3/2$$

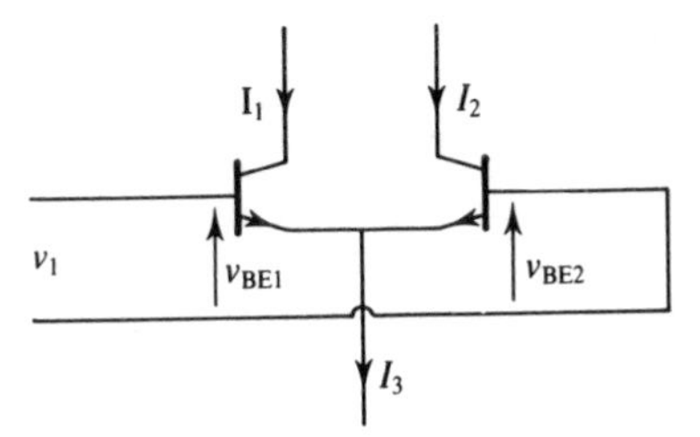

Figure 8.11 Basic long-tailed pair

Now apply a small input voltage, v_1. Because of the identical characteristics, this will divide equally across the two base–emitter junctions, producing the same value but of opposite signs. So if v_{BE1} changes by v_{be1} and v_{BE2} by v_{be2}, the I_1 and I_2 change by equal and opposite amounts i_1 and i_2 where

$$v_{be1} = -v_{be2} = v_1/2$$

$$i_1 = -i_2$$

also

$$i_1 = g_m v_{be1} = \tfrac{1}{2}40 I_1 v_1 = 10 I_3 v_1$$

That is

$$\boxed{i_1 = 10 v_1 I_3 = -i_2}$$

This means that the change of current in each transistor is proportional to the product $I_3 v_1$. This assumes small changes only in i_1, i_2, and v_1. For v_1 this criterion means $|v_1| < v_T$.

The next step towards obtaining product modulation is to arrange for a second input voltage v_2 to control the two collector currents I_1 and I_2. This is done by making I_1 and I_2 the tail currents of two more long-tailed pairs having the same differential input, v_2, to each. This arrangement is that of a **quad differential amplifier**. All six transistors must have identical characteristics. The circuit is now as shown in Figure 8.12.

A similar analysis to that for v_1 will show that, for a small input v_2, small current

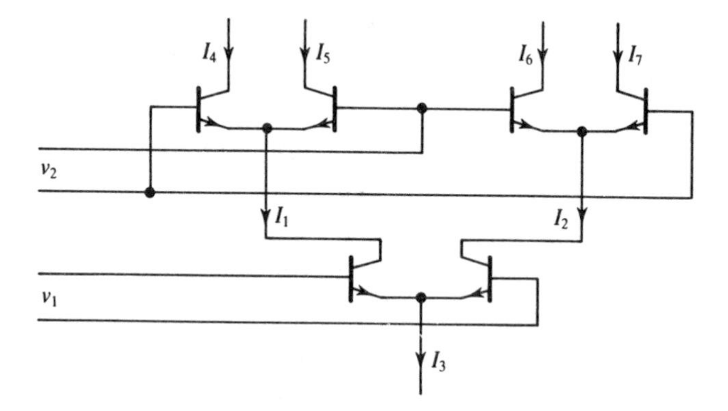

Figure 8.12 Basic multiplier circuit

changes occur in the collector currents of the quad amplifier (i_1, i_2, i_3, and i_4) such that

$$i_4 = 10v_2 I_1 = -i_5$$
$$i_6 = 10v_2 I_2 = -i_7$$

Therefore

$$I_4 = I_1/2 + i_4 = I_1(\tfrac{1}{2} + 10v_2)$$
$$= I_3(\tfrac{1}{2} + 10v_1)(\tfrac{1}{2} + 10v_2)$$

or

$$I_4 = I_3(\tfrac{1}{4} + 5v_1 + 5v_2 + 100v_1 v_2)$$

similarly

$$I_5 = I_3(\tfrac{1}{4} + 5v_1 - 5v_2 - 100v_1 v_2)$$

and

$$I_6 = I_3(\tfrac{1}{4} - 5v_1 + 5v_2 - 100v_1 v_2)$$

and

$$I_7 = I_3(\tfrac{1}{4} - 5v_1 - 5v_2 + 100v_1 v_2)$$

These currents contain terms in d.c., carrier, baseband, and, most importantly, the wanted modulation product, $v_1 v_2$. The unwanted terms are removed by adding pairs of currents and subtracting them according to the equation

$$\boxed{(I_4 + I_7) - (I_5 + I_6) = 400 I_3(v_1 v_2)}$$

You may care to check the algebra for yourself. Strictly the factor 400 is $\frac{1}{4}(v_T^{-2})$ where $v_T = kT/e \simeq 1/40$ at room temperature. As we only want an output proportional to the product $v_1 v_2$ the exact value of this constant does not matter.

The actual output is obtained as a differential voltage between the live ends of two 3.9 kΩ loads through which the current pairs $((I_4 + I_7)$ and $(I_5 + I_6))$ flow, as shown on the manufacturer's circuit schematics in Figure 8.13.

There is also provision for altering I_3 (using bias) which alters the gain, and for adjusting the balance to remove the carrier from the output. The carrier is always v_2 and the baseband signal, v_1.

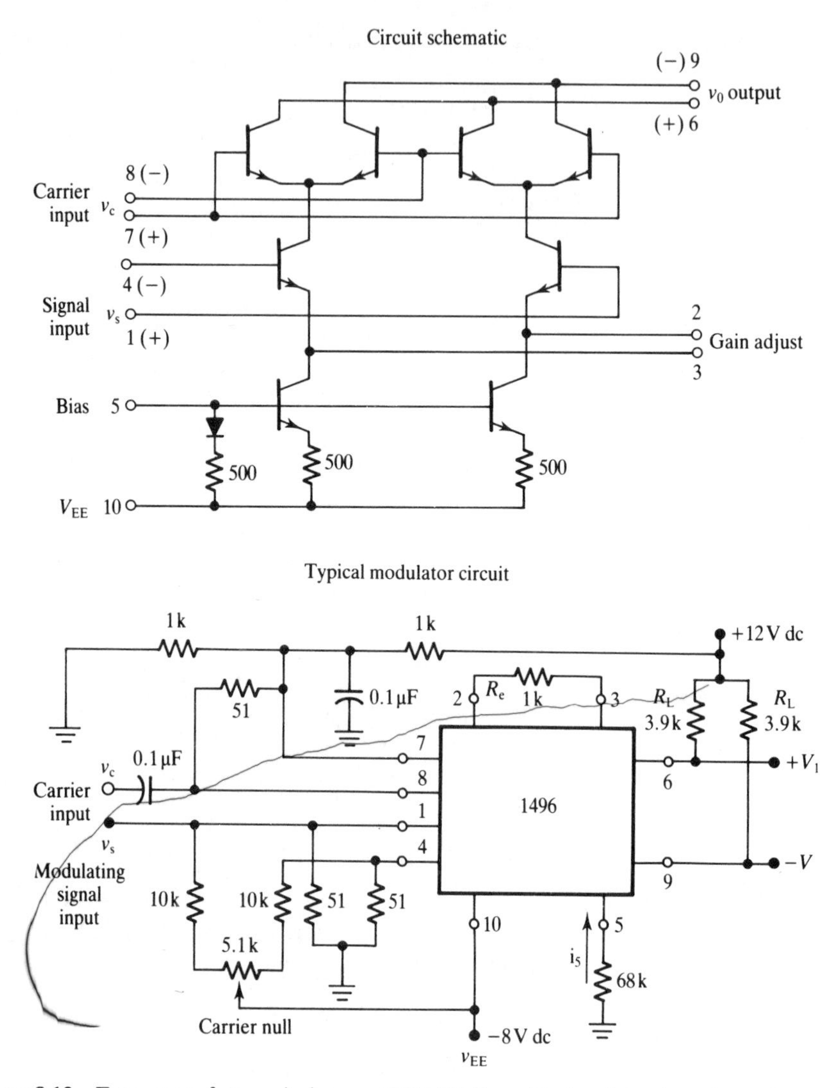

Figure 8.13 From manufacturer's data on MC 1496 (from National Semiconductor Data Sheet)

Using circuits very similar to that in the bottom half of Figure 8.13 a 1496 can be used to obtain both DSBSC (directly) and full AM (by unbalancing the carrier). SSB can also be obtained fairly easily by adding a filter to remove one of the sidebands from a DSBSC signal. Thus it is a very useful i.c. indeed. Then, in addition, it can be used as a mixer to reduce a carrier to an intermediate frequency in many types of radio receiver and also to demodulate it coherently as we shall see in Chapters 9 and 16. For larger signals this circuit's operation can also be explained as that of a chopper modulator.

8.7 • Modulation using balanced modulators

Here is a summary of the ways in which double balanced modulators can be used to obtain all three types of amplitude modulation (Figure 8.14). The balanced modulator itself basically produces a DSBSC output.

The fourth (**filter**) method for SSB is self-explanatory, but the first needs more explanation. This **phase shift method** obtains SSB by synthesizing it from the equation

$$\begin{aligned}\text{LSB} &= \cos(\omega_c - \omega_m)t = \cos\omega_c t \cos\omega_m t + \sin\omega_c t \sin\omega_m t \\ &= \cos\omega_c t \cos\omega_m t + \cos(\omega_c t - \pi/2)\cos(\omega_m t - \pi/2)\end{aligned}$$

This is just the sum of two products, one of the carrier and the baseband, and the

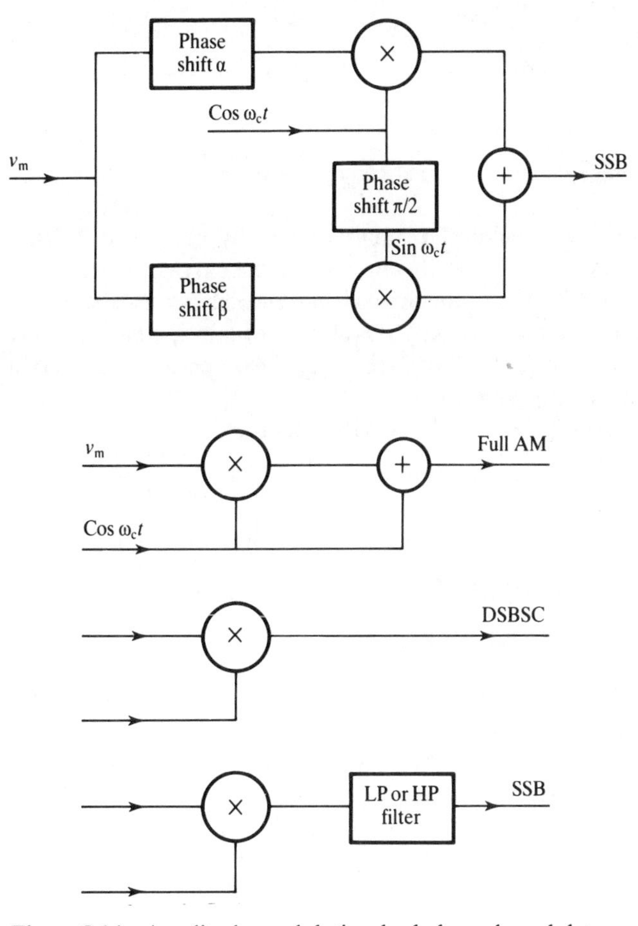

Figure 8.14 Amplitude modulation by balanced modulator

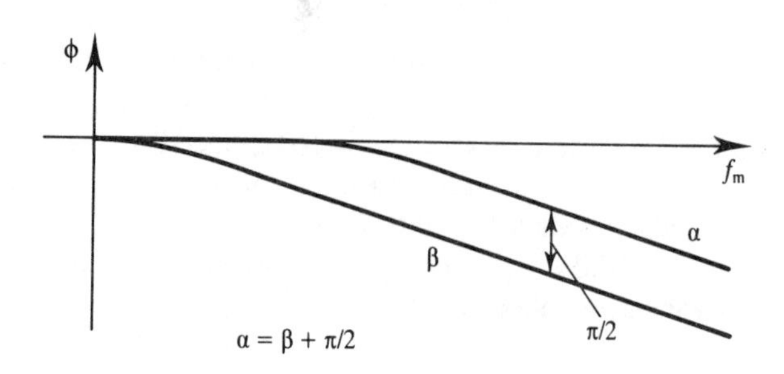

Figure 8.15 Obtaining constant phase shift over a band

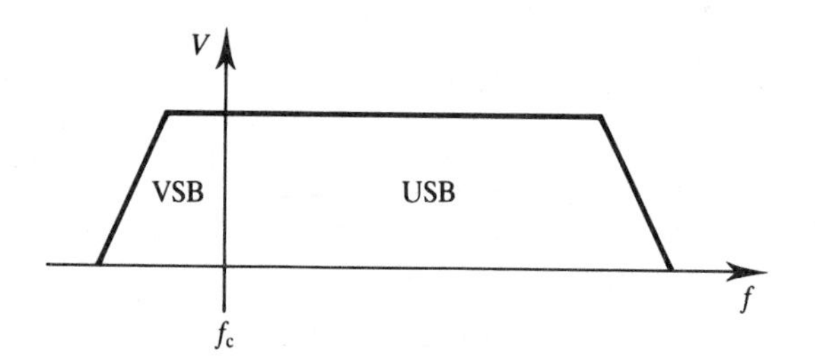

Figure 8.16 TV signal spectrum showing vestigial sideband

other of the same two signals but phase shifted by $\pi/2$ (their Hilbert transforms). This phase shift can be done easily enough for the carrier, which is at a single fixed frequency, but is much harder for the baseband which nearly always contains a band of frequencies. In practice it is only possible to obtain a constant phase shift over a band by using two phase shift networks whose **difference** of shift is constant (at $\pi/2$ here) as illustrated in Figure 8.15.

The LSB sum of products above then becomes

$$\begin{aligned}
&\cos \omega_c t \cos(\omega_m t + \alpha) + \cos(\omega_c t - \pi/2) \cos(\omega_m t + \beta) \\
&= \cos \omega_c t \cos(\omega_m t + \alpha) + \cos(\omega_c t - \pi/2) \cos(\omega_m t + \alpha - \pi/2) \\
&= \cos \omega_c t \cos(\omega_m t + \alpha) + \sin \omega_c t \sin(\omega_c t + \alpha) \\
&= \cos[(\omega_c - \omega_m)t - \alpha]
\end{aligned}$$

Thus LSB is obtained. The phase shift, α, varies linearly with signal frequencies and has no effect on most signals.

There is also a third method of **Weaver and Barber** which is more complicated but avoids the phase shift problem.

There are two variations of SSB which need mentioning. One is **vestigial sideband** (VSB) used in the transmission of TV signals. Most of the video baseband is sent as spectral lines within a USB modulation envelope above the vision carrier, but some (a vestige?) of the lower sideband is also sent as in Figure 8.16.

This allows the d.c. component to be received as well as preventing phase distortion of the low-frequency components. The modulation bandwidth only increases by 1.25 MHz. The second is the use of **independent sideband**. This is an apparently double sideband signal but with each sideband carrying a different baseband. It puts two information signals on one (suppressed) carrier with no increase of bandwidth.

8.8 • Modulation at high powers

Full AM can be produced directly by modulating E_c and this method is commonly used in power situations such as radio transmitters. The modulation is done at the last (class C) stage of amplification of the r.f. carrier as in Figure 8.17 where the circuit is operating in a highly non-linear state.

The modulation is achieved by using the baseband signal to alter the instantaneous power supply voltage to the output stage. Figure 8.18 shows a circuit which does this.

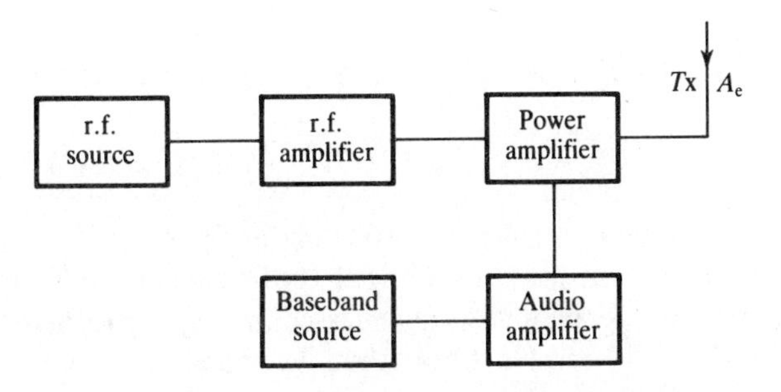

Figure 8.17 Direct modulation at output stage

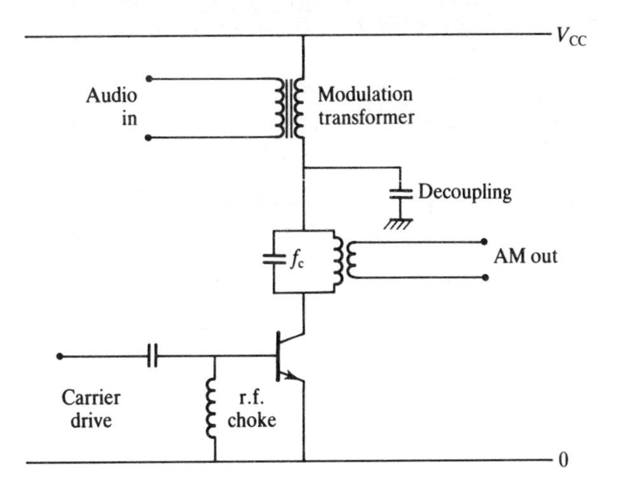

Figure 8.18 Full AM by modulating V_{cc} direct

8.9 • Summary

To achieve modulation, signals have to interact in some form of non-linear circuit. The non-linearity can be abrupt (e.g. chopper modulators) or continuously variable (e.g. "square-law" diodes). Several types of each are described leading to present-day i.c. forms. SSB also uses synthesis methods.

The other important idea in this chapter is that of balanced modulators in which unwanted modulation products are removed by cancellation.

8.10 • Conclusions

Naturally, having discovered how to modulate a carrier, we now need to know how to demodulate it to recover the baseband signal. In some ways this is a much more straightforward business, as we shall see in the next chapter.

8.11 • Problems

8.1 A non-linear circuit has an input voltage, v_i, and an output voltage, v_o, where

$$v_o = 0.5v_i + 0.1v_i^2$$

v_i contains a carrier of 1.0 V p–p at 100 kHz, and a tone of 0.7 V p–p at 5 kHz. Write down an expression containing all the frequencies and their amplitudes in v_o and draw its spectrum. v_o is now passed through an ideal low-pass filter of passband 65 to 135 kHz. Show that the signal spectrum is now a full AM one with $m = 0.28$.

Now find the largest values of modulation tone and frequency which can be used without producing distortion or interference in the filter output.

8.2 Figure 8.19 shows a non-linear modulation system and its characteristics. The input, v_i, consists of an 80 kHz carrier of 20 V p–p amplitude and a 10 kHz

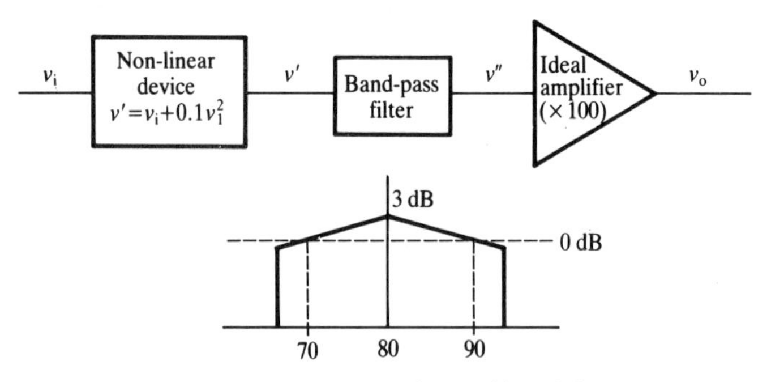

Figure 8.19 System in Problem 8.2

signal of 2.0 V p–p amplitude. Determine the amplitude and frequencies of all components appearing in the filter output. Give the values of the modulation factor before and after the filter. Calculate the ratio of sideband power to total power in the output of the ideal amplifier.

8.3 A system similar to that in Figure 8.19 has an ideal bandpass filter of limits 70 and 90 kHz. The characteristic of the modulator is now given by

$$v_o = 0.1v_i + bv_i^2 + cv_i^3$$

and the input is given by

$$v_i = 10\cos(2\pi 80 \times 10^3 t) + 1.0\cos(2\pi \times 10^4 t)$$

Show that the full AM output has $m = 1$ if $b = 0.05$ and $c = 0$. For this value of b, show that the value of c which makes the highest harmonic of f_m has only one-tenth the amplitude of f_m itself, is 6.8×10^{-3}.

8.4 A diode has the ideal characteristic of

$$I = I_o[\exp(40V) - 1]$$

(i) What value of V produces a 1% distortion in I?
(ii) Expand I as a power series up to V^3.
(iii) If $V = 0.1\cos\omega_1 t + 0.1\cos\omega_2 t$, list the frequencies appearing in I and their relative magnitudes using the model in (ii).

8.5 Two identical non-linear modulators are used to produce DSBSC in the system shown in Figure 8.20. Each has a characteristic of

$$v_{OUT} = 0.5v_{IN} + 0.2v_{IN}^2 + 0.1v_{IN}^3$$

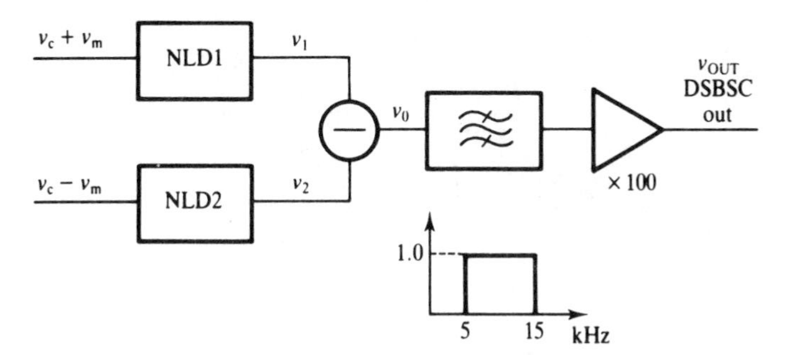

Figure 8.20 Circuit for Problem 8.5

The inputs are given by

$$v_c = 2.0\cos(2\pi \times 10^4 t) \quad \text{and} \quad v_m = 0.2\cos(2\pi \times 10^3 t)$$

Sketch the spectrum of the DSBSC output and calculate its available power (into $1\,\Omega$).

Now let f_m increase to the largest value it can have without causing distortion or interference in the output.

8.6 Explain how modulation occurs in the piecewise linear modulator in Figure 8.7. If the carrier is exactly half wave rectified when $v_m = 0$ estimate m when $E_m = E_c/10$. [Assume product modulation by unchanged f_m of Fourier harmonics of rectified f_c.]

8.7 Product modulators are sometimes called "balanced modulators".

(i) What is meant by the term "balanced" here?
(ii) Why is balanced used?
(iii) What is a double balanced modulator?
(iv) What features are required for circuits to produce balanced modulation?
(v) How are these requirements achieved in i.c. product modulators?

8.8 Explain the principle behind the product modulation of two signals using the principle of variable transconductance from the circuits in Figure 8.13. Calculate the value of "I_3" (see Fig. 18.12) and hence find the value of the output voltage when the two inputs each have 10 mV p–p amplitude.

8.9 Estimate the order of the filter required to separate the lower sideband from the upper one when a standard telephone channel is product modulated on to a 64 kHz carrier.

•9•

Amplitude demodulation

9.1 • Introduction

Demodulation is the process of recovering the original baseband from the modulated carrier when it has reached the receiving end of the transmission system. For historical reasons it is often called "detection" and both terms will be used interchangeably.

For a.m. signals there are **two** basic types of demodulator: envelope detectors and coherent demodulators. Each has its particular applications and both are described below.

9.2 • Envelope detection

This is a very common method with a simple basic circuit and many different applications. It is restricted to signals where the baseband waveform is the envelope of the modulated carrier voltage excursions (i.e. full AM); hence the name of **envelope detector**. An obvious application is in ordinary broadcast radio receivers where full AM signals are used, but they also occur as integral parts of some detection systems in FM receivers at VHF. All this makes them very common indeed – yet it is often hard to find the envelope detector in a circuit because it has so few components.

The **basic circuit** is a diode rectifier, with a CR time constant, followed by a d.c. blocking high-pass CR filter part of which is the load. A representative arrangement is shown in Figure 9.1.

The product CR is chosen so that the rectified signal follows the modulation envelope closely but not the much more rapid carrier changes. A useful **rule of thumb** is that $\omega CR = 1$, where ω is the highest baseband frequency being detected. C' must be much larger than C so that it allows all baseband frequencies through and merely blocks out the d.c. component. R' has to be much larger than R so that it does not load CR and alter its effective value. More details are given later in Section 9.3. The only external condition is that the input voltage must be large enough so that the envelope voltage is much greater than the diode threshold voltage (0.3 V for Ge; 0.6 V for Si) at all times, otherwise distortion will occur.

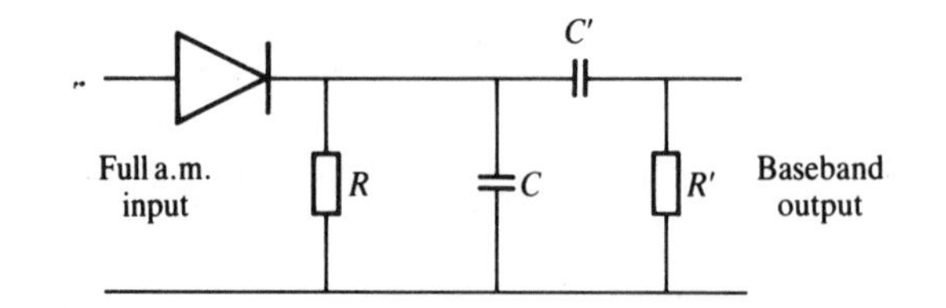

Figure 9.1 Basic envelope detector circuit

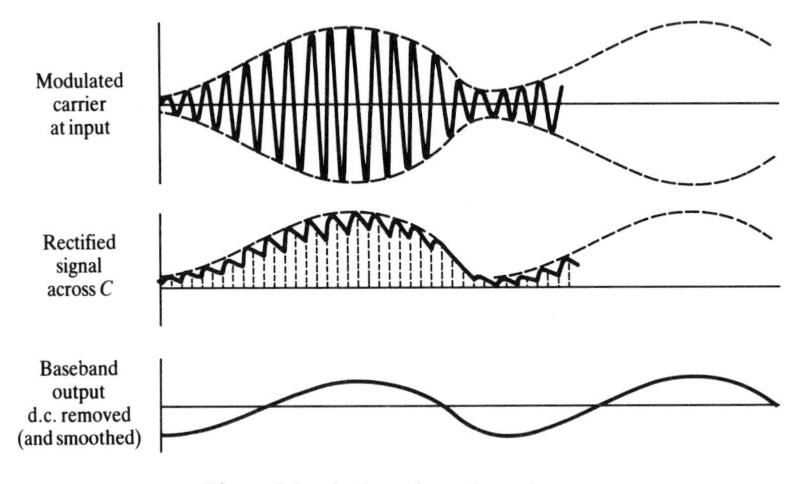

Figure 9.2 Action of envelope detector

The detector action is shown in Figure 9.2.

The diode allows through only the positive half cycles of the carrier. The effect of this is to charge up C to the carrier peak voltage each time. But because C discharges comparatively slowly through R, its voltage does not follow the carrier voltage down to zero between peaks. Instead it falls a small amount before being charged up again by the next positive half cycle of the carrier voltage. The diagrams above show some distortion (the "saw-tooth" line) but this becomes negligible with larger relative carrier frequencies when the capacitor voltage has much less time to fall between peaks. In practice f_c is made greater than $100f_m$.

Finally C' blocks out the d.c. component and the original baseband waveform is obtained at the output.

The trick is to arrange for the rate of discharge of C to be slow enough for it not to fall much between carrier peaks, yet large enough for it to be able to follow the fastest rate of decrease of the envelope voltage. This occurs on the forward sides of the baseband waveform as you can see. If the capacitor voltage falls too slowly it cannot follow the envelope troughs which then become distorted. If it falls too quickly it tends merely to follow the carrier voltage and the baseband is still not isolated. It is an interesting practical exercise to set up a working detector and alter the values of C (or R) to see these two effects.

9.3 • Deciding component values

The considerations above lead to a theoretical method of deciding the values of time constant (product CR) to use. The answer turns out to depend on the modulation factor, m, and can be derived as follows for a single sinusoid baseband:

take a carrier	$v_c = V_c \cos \omega_c t$
and modulation	$v_m = V_m \cos \omega_m t$
in full AM form	$v_{AM} = V_c(1 + m \cos \omega_m t) \cos \omega_c t$
with modulation envelope	$v_{ENV} = V_c(1 + m \cos \omega_m t)$

We want to find the value for the time constant (product CR) at which the capacitor discharge rate is just equal to the fastest rate of decrease of the envelope voltage. First equate these rates in general terms:

$$\frac{dv_{ENV}}{dt} = -V_c m \omega_m \sin \omega_m t$$

$$\frac{dv_{CAP}}{dt} \simeq \frac{-v_{ENV}}{CR} = \frac{-V_c}{CR}(1 + m \cos \omega_m t)$$

The approximation is valid because we are only concerned with a small fall in v_c when charged to the envelope voltage at a carrier peak. The discharge can then be regarded as linear not exponential. Derive the expression yourself if you are not sure.

Now equate the two rates of discharge. That is

$$-V_c m \omega_m \sin \omega_m t = \frac{-V_c}{CR}(1 + m \cos \omega_m t)$$

or

$$CR = \frac{1 + m \cos \omega_m t}{m \omega_m \sin \omega_m t}$$

Obviously this can have a wide range of values up to ∞ depending on time (i.e. on values of $\sin \omega_m t$ and $\cos \omega_m t$). But we want to know the worst case which restricts the value of CR the most. This occurs when the proportionate rate of discharge is greatest, which is just **after** the mean envelope voltage as shown in Figure 9.3.

The exact point of maximum discharge rate will depend on m and we need to find the time (or $\cos \omega_m t$ value) at which this occurs. This is done by differentiation to find the point at which the rate of change of CR with time is zero – standard calculus of minima.

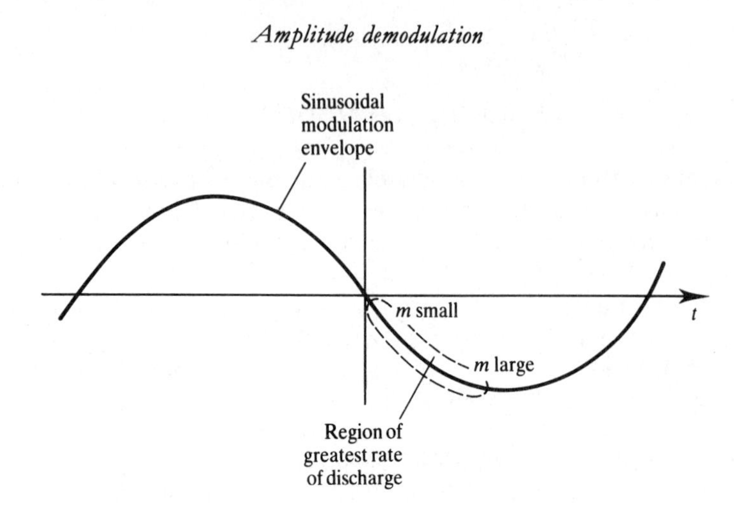

Figure 9.3 Point of greatest rate of discharge

That is, we want CR value when $\mathrm{d}(CR)/\mathrm{d}t = 0$:

$$\frac{\mathrm{d}(CR)}{\mathrm{d}t} = \frac{\mathrm{d}}{\mathrm{d}t}\left(\frac{1 + m\cos\omega_{\mathrm{m}}t}{m\omega_{\mathrm{m}}\sin\omega_{\mathrm{m}}t}\right)$$

$$= -\frac{m\omega_{\mathrm{m}}\sin\omega_{\mathrm{m}}t}{m\omega_{\mathrm{m}}\sin\omega_{\mathrm{m}}t} - \frac{\omega_{\mathrm{m}}\cos\omega_{\mathrm{m}}t}{m\omega_{\mathrm{m}}\sin^2\omega_{\mathrm{m}}t}(1 + m\cos\omega_{\mathrm{m}}t)$$

$$= -1 - \frac{\cos\omega_{\mathrm{m}}t}{m\sin^2\omega_{\mathrm{m}}t} - \frac{\cos^2\omega_{\mathrm{m}}t}{\sin^2\omega_{\mathrm{m}}t}$$

This equals zero when

$$1 = -\frac{\cos\omega_{\mathrm{m}}t}{\sin^2\omega_{\mathrm{m}}t}\left(\frac{1}{m} + \cos\omega_{\mathrm{m}}t\right)$$

or

$$\boxed{m = -\cos\omega_{\mathrm{m}}t}$$

that is

$$CR_{\mathrm{MAX}} = \frac{1 - m^2}{m\omega_{\mathrm{m}}(1 - m^2)^{1/2}}$$

or

$$CR \leqslant \frac{(1 - m^2)^{1/2}}{m\omega_m}$$

[For example, if $f_m = 1$ kHz and $m = 0.5$ then $CR = 2.76 \times 10^{-4}$, so a choice of $C = 10$ nF would require $R = 28$ kΩ.]

Of course this does not fix the exact value of C or R but only their product. A sensible choice of one of them has to be made for other reasons. Some useful ideas are as follows:

- R must not be too small or it overloads the signal source.
- R must not be too large or it requires too large an R'. (These two often put R around 1 or 10 kΩ.)
- Fix C first (as fewer preferred values) which also fixes R.
- Then take $R' \geqslant 10R$ and $C' \geqslant 10C$.
- Remember C' can be quite large as it only has to block d.c.
- R' may often be built in already in the input of the next stage.
- Use the rough rule of $CR \simeq 1/\omega_m$ first.

At first sight this criterion does not seem to depend on f_c at all. But it actually does so in a hidden way because we have assumed small linear capacitor discharge between carrier peaks. To be safe this requires $f_c = 100 f_m$ at least. The effect of not achieving this has already been mentioned above – the saw-tooth appearance of the demodulated waveform becomes quite large and takes the average line position away from the correct one, so introducing error.

Also, envelopes near maximum (100%) modulation cannot be detected accurately when near the zero voltage parts of the envelope. At first sight this seems to be because one must have $CR = 0$ from the equation above. But a more likely reason is that the voltage is below the diode threshold there.

The simple rule of taking $CR = 1/\omega_m$ is not far from correct for most practical modulation levels and is actually the correct value for $m = 0.7$. This gives $CR = 1.6 \times 10^{-4}$ for the example above.

Envelope detectors can be used equally well with the diode the other way round. The only effect is to detect the baseband from the negative part of the modulated carrier.

9.4 • Square-law diode detection

For small input signal levels a diode characteristic can be approximated to a square law as already explained in Chapter 8. Its effect is to mix the signal with itself (v^2)

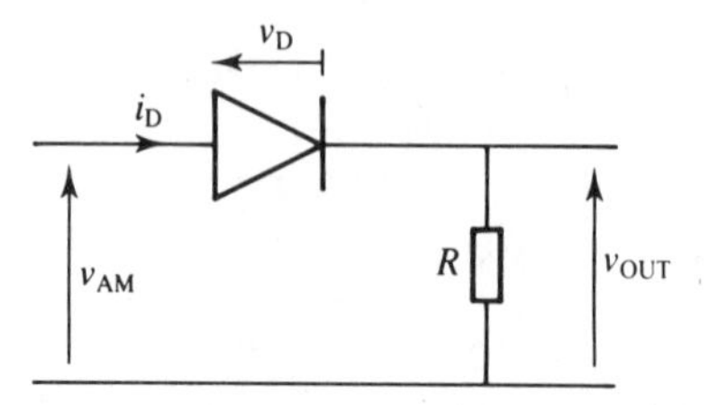

Figure 9.4 Square-law demodulation

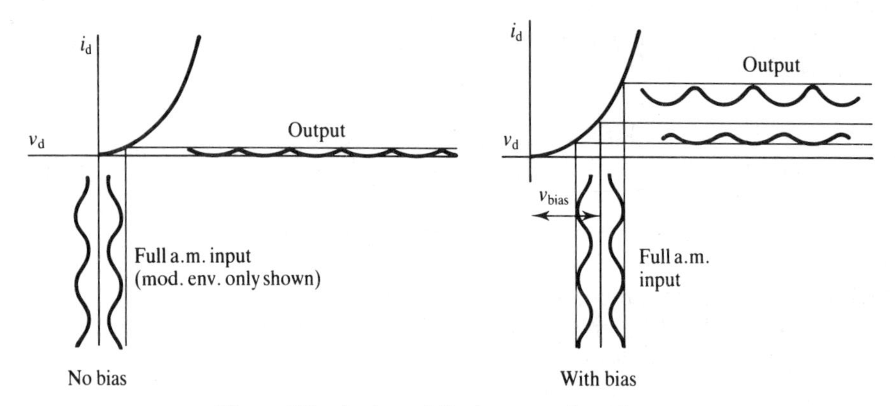

Figure 9.5 Action of diode square law detector

producing intermodulation products among which will be the original baseband itself. A simple circuit like that in Figure 9.4 will carry out this process but would need to be followed by a low-pass filter to isolate the baseband.

For a full AM signal the action can be described mathematically as follows:

$$v_{AM} = V_0(1 + m \cos \omega_m t) \cos \omega_c t$$
$$= V_0 \cos \omega_c t + \tfrac{1}{2}V_0 m \cos(\omega_c - \omega_m)t + \tfrac{1}{2}V_0 m \cos(\omega_c + \omega_m)t$$

Now

$$v_O = Ri_D$$
$$= R(av_D + bv_D^2)$$

then, assuming $v_O \leqslant v_D$,

$$v_O = R(av_{AM} + bv_{AM}^2)$$

Completing the analysis will show that the first term is just the carrier and its sidebands and that the other term produces frequencies at f_m (the required baseband), $2f_m$, $2(f_c - f_m)$, $2f_c$, and $2(f_c + f_m)$ as well as a d.c. term. Try it for yourself if you are in any doubt.

The detector can be used with or without d.c. bias to the diode and the action for both cases is best seen diagrammatically as in Figure 9.5.

The principal disadvantage of this method is the presence of $2\omega_m$ in the output which makes it useless for wideband modulation. One common use is in microwave detectors where a 1 kHz modulation tone is used to detect the gigahertz carrier. A tuned amplifier then selects this tone only. This method uses a bias current of a few milliamps with a specially mounted "coaxial" diode. One especially useful aspect is that the level of the detected 1 kHz output voltage is proportional to the microwave signal power level because of the two square laws involved.

9.5 • Coherent demodulation

Neither DSBSC nor SSB can be demodulated by an envelope detector because they do not have an "envelope" of the right kind. DSBSC does appear to have one but it crosses the axis and cannot be recovered by rectification.

Instead an entirely different method has to be used based on standard mixing techniques already described for product modulation. The only new features are the types of input. Now these are the signal to be demodulated and a locally generated carrier which forms the "local oscillator" as shown in Figure 9.6.

The name arises because the carrier signal used, whilst being generated at the receiver, must be of exactly the same frequency and phase as the carrier sent from the transmitter in the modulated signal. That is, it must be coherent with the incoming carrier, and hence it is called a **coherent detector**. As you can imagine this requirement leads to problems.

But first let us carry out the basic analysis. This will be done only for **DSBSC** because it automatically includes both types of SSB. Essentially the mixer gives the product $v_c . v_{MOD}$ and we need to show that this produces the original baseband. For simplicity we take a sinusoidal baseband, but the principle applies to all frequencies in a baseband. Amplitudes are omitted (i.e. equal to one):

$$\begin{aligned} v_{DSBSC} \cdot v_c &= [\cos(\omega_c - \omega_m)t + \cos(\omega_c + \omega_m)t]\cos\omega_c t \\ &= \cos(\omega_c - \omega_m)t \cdot \cos\omega_c t + \cos(\omega_c + \omega_m)t \cdot \cos\omega_c t \end{aligned}$$

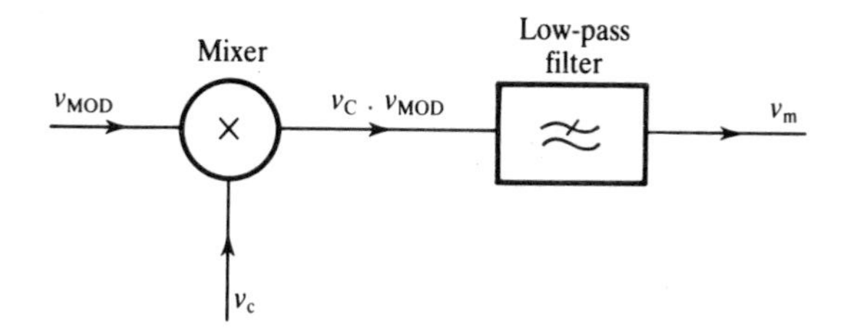

Figure 9.6 Principle of coherent demodulation

$$= \tfrac{1}{2}[\cos(-\omega_m)t + \cos(2\omega_c - \omega_m)t] + \tfrac{1}{2}[\cos \omega_m t + \cos(2\omega_c + w_m)t]$$
$$= \cos \omega_m t + \tfrac{1}{2}[\cos(2\omega_c - \omega_m)t + \cos(2\omega_c + \omega_m)t]$$

The $2\omega_c$ terms are filtered off leaving only the original baseband at ω_m. Thus the signal has been successfully demodulated. Obviously each frequency within a broader baseband can be recovered in this way leading to faithful reproduction of v_m at the receiver.

The analysis of **SSB** is similar, being just the two halves of the DSBSC analysis above. Each produces v_m separately.

Neither DSBSC nor SSB can be demodulated in any other way and SSB, in particular, is such an important modulation technique, because of its economy of power and bandwidth, that it is worth tolerating the extra complexity of circuitry (and cost) to use it. The big problem is the coherence requirement, and two classes of method are used to achieve this – pilot tone and carrier recovery. For example, VHF FM stereo broadcasting uses an intermediate stage of DSBSC modulation on to a subcarrier of 38 kHz transmitting this as a 19 kHz pilot tone. In microwave radio communication links the subcarrier is often recovered at the receiver by squaring the received signal – a carrier recovery circuit.

9.6 • Effect of phase and frequency errors in coherent demodulation

To recover the original baseband accurately, the locally generated carrier has to have the same frequency and phase as that suppressed from the signal. If this is not so then there will be errors in the demodulated signal which will distort it in a more or less acceptable fashion depending on circumstances. We now look at these effects analytically for an SSB-modulated signal which is by far the most common suppressed carrier modulation.

Take a USB signal of unity amplitude

$$v_{USB} = \cos(\omega_c + \omega_m)t$$

and coherently demodulate it with a locally generated carrier (v_c) having either a phase error ($\Delta\phi$) or a frequency error ($\Delta\omega$).

For **phase error** use

$$v_c = \cos(\omega_c t + \Delta\phi)$$

then

$$\begin{aligned} v_{\mathrm{DEMOD}} &= v_{\mathrm{SSB}} \cdot v_{\mathrm{c}} \\ &= \cos(\omega_{\mathrm{c}} + \omega_{\mathrm{m}})t \,.\, \cos(\omega_{\mathrm{c}}t + \Delta\phi) \\ &= \tfrac{1}{2}\cos(2\omega_{\mathrm{c}}t + \omega_{\mathrm{m}}t + \Delta\phi) + \tfrac{1}{2}\cos(\omega_{\mathrm{m}} - \Delta\phi) \end{aligned}$$

The first term is removed by the low-pass filter in the demodulator leaving the baseband correctly recovered except for a phase error. On a voice signal this will obviously have little noticeable effect.

For **frequency error** use

$$v_{\mathrm{c}} = \cos(\omega_{\mathrm{c}} + \Delta\omega)t$$

then

$$\begin{aligned} v_{\mathrm{DEMOD}} &= v_{\mathrm{SSB}} \cdot v_{\mathrm{c}} \\ &= \cos(\omega_{\mathrm{c}} + \omega_{\mathrm{m}})t \,.\, \cos(\omega_{\mathrm{c}} + \Delta\omega)t \\ &= \tfrac{1}{2}\cos(2\omega_{\mathrm{c}} + \omega_{\mathrm{m}} + \Delta\omega)t + \tfrac{1}{2}\cos(\omega_{\mathrm{m}} - \Delta\omega)t \end{aligned}$$

Again the $2f_{\mathrm{c}}$ term is removed by the LPF leaving the required baseband but with a frequency shift. For the most likely $\Delta\omega$ of only a few kilohertz this error could be subjectively serious, sounding as an increase (or decrease) in voice pitch with the meaning possibly unintelligible.

For DSBSC-modulated signal the corresponding results (which you may care to confirm for yourself) are

phase error	$\cos\omega_{\mathrm{m}}t \,.\, \cos\Delta\phi$
frequency error	$\cos\omega_{\mathrm{m}}t \,.\, \cos\Delta\omega$

The frequency error effect is similar to that for SSB but now with two shifts (one sideband up, the other down). The phase effect looks terrible (as $\cos\Delta\phi$ can easily be zero) but is not a serious matter in practice.

The general lesson is to avoid these errors, which can be done easily enough – if necessary with modern carrier recovery circuits.

Finally note that full AM (with carrier) can also be coherently demodulated. Try proving this for yourself by the methods above. Is it of any use?

9.7 • Summary

Two methods to demodulate AM signals are in common use.

Envelope detection uses a diode as a rectifier in a filter circuit (Figure 9.1) for full AM only. The principle of operation is obvious but hard to understand in detail.

Its component values are given by

$$CR \leqslant \frac{(1 - m^2)^{1/2}}{m\omega_m} \simeq \frac{1}{\omega_m}$$

for normal m values.

Coherent demodulation mixes the modulated signal with the original carrier (Figure 9.6) for DSBSC and SSB. The basic equation of operation is

$$v_{MOD} \cdot v_c \rightarrow v_m + \text{terms near } 2\omega_c$$

The carrier signal used for demodulation **must** be exactly coherent with that in the modulated signal otherwise errors occur – see Section 9.6. This requires either pilot tones or carrier recovery circuits.

9.8 • Conclusion

In the last three chapters we have discussed the basic ideas and methods of communicating by amplitude modulation. Next we must look at the same things for frequency modulation – it is much harder to understand at first but its advantages of dynamic range and noise suppression make the effort worth while.

9.9 • Problems

9.1 For an envelope detector, with a full AM input, answer the following questions:

(i) What is the maximum CR value at $f_m = 1.0$ kHz for $m = 1, 0.3$, and 0.9?

(ii) Find values of m for which $CR = 1/\omega_m$ and 0.

(iii) CR values seem to be independent of f_c. What assumption has been made to cause this to be true and under what circumstances is it invalid? What happens then?

(iv) For the results in (i), assign actual values for C, R, C', and R' explaining your choice of values.

(v) For $m = 0.3$ sketch the envelope detector output waveforms for input peak voltages of 10, 1, and 0.1 V.

(vi) What would be the effect of having the diode the other way round on the output waveforms in (vi)?

(vii) If any of the output waveforms in (vi) is distorted, how can you improve them?

(viii) Assign C, R, C', and R' values for a suitable mid-range value of CR in (iv).

9.2 A coherent detector mixes a carrier, f_c, with an AM input signal. Give the frequencies in the detector output (with relative amplitudes where appropriate) for the inputs below. In each case say also how the baseband would then be isolated:

(i) SSB (USB): $f_i = 101$ kHz, $f_o = 100$ kHz. [f_i is the "instantaneous" frequency of the incoming signal; f_o is the "local oscillator" frequency.]
(ii) DSBSC: same f_0 as in (i), f_1 now 99 kHz and 101 kHz.
(iii) Full AM: same bandwidth and f_m as in (ii).
(iv) As in (i) and (ii) but $f_o \pi/2$ out of phase with the original carrier, f_c; $f_o = f_c$.
(v) As (i) and (ii) but $f_o = 1010$ kHz; f_c unchanged.
(vi) SSB in band 990–999 kHz with $f_o = 1$ MHz.
(vii) As (vi) but DSBSC in band 990–1010 kHz.
(viii) As (vii) but $f_o = 1010$ kHz.
(ix) What output would you expect from an envelope detector having (ii) as its input?

9.3 Draw the circuit diagram of an envelope detector and explain its principle of operation assuming it is fed by a full AM signal produced by a single modulating tone.

9.4 Show how a diode can be used to detect the baseband in a full AM signal. Explain carefully all the conditions which must be met for the detector to operate successfully. In particular show that the time constant, τ, of your circuit is related to the modulation factor, m, for a single sinusoid baseband of angular frequency, ω_m, by

$$\tau = \frac{(1 - m^2)^{1/2}}{m\omega_m}$$

Give suitable values of components for a detector to recover a baseband having frequencies in the range up to 15 kHz with $m = 0.5$.

9.5 The detector circuit shown in Figure 9.7 receives a full AM signal at 198 kHz carrier frequency carrying a baseband limited to 7500 Hz.

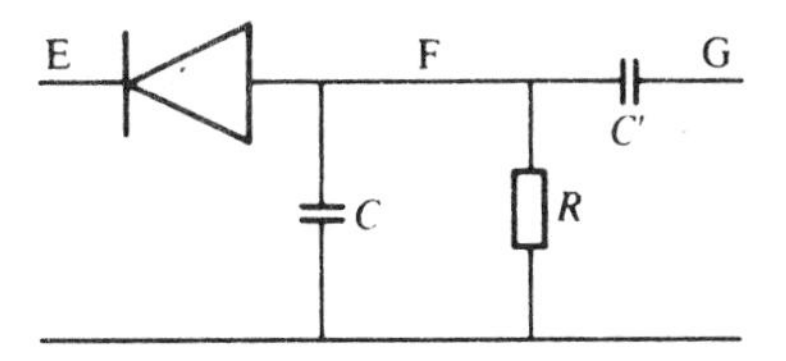

Figure 9.7 Circuit for Problem 9.5

Choose a sensible value for m then calculate values for the components, explaining any assumptions you need to make. Draw rough sketches showing the main features of the waveforms at the points marked E, F, and G.

9.6 Design an envelope detector to demodulate an AM signal having $m = 0.6$ with the standard broadcast baseband bandwidth and intermediate frequency. Give all component values and explain how these were obtained, deriving any expressions used. Also give representative voltage values with reasons.

9.7 Explain how an envelope detector recovers the baseband from a fully AM signal.

Discuss the choice of component values for recovering a 1 kHz sine wave from a carrier of 100 kHz.

9.8 Design an envelope detector to demodulate a steady 5 MHz signal modulated on to a 1 MHz carrier at $m = 0.6$. You must explain your procedure in full but there is no need to derive equations in detail.

Give and explain the signal conditions required for accurate recovery of the baseband. State which of these conditions are already met and give any extra requirements needed to meet the others.

• 10 •

Frequency modulation theory

10.1 • Introduction

In Chapter 6 (Section 6.6) we saw how the general expression for a carrier waveform

$$v_c = E_c \cos \theta_c = E_c \cos(\omega_c t + \phi_c)$$

ANGLE MOD. (θ_c) AM (E_c) FM (ω_c) PM (ϕ_c)

leads to the three types of **analogue modulation**.

Changes in E_c lead to amplitude modulation which has already been dealt with. The other two types occur when θ_c is altered by a baseband and come under the general heading of **angle modulation**. Changing ω_c produces **frequency modulation** whereas changes in ϕ_c cause **phase modulation**. These last two often seem very similar in action (after all, frequency is merely rate of change of phase) and have similar analytical treatments, although their effect is quite different in practice (and in application). This difference is seen clearly in Figure 10.1 for a square wave baseband but is much harder to see for a continuously varying baseband. Because they are actually very different, FM is dealt with here and PM is treated separately in Chapter 13.

10.2 • General basis for FM

The basic principle of frequency modulation is that a baseband voltage (v_m) **changes** the carrier frequency **linearly** by a **small amount** (i.e. keeping $\delta f_c \ll f_c$). In this way frequency changes carry the information which can then be recovered by the receiver.

Thus the basic equation is

$$\delta\omega \propto v_m$$

or

$$\delta\omega_c = K v_m$$

Basis for FM

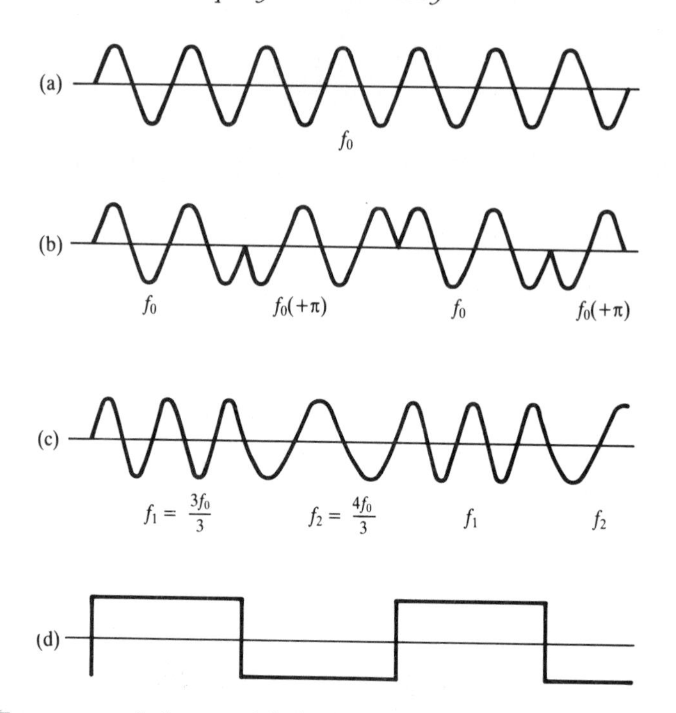

Figure 10.1 Frequency and phase modulation by a square wave baseband: (a) unmodulated carrier waveform, ω_c and ϕ_c unvarying; (b) modulated by step changes of phase, $\Delta\phi_c = 180°$; (c) modulated by step changes of frequency, $f_1 = 3f_c/4$, $f_2 = 4f_c/3$; (d) the square wave baseband

K is the **modulation sensitivity** having units of rad s^{-1} V^{-1}, although more likely to be expressed in practice as kHz mV^{-1}. K is often written K_P to distinguish it from K_F used for phase modulation.

$\delta\omega$ is the amount by which the baseband has caused the instantaneous carrier frequency, ω_i, of the carrier to **deviate** from the unmodulated frequency, ω_c. That is

$$\omega_i = \omega_c + \delta\omega$$

At first sight you might think you could now write simply that

$$v_{FM} = E_c \cos \omega_i t \qquad \textbf{WRONG}$$

but this would not be correct because of the previous history of ω_i which would introduce an unknown additional phase term. Instead you have to write that

$$\boxed{v_{FM} = E_c \cos \theta_i}$$

where θ_i is the instantaneous phase angle of the carrier, and then go back to the basic idea that frequency is the rate of change of phase so that $\omega_i = d\theta_i/dt$.

The correct expression for θ_i can now be found by integrating ω_i from time zero to the present instant, t. That is

$$d\theta_i = \omega_i \, dt$$

Therefore

$$\int_0^{\theta_i} d\theta_i = \int_0^t \omega_i \, dt$$

$$\theta_i = \int_0^t (\omega_c + \delta\omega) \, dt$$

$$= \int_0^t (\omega_c + Kv_m) \, dt$$

$$= \int_0^t \omega_c \, dt + \int_0^t Kv_m \, dt$$

Thus

$$\theta_i = \omega_c t + K \int_0^t v_m \, dt$$

giving

$$v_{FM} = E_c \cos\left(\omega_c t + K \int_0^t v_m \, dt\right)$$

General expression for f.m. waveform

This is a completely general expression for any baseband signal frequency modulated onto a carrier.

10.3 • Frequency modulation by a single sinusoid baseband

Obviously the expression above does not tell you much. In particular it gives no idea of the spectrum produced in the carrier by the act of frequency modulating it. To see this it is necessary, as for AM, to restrict the analysis to a baseband consisting of a single frequency. This is not as restricting as it seems because any real signal is made up of a number of such single frequency signals as was shown by Fourier analysis in Chapters 3 and 4. Thus it is both valid and useful to use

$$\boxed{v_{\mathrm{m}} = E_{\mathrm{m}} \cos \omega_{\mathrm{m}} t}$$

which gives

$$\begin{aligned}\delta\omega &= KE_{\mathrm{m}} \cos \omega_{\mathrm{m}} t\\ &= \Delta\omega \cos \omega_{\mathrm{m}} t\\ &= 2\pi\Delta f \cos \omega_{\mathrm{m}} t\end{aligned}$$

where $\Delta\omega$ and Δf are the **maximum** deviations which occur (when $\cos \omega_{\mathrm{m}} t = \pm 1$). Both quantities are actually referred to merely as the **deviation**. They are important parameters of any f.m. system (e.g. ± 75 kHz for VHF):

$$\begin{aligned}\Delta\omega &= KE_{\mathrm{m}}\\ \Delta f &= KE_{\mathrm{m}}/2\pi\end{aligned} \qquad \text{DEVIATIONS}$$

Putting v_{m} into the general expression for v_{FM} at the end of the last section gives

$$\begin{aligned}v_{\mathrm{FM}} &= E_{\mathrm{c}} \cos\left(\omega_{\mathrm{c}} t + K\int_0^t E_{\mathrm{m}} \cos \omega_{\mathrm{m}} t \, \mathrm{d}t\right)\\ &= E_{\mathrm{c}} \cos\left(\omega_{\mathrm{c}} t + KE_{\mathrm{m}}\int_0^t \cos \omega_{\mathrm{m}} t \, \mathrm{d}t\right)\\ &= E_{\mathrm{c}} \cos\left(\omega_{\mathrm{c}} t + \Delta\omega\int_0^t \cos \omega_{\mathrm{m}} t \, \mathrm{d}t\right)\\ &= E_{\mathrm{c}} \cos\left(\omega_{\mathrm{c}} t + \frac{\Delta\omega}{\omega_{\mathrm{m}}} \sin \omega_{\mathrm{m}} t\right)\end{aligned}$$

Now write the quantity $\Delta\omega/\omega_{\mathrm{m}}$ as a single constant, β, giving

$$v_{\mathrm{FM}} = E_{\mathrm{c}} \cos(\omega_{\mathrm{c}} t + \beta \sin \omega_{\mathrm{m}} t)$$

where β is the **modulation index** and is another important quantity in any f.m. communication system. Here it is given by

$$\beta = \frac{\Delta\omega}{\omega_{\mathrm{m}}} = \frac{\Delta f}{f_{\mathrm{m}}} \qquad \text{MODULATION INDEX}$$

β values obviously change with f_m, so β will vary over a band of signal frequencies and can be very large. For example, 300 Hz audio with 75 kHz deviation gives a value of β of 250 whereas 15 kHz audio gives only 5. This is considered more in Section 10.5.

Now the general expression for v_{FM} can be analysed one stage further to give

$$v_{FM} = E_c \cos \omega_c t \cos(\beta \sin \omega_m t) - E_c \sin \omega_c t \sin(\beta \sin \omega_m t)$$

It is this expression which can be analysed to give the spectral components of an f.m. signal. There are two ways of doing this, one requiring simplifying assumptions and the other not.

10.4 • Narrow-band frequency modulation (NBFM)

Here a simplifying assumption is made that β is small – so small in fact that it makes $\beta \sin \omega_m t$ also small, allowing the following further assumptions to be made:

$$\cos(\beta \sin \omega_m t) = 1$$

$$\sin(\beta \sin \omega_m t) = \beta \sin \omega_m t$$

The usual criterion for these assumptions to be valid is that $\beta \leqslant 0.2$ rad. [Check that $\cos(0.2) = 0.980$ and $\sin(0.2) = 0.199$ – good enough?]

Now the expression for v_{FM} can be written in a much simplified form as

$$\begin{aligned} v_{FM} &= E_c \cos \omega_c t \,.\, 1 - E_c \cos \omega_c t(\beta \sin \omega_m t) \\ &= E_c \cos \omega_c t - \beta E_c \cos \omega_c t \sin \omega_m t \end{aligned}$$

giving

$$v_{NBFM} = E_c \cos \omega_c t - \tfrac{1}{2}\beta E_c \cos(\omega_c - \omega_m)t + \tfrac{1}{2}\beta E_c \cos(\omega_c + \omega_m)t$$

This is a very simple expression, quite easy to understand. It consists of a carrier and one sideband pair similar to full AM, the only differences being that β is used instead of m and the lower sideband is negative. The two spectra are compared in Figure 10.2. The value of β can be measured from this spectrum just as for m (p. 96).

The fact that the lower sideband is negative is crucial and makes the modulated waveforms quite different as can be seen in Figure 10.3.

The frequency shift in NBFM at $\beta = 0.2$ is so small (βf_m compared with f_c) that it cannot be illustrated visibly, unlike the envelope variations for $m = 0.2$. The effect of this negative lower sideband can be seen much more graphically by considering the **stationary phasor** diagrams in Figure 10.4.

Here you can see very clearly how the change in sign of the lower sideband causes changes in phase angle for NBFM but in amplitude only for full AM. It also

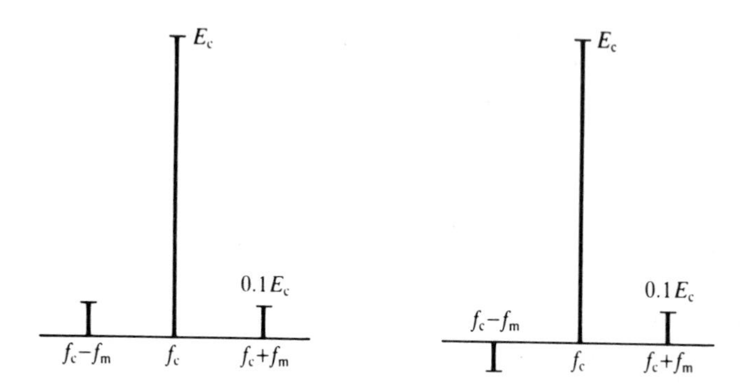

Figure 10.2 NBFM (right) and full AM (left) spectra compared (for $\beta = m = 0.2$)

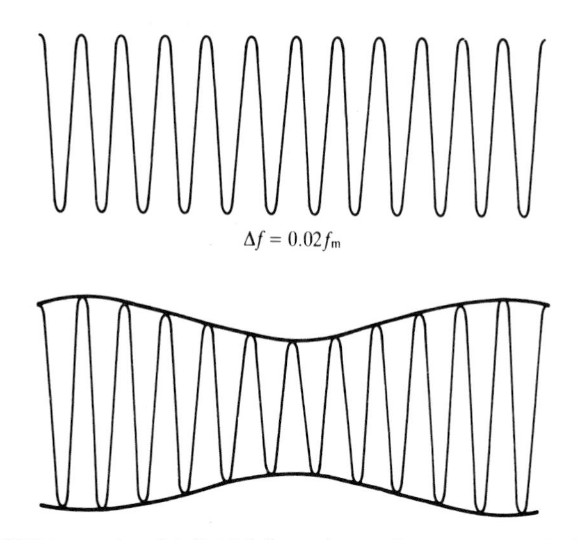

Figure 10.3 NBFM (upper) and full AM (lower) waveforms compared (for $\beta = m = 0.2$)

shows how FM produces a phase modulation type of effect ($\Delta\phi_{FM} = \beta = \Delta f/f_m$). Note also that there is a small amount of amplitude variation in NBFM (about 3% maximum) but this does not matter as an f.m. demodulator will ignore it.

From Figure 10.3 you can see that the **bandwidth** of an NBFM signal is the same as that for full AM

$$B = 2f_m \qquad \text{NBFM BANDWIDTH}$$

Hence the name **narrow** band – compared with WBFM.

NBFM is not just a curiosity but forms a vital part of many f.m. systems – as we shall see later in Chapter 11. This is partly because of its narrow bandwidth, but also because the deviation is so small that it is easy to make it very linear.

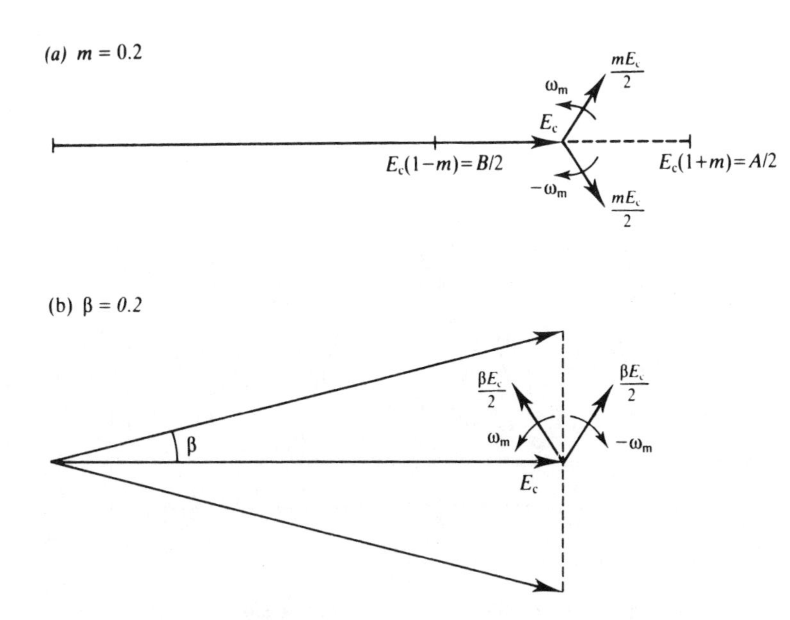

Figure 10.4 Stationary phasor diagrams for (a) full AM and (b) NBFM

10.5 • Wideband frequency modulation (WBFM)

For values of β larger than 0.2, the general expression at the end of Section 10.3 cannot be simplified any further but must be used as it is. Luckily the awkward parts can be expanded by using the following relationships (which are merely stated here and not derived):

$$\cos(\beta \sin \omega_m t) = J_0(\beta) + 2J_2(\beta) \cos 2\omega_m t + 2J_4(\beta) \cos 4\omega_m t + \cdots$$

$$\sin(\beta \sin \omega_m t) = 2J_1(\beta) \sin \omega_m t + 2J_3(\beta) \sin 3\omega_m t + 2J_5(\beta) \sin 5\omega_m t + \cdots$$

where $J_n(\beta)$ etc. are Bessel functions of β of order n.

To use these expressions it is not necessary to understand Bessel functions in full but a certain familiarity with them is useful. They come from solving a particular type of differential equation which often arises in actual physical situations. Bessel function values all have the form of pseudo-periodic functions of β, "decaying" as β increases, and of "period" increasing with n. They all start at zero for $\beta = 0$ except for J_0 which starts at one. All this is shown in Figure 10.5.

Also, to use Bessel functions quantitatively it is necessary to be able to look up their values in tables such as those given in Tables 10.1 and 10.2.

Now let us see the effect of these relationships on the general expression for a

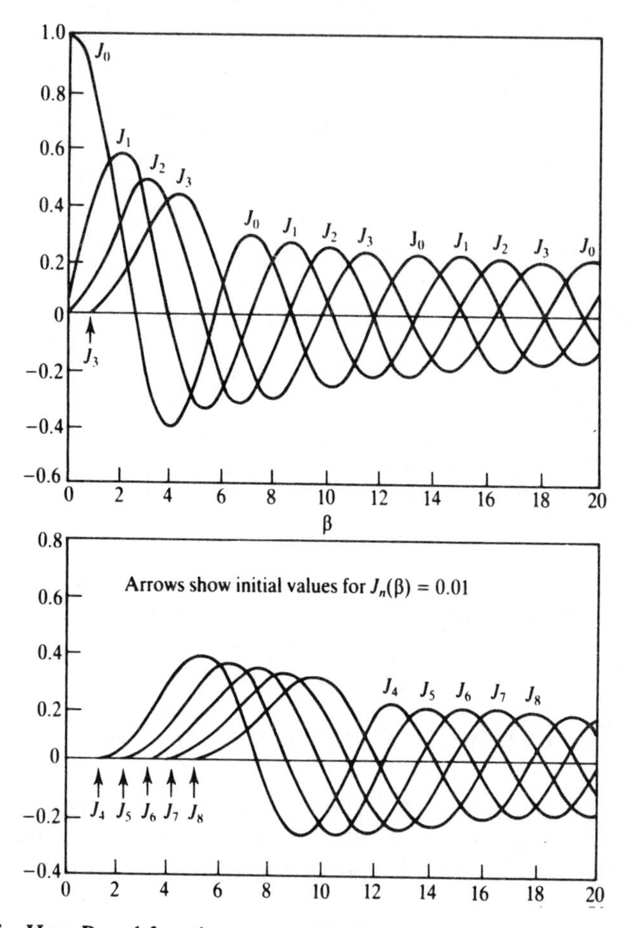

Figure 10.5 How Bessel functions vary with β and n (after Brown and Glazier, 1964)

carrier modulated by a sinusoidal baseband in WBFM:

$$
\begin{aligned}
v_{\mathrm{WBFM}} &= E_c \cos \omega_c t \cos(\beta \sin \omega_m t) - E_c \sin \omega_c \sin(\beta \sin \omega_m t) \\
&= E_c \cos \omega_c t[J_0(\beta) + 2J_2(\beta) \cos 2\omega_m t + \cdots] \\
&\quad - E_c \sin \omega_c t[2J_1(\beta) \sin \omega_m t + 2J_3(\beta) \sin 3\omega_m t + \cdots] \\
&= J_0(\beta) E_c \cos \omega_c t + 2J_2(\beta) E_c \cos \omega_c t \cos 2\omega_m t + \cdots \\
&\quad - 2J_1(\beta) E_c \sin \omega_c t \sin \omega_m t - 2J_3(\beta) E_c \sin \omega_c t \sin 3\omega_m t - \cdots \\
&= J_0(\beta) E_c \cos \omega_c t + J_2(\beta) E_c \cos(\omega_c - 2\omega_m)t \\
&\quad + J_2(\beta) E_c \cos(\omega_c + 2\omega_m)t + \cdots \\
&\quad - J_1(\beta) E_c \cos(\omega_c - \omega_m)t + J_1(\beta) E_c \cos(\omega_c + \omega_m)t \\
&\quad - J_3(\beta) E_c \cos(\omega_c - 3\omega_m)t + J_3(\beta) E_c \cos(\omega_c + 3\omega_m)t - \cdots
\end{aligned}
$$

Table 10.1 Bessel function values, $\beta = 0\text{–}2.5$ (0.2 steps) and $n = 0\text{–}8$ (after Betts, 1970)

n	$J_n(0.2)$	$J_n(0.4)$	$J_n(0.6)$	$J_n(0.8)$	$J_n(1.0)$	$J_n(1.25)$	$J_n(1.5)$	$J_n(1.75)$	$J_n(2.0)$	$J_n(2.5)$
0	+0.9900	+0.9604	+0.9120	+0.8463	+0.7652	+0.6459	+0.5118	+0.3690	+0.2239	−0.0484
1	+0.0995	+0.1960	+0.2867	+0.3688	+0.4401	+0.5106	+0.5579	+0.5802	+0.5767	+0.4971
2	+0.0050	+0.0197	+0.0437	+0.0758	+0.1149	+0.1711	+0.2321	+0.2940	+0.3528	+0.4461
3	+0.0002	+0.0013	+0.0044	+0.0102	+0.0196	+0.0369	+0.0610	+0.0918	+0.1289	+0.2166
4			+0.0003	+0.0010	+0.0025	+0.0059	+0.0118	+0.0209	+0.0340	+0.0738
5					+0.0002	+0.0007	+0.0018	+0.0038	+0.0070	+0.0195
6							+0.0002	+0.0006	+0.0012	+0.0042
7								+0.0001	+0.0002	+0.0008
8										+0.0001

Table 10.2 Bessel function values, $\beta = 0\text{–}20$ (1.0 steps) and $n = 0\text{–}25$ (after Betts, 1970)

n	$J_n(1)$	$J_n(2)$	$J_n(3)$	$J_n(4)$	$J_n(5)$	$J_n(6)$	$J_n(7)$	$J_n(8)$	$J_n(9)$	$J_n(10)$
0	+0.7652	+0.2239	−0.2601	−0.3971	−0.1776	+0.1506	+0.3001	+0.1717	−0.0903	−0.2459
1	+0.4400	+0.5767	+0.3391	−0.0660	−0.3276	−0.2767	−0.0047	+0.2346	+0.2453	+0.0435
2	+0.1149	+0.3528	+0.4861	+0.3641	+0.0466	−0.2429	−0.3014	−0.1130	+0.1448	+0.2546
3	+0.0196	+0.1289	+0.3091	+0.4302	+0.3648	+0.1148	−0.1676	−0.2911	−0.1809	+0.0584
4	+0.0025	+0.0340	+0.1320	+0.2811	+0.3912	+0.3576	+0.1578	−0.1054	−0.2655	−0.2196
5	+0.0002	+0.0070	+0.0430	+0.1321	+0.2611	+0.3621	+0.3479	+0.1858	−0.0550	−0.2341
6		+0.0012	+0.0114	+0.0491	+0.1310	+0.2458	+0.3392	+0.3376	+0.2043	−0.0145
7		+0.0002	+0.0025	+0.0152	+0.0534	+0.1296	+0.2336	+0.3206	+0.3275	+0.2167
8			+0.0005	+0.0040	+0.0184	+0.0565	+0.1280	+0.2235	+0.3051	+0.3179
9				+0.0009	+0.0055	+0.0212	+0.0589	+0.1263	+0.2149	+0.2919
10				+0.0002	+0.0015	+0.0070	+0.0235	+0.0608	+0.1247	+0.2075
11					+0.0004	+0.0020	+0.0083	+0.0256	+0.0622	+0.1231
12						+0.0005	+0.0027	+0.0096	+0.0274	+0.0634
13						+0.0001	+0.0008	+0.0033	+0.0108	+0.0290
14							+0.0002	+0.0010	+0.0039	+0.0120
15								+0.0003	+0.0013	+0.0045
16									+0.0004	+0.0016
17									+0.0001	+0.0005
18										+0.0002
19										
20										
21										
22										
23										
24										
25										

Table 10.2 (continued)

n	$J_n(11)$	$J_n(12)$	$J_n(13)$	$J_n(14)$	$J_n(15)$	$J_n(16)$	$J_n(17)$	$J_n(18)$	$J_n(19)$	$J_n(20)$
0	−0.1712	+0.0477	+0.2069	+0.1711	−0.0142	−0.1749	−0.1699	−0.0134	+0.1466	+0.1670
1	−0.1768	−0.2234	−0.0703	+0.1334	+0.2051	+0.0904	−0.0977	−0.1880	−0.1057	+0.0668
2	+0.1390	−0.0849	−0.2177	−0.1520	+0.0416	+0.1862	+0.1584	−0.0075	−0.1578	−0.1603
3	+0.2273	+0.1951	+0.0033	−0.1768	−0.1940	−0.0438	+0.1349	+0.1863	+0.0725	−0.0989
4	−0.0150	+0.1825	+0.2193	+0.0762	−0.1192	−0.2026	−0.1107	+0.0696	+0.1806	+0.1307
5	−0.2383	−0.0735	+0.1316	+0.2204	+0.1305	−0.0575	−0.1870	−0.1554	+0.0036	+0.1512
6	−0.2016	−0.2437	−0.1180	+0.0812	+0.2061	+0.1667	+0.0007	−0.1560	−0.1788	−0.0551
7	+0.0184	−0.1703	−0.2406	−0.1508	+0.0345	+0.1825	+0.1875	+0.0514	−0.1165	−0.1842
8	+0.2250	+0.0451	−0.1410	−0.2320	−0.1740	−0.0070	+0.1537	+0.1959	+0.0929	−0.0739
9	+0.3089	+0.2304	+0.0670	−0.1143	−0.2200	−0.1895	−0.0429	+0.1228	+0.1947	+0.1251
10	+0.2804	+0.3005	+0.2338	+0.0850	−0.0901	−0.2062	−0.1991	−0.0732	+0.0916	+0.1865
11	+0.2010	+0.2704	+0.2927	+0.2357	−0.1000	−0.0682	−0.1914	−0.2041	−0.0984	+0.0614
12	+0.1216	+0.1953	+0.2615	+0.2855	+0.2367	+0.1124	−0.0486	−0.1762	−0.2055	−0.1190
13	+0.0643	+0.1201	+0.1901	+0.2536	+0.2787	+0.2368	+0.1228	−0.0309	−0.1612	−0.2041
14	+0.0304	+0.0650	+0.1188	+0.1855	+0.2464	+0.2724	+0.2364	+0.1316	−0.0151	−0.1464
15	+0.0130	+0.0316	+0.0656	+0.1174	+0.1813	+0.2399	+0.2666	+0.2356	+0.1389	−0.0008
16	+0.0051	+0.0140	+0.0327	+0.0661	+0.1162	+0.1775	+0.2340	+0.2611	+0.2345	+0.1452
17	+0.0019	+0.0057	+0.0149	+0.0337	+0.0665	+0.1150	+0.1739	+0.2286	+0.2559	+0.2331
18	+0.0006	+0.0022	+0.0063	+0.0158	+0.0346	+0.0668	+0.1138	+0.1706	+0.2235	+0.2511
19	+0.0002	+0.0008	+0.0025	+0.0068	+0.0166	+0.0354	+0.0671	+0.1127	+0.1676	+0.2189
20		+0.0003	+0.0009	+0.0028	+0.0074	+0.0173	+0.0362	+0.0673	+0.1116	+0.1647
21			+0.0003	+0.0010	+0.0031	+0.0079	+0.0180	+0.0369	+0.0675	+0.1106
22			+0.0001	+0.0004	+0.0012	+0.0034	+0.0084	+0.0187	+0.0375	+0.0676
23				+0.0001	+0.0004	+0.0013	+0.0037	+0.0089	+0.0193	+0.0381
24						+0.0005	+0.0015	+0.0039	+0.0093	+0.0199
25						+0.0002	+0.0006	+0.0017	+0.0042	+0.0098

$$\begin{aligned} v_{\text{WBFM}} = E_c[\cdots &- J_3(\beta)\cos(\omega_c - 3\omega_m)t \\ &+ J_2(\beta)\cos(\omega_c - 2\omega_m)t - J_1(\beta)\cos(\omega_c - \omega_m)t \\ &+ J_0(\beta)\cos\omega_c t + J_1(\beta)\cos(\omega_c + \omega_m)t \\ &+ J_2(\beta)\cos(\omega_c + 2\omega_m)t + J_3(\beta)\cos(\omega_c + 3\omega_m)t + \cdots] \end{aligned}$$

Note that now, unlike NBFM, there are many more sideband pairs (up to $\beta + 1$ of them). It also appears that there is a pattern about their signs – all positive except the odd lower ones – but this neat pattern is distorted in reality by the signs of the Bessel functions which are just as likely to be negative as positive as shown in Figure 10.6.

Of much more significance are the magnitudes of the spectral lines which are the same for both upper and lower components of each sideband pair producing

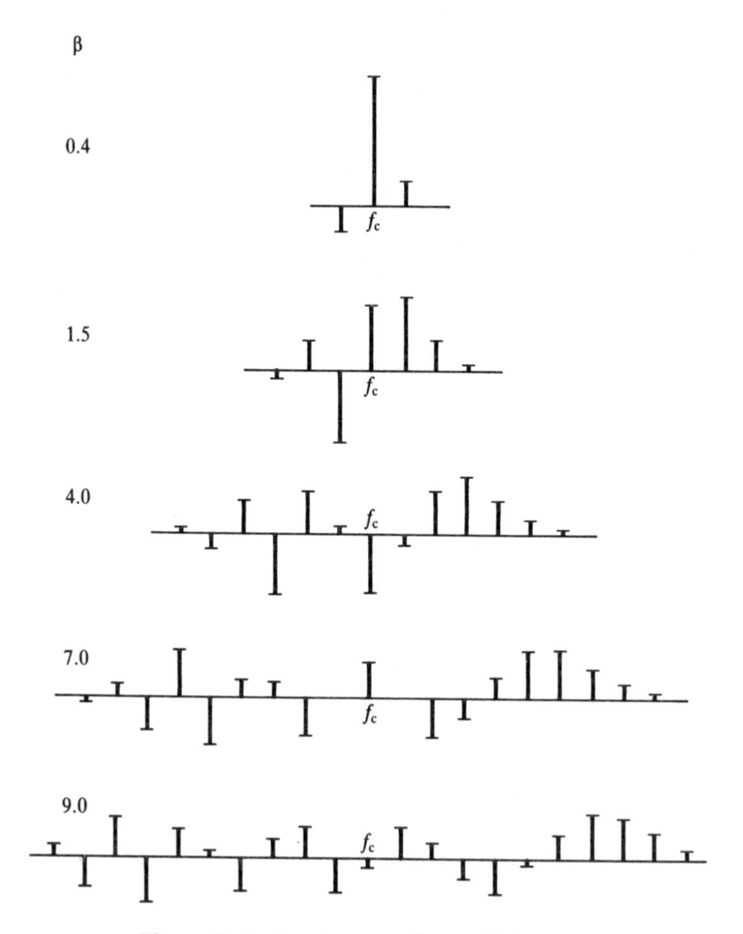

Figure 10.6 Total spectra for small β values

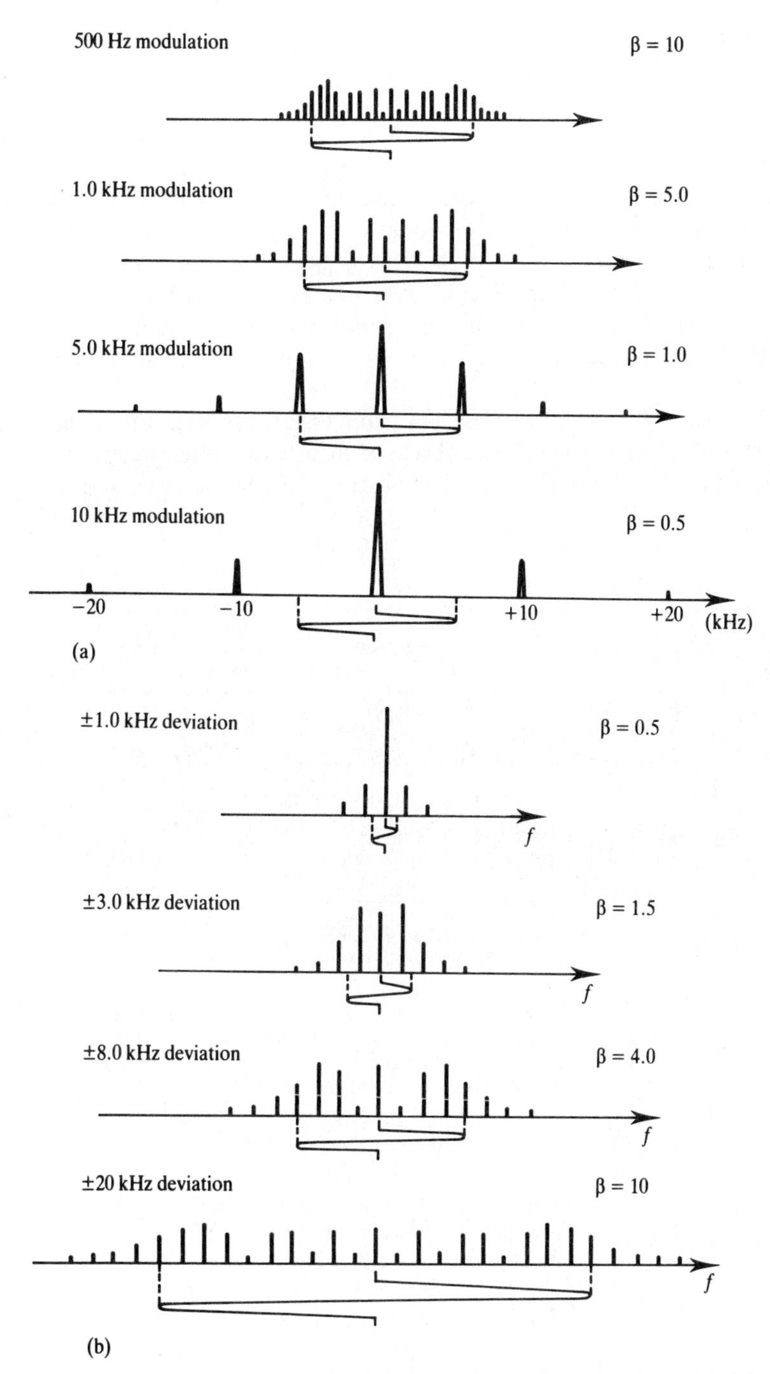

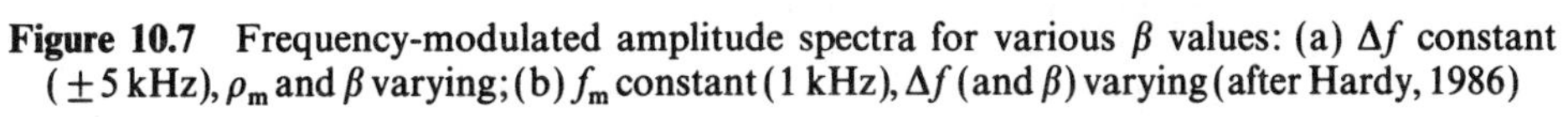

Figure 10.7 Frequency-modulated amplitude spectra for various β values: (a) Δf constant (± 5 kHz), ρ_m and β varying; (b) f_m constant (1 kHz), Δf (and β) varying (after Hardy, 1986)

characteristic patterns of amplitude symmetrical about the carrier position. The carrier itself is, otherwise, of no special importance and may be very small or even zero in value. All this is shown in Figure 10.7 for a wide range of β and Δf.

Note how the line amplitudes fall off rapidly at frequencies beyond its deviation. This is related to bandwidth which is discussed later (Section 10.6).

Note also how these amplitude spectra peak just before they fall into insignificance. The general rule is that the lines are strongest for deviations at which the baseband waveform lingers for the longest time. This is near the peaks for a sinusoid and produces the characteristic "eared" envelope shape as in Figure 10.7. Other waveforms will have their spectral peaks at different places and, for example, there will be an even greater concentration at the extremes for a square wave. An example is given in Figure 10.8.

Another way in which the Bessel function values are useful is to help identify the β value for certain spectra. This is because their zero values give zero amplitudes for the spectral lines for which they are a multiplier. Some examples are given in Table 10.3.

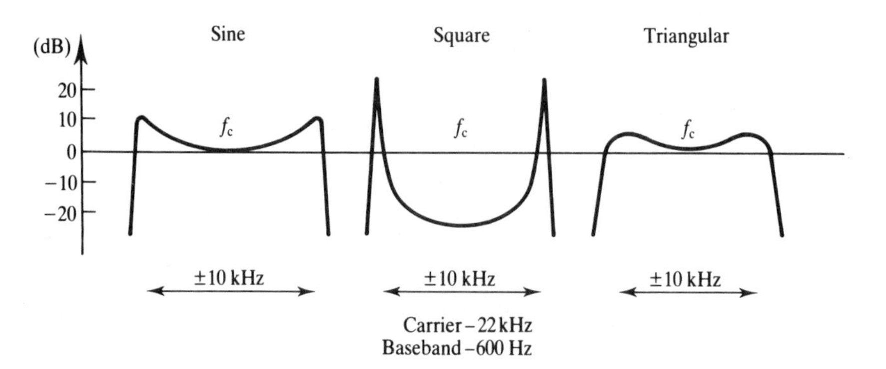

Figure 10.8 Spectral concentration near deviation "pauses"

Table 10.3 Spectral line zeros

At	$\beta = 2.4$	the carrier is zero
At	$\beta = 3.8$	the first pair are zero
At	$\beta = 5.2$	the second pair are zero
At	$\beta = 5.5$	the carrier is again zero
At	$\beta = 6.4$	the third pair are zero
At	$\beta = 7.0$	the first pair are again zero
At	$\beta = 7.6$	the fourth pair are zero

and so on (e.g. the carrier is again zero at 8.7, 11.9, 15.0 and 18.0)

It is very illustrative to set up a simple FM system (e.g. using the VCO input to a function generator) and vary v_m, checking β values from the zeros above and also by measuring bandwidth as below. This forms an excellent teaching experiment in the laboratory.

10.6 • WBFM bandwidth

As you can see from the spectra in Figure 10.7, there is no clearcut limit on bandwidth for WBFM as there is for NBFM and full AM. On the other hand, spectral line magnitude does fall off very rapidly at deviations greater than $\pm\Delta f$ so that the uncertainty lies within fairly narrow limits, as is also clear from Figure 10.7. The value of $2\Delta f$ is a good approximation known as the **nominal bandwidth**, especially at larger β values, but, for general use, more quantitative definitions are required.

The most common of these is the **Carson bandwidth** defined as that bandwidth which will let through at least 98 % of the signal power. The usefulness of this definition is that it always requires $\beta + 1$ sideband pairs at any value of β. Since the spectral lines are f_m apart, this means that the bandwidth is B where

$$B = 2(\beta + 1)f_m$$

which can also be written $B = 2(\Delta f + f_m)$ CARSON BANDWIDTH

Note that this includes one more sideband pair than the nominal bandwidth but that that extra width is negligible if β is large.

Another definition is the **1% bandwidth** defined as that bandwidth lying inside the sideband pair beyond which all spectral amplitudes are smaller than 1% of the largest line amplitude. This is usually wider than the Carson bandwidth.

For $\beta = 1$ the Carson bandwidth gives the following values:

$$v_{FM} = E_c \cos(\omega_c t + \sin \omega_m t)$$

giving normalised power (into 1 Ω) of $P = \frac{1}{2}E_c^2$. Also

$$\begin{aligned} v_{FM} &= J_0(1)\cos\omega_c t - J_1(1)\cos(\omega_c - \omega_m)t + J_1(1)\cos(\omega_c + \omega_m)t \\ &\quad + J_2(1)\cos(\omega_c - 2\omega_m)t + J_2(1)\cos(\omega_c + 2\omega_m)t \\ &= 0.765E_c - 0.440E_c + 0.440E_c + 0.115RE_c + 0.115E_c \end{aligned}$$

Therefore

$$\begin{aligned} P &= [0.765^2 + 2(0.440)^2 + 2(0.115)^2]E_c^2 \\ &= [0.4997]E_c^2 \end{aligned}$$

This is 99.93% of the total power – well within the Carson definition. The third sideband pair with $J_3(1)$ of 0.0196 add very little – calculate them to check.

10.7 • Modulation by a band of frequencies

So far we have only considered modulation by a single sinusoid signal. What happens when, as is usually the case, the signal is a band of frequencies – a true base**band**? An example of such a band is a single telephone channel of 0–4 kHz nominal (300–3400 Hz usable). Obviously the first step is to consider this band as the sum of a whole lot of frequencies, and then see what they have in common and where they differ.

The main **common factor** is Δf which is the **same** for all f_m values because it depends on signal amplitude only, not frequency. On the other hand, β is **less useful** because it varies inversely with f_m ($\beta = \Delta f / f_m$). The baseband **waveform** itself will still be periodic but non-sinusoidal and also varying continuously in shape as E_m and f_m change rapidly. Thus, this is not particularly useful except to give an instantaneous idea of what the signal might look like. It does give a very qualitative picture of the range of wavelengths involved. The **spectrum** will still consist of separate lines. There will be many more of them, closer together, and they will be continuously varying in position as the spectral content of v_m changes. Instantaneously there will still be symmetry in the amplitude spectrum but this is of little use.

The **bandwidth** will depend partially on f_m which must therefore be large enough to allow for the largest f_m value. For example, a telephone channel requires

$$B = 2(\Delta f + 3400)\ \text{Hz}$$

From all these considerations it can be seen that the only effectively **useful quantities** are Δf and B, and this is recognized in **practical specifications**. For example, the chief quantities specified for VHF FM radio broadcasting in UK are

$$\Delta f = \pm 75\ \text{kHz} \quad \text{and} \quad B = 225\ \text{kHz}$$

10.8 • Summary of expressions

Basis of FM:

$$\delta\omega = K v_m$$

General expression for any baseband:

$$v_{FM} = E_c \cos\left(\omega_c t + K \int_0^t v_m \, dt\right)$$

General expression for a single sinusoid baseband:

$$v_{\mathrm{FM}} = E_{\mathrm{c}} \cos[\omega_{\mathrm{c}} t + \beta \sin \omega_{\mathrm{m}} t]$$

NBFM:

$$v_{\mathrm{NBFM}} = E_{\mathrm{c}} \cos \omega_{\mathrm{c}} t - \tfrac{1}{2}\beta E_{\mathrm{c}} \cos(\omega_{\mathrm{c}} - \omega_{\mathrm{m}})t + \tfrac{1}{2}\beta E_{\mathrm{c}} \cos(\omega_{\mathrm{c}} + \omega_{\mathrm{m}})t$$

WBFM:

$$\begin{aligned} v_{\mathrm{WBFM}} &= E_{\mathrm{c}} \cos \omega_{\mathrm{c}} t \cos(\beta \sin \omega_{\mathrm{m}} t) - E_{\mathrm{c}} \sin \omega_{\mathrm{c}} t \sin(\beta \sin \omega_{\mathrm{m}} t) \\ &= \cdots - J_3(\beta) \cos(\omega_{\mathrm{c}} - 3\omega_{\mathrm{m}})t + J_2(\beta) \cos(\omega_{\mathrm{c}} - 2\omega_{\mathrm{m}})t \\ &\quad - J_1(\beta) \cos(\omega_{\mathrm{c}} - \omega_{\mathrm{m}})t + J_0(\beta) \cos \omega_{\mathrm{c}} t + J_1(\beta) \cos(\omega_{\mathrm{c}} + \omega_{\mathrm{m}})t \\ &\quad + J_2(\beta) \cos(\omega_{\mathrm{c}} + 2\omega_{\mathrm{m}})t + J_3(\beta) \cos(\omega_{\mathrm{c}} + 3\omega_{\mathrm{m}})t + \cdots \end{aligned}$$

Definitions

(i) Modulation sensitivity (K) $\quad \delta\omega = K v_{\mathrm{m}}$
(ii) Deviation ($\Delta\omega$) $\quad \Delta\omega = K E_{\mathrm{m}}$
(iii) Modulation index (β) $\quad \beta = \Delta\omega / \omega_{\mathrm{m}} = \Delta f / f_{\mathrm{m}}$
(iv) Carson bandwidth (B) $\quad B = 2(\Delta f + f_{\mathrm{m}})$ $= 2(\beta + 1) f_{\mathrm{m}}$
(v) Nominal bandwidth $\quad B = 2\Delta f$

10.9 • Conclusion

Frequency modulation is capable of both high linearity and large dynamic range. Also, it is much less susceptible to amplitude noise corruption than amplitude modulation provided the signal power is large enough, although the relevant analysis has not been done here. By using higher carrier frequencies, larger baseband bandwidths can be used enabling quality stereo sound broadcasting.

True NBFM (Δf of the order of Hertz) is used in broadcasting to provide highly linear modulation which is then converted to WBFM for transmission. Low Δf values (e.g. ± 2.5 kHz) are also used in communications applications such as mobile radio, CB, and amateur radio. These give noise advantages whilst keeping within small channel bandwidths.

Now let us see how FM is achieved and then recovered.

10.10 • Problems

10.1 For $f_m = 3$ kHz, $E_m = 3.0$ V, $\beta = 4.0$, $E_c = 2.0$ V, and $f_c = 200$ kHz calculate:
(i) The frequency deviation (Δf).
(ii) The frequency modulation sensitivity (K).
(iii) The maximum and minimum instantaneous frequencies (f_i).
(iv) The Carson bandwidth (B).
(v) The f.m. side frequencies and amplitudes.

10.2 Sketch the spectrum for 10.1(v) above, showing the bandwidth.

10.3 Repeat Problems 10.1 and 10.2 for $\beta = 0.04$ and $\beta = 40$ [you will need plenty of time].

10.4 For $\Delta f = 75$ kHz, $f_c = 100$ MHz, and an audio baseband extending from 300 Hz to 3400 Hz calculate:
(i) The r.f. bandwidth.
(ii) The maximum and minimum modulation indices.
(iii) The highest baseband frequency which could be sent without using NBFM.
(iv) The modulation index for the stereo pilot tone at 19 kHz.

10.5 A square wave baseband at 2.5 kHz repetition frequency and ± 6 V magnitude is frequency modulated on to a carrier at 25 kHz with modulation sensitivity of 2500 rads^{-1} V^{-1}:
(i) What is the modulation index?
(ii) What is the frequency deviation?
(iii) What is the bandwidth?
(iv) Sketch the modulated waveform.
(v) Draw the instantaneous amplitude spectrum at the two extremes of high voltage.
(vi) Draw the f.m. spectrum and compare with that in (v).

10.6 Draw the phasor diagram for $\beta = 0.2$ showing especially:
(i) maximum and minimum phase deviations (compare with calculated values)
(ii) the phasor position when $f_i = f_c$; at min f_i; at max f_i; and v_m is 45° from v_c.
(iii) Estimate how much AM occurs (calculate equivalent m).

10.7 Explain why FM is used for high-quality sound broadcasting. Also explain why it is *not* used for medium-wave broadcasting.

10.8 Find the value of the modulation index (β) for:
(i) $v_{PM} = E_c \cos(\omega_c t + \int K \cos \omega_m t \, dt)$.
(ii) $v_{FM} = 1.0 \cos(10^5 t + 10 \sin 10^3 t)$.

(iii) $v_{FM} = 5.0 \cos(2\pi \times 10^6 t + 0.5 \sin 2\pi \times 10^4 t)$.
(iv) $1.0 \cos 100t$ frequency modulated on to $1.0 \cos 10^5 t$ with $K = 10^4$.
(v) As (iv) but phase modulated with $\beta_P = 1$ (see Chapter 13).

10.9 Find the frequency separation for the first pair of sidebands for each of the waveforms in Problem 10.8.

10.10 Work out the relative sideband amplitudes for f.m. waveforms for which $\beta = 0.1$, 2.0, 20, and 100. Go up to the $(n + 1)$th pair.

10.11 Sketch the phasor diagrams for both full AM and NBFM ($m = \beta = 0.2$) with sidebands (LSB and USB) at the following relationships to the carrier:
(i) LSB in phase with the carrier.
(ii) LSB in anti-phase with the carrier.
(iii) LSB leads carrier by $\pi/2$.
(iv) LSB lags carrier by $\pi/2$.

10.12 For the situations in Problem 10.11 work out:
(i) Amplitude modulation depth.
(ii) Amount of AM in the NBFM signal.
(iii) Maximum phase deviation in the NBFM signal.
(iv) The greatest rate of change of phase in NBFM if $f_m = 1$ kHz.
(v) The instantaneous frequency at the points of maximum phase deviation for $f_c = 100$ kHz.

10.13 Find the percentage power transmitted within the Carson bandwidth for $\beta = 0.4$, 2.0, 10, and 50. Then calculate the percentage power lost if the outermost sideband pair is not sent.

10.14 For VHF FM radio find the highest modulating signal frequency for which Δf meets Carson's rule.

10.15 A carrier signal whose voltage is given by

$$v_c = 10 \cos(2\pi \times 10^6)t$$

is frequency modulated by a single frequency tone whose voltage is

$$v_m = 2.0 \cos(2\pi \times 10^4)t$$

using a modulation index of 5.

Write down an expression for the modulated signal showing clearly how frequency change is related to modulation amplitude. From this obtain an expression showing the amplitudes and frequencies of all components of the spectrum of the modulated signal which fall within the Carson bandwidth. Sketch the amplitude spectrum you obtain.

Explain the criteria for deciding on the bandwidth you use and verify if it is correct in this case.

10.16 The circuit shown in Figure 10.9 has a frequency-modulated input signal whose waveform is given by the expression

$$v_c = E_c \cos\left(\omega_c t + \int_0^t K v_m \, dt\right)$$

where v_m is the waveform of the original baseband, K is the modulation sensitivity, and E_c is the carrier amplitude.

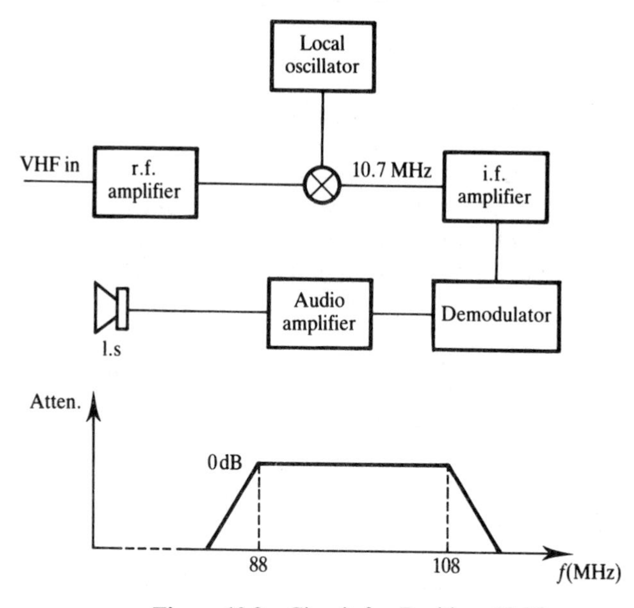

Figure 10.9 Circuit for Problem 10.16

From this expression obtain a relationship giving the modulation index in terms of frequency deviation and baseband frequency for a single sinusoid signal.

Given that: the centre frequency of the carrier is 90 MHz
the centre frequency of the i.f. signal is 10.7 MHz
the frequency deviation is ± 75 kHz
the baseband is from 30 Hz to 15 kHz

calculate: (i) the extreme values of modulation index
(ii) the maximum Carson bandwidth

(iii) the local oscillator frequency
(iv) the probable practical i.f. bandwidth.

Derive any additional expression you use. [Note: this question requires some understanding of the organization and operation of a radio receiver.]

10.17 Define the modulation index of a frequency-modulated signal. A carrier has a waveform given by $v_c = 5.0 \cos(2\pi \times 10^8)t$. It is frequency modulated by a baseband signal whose waveform is given by $v_m = 1.0 \cos \omega_m t$, where $\omega_m = 2\pi f_m$ and f_m is in the range 200 Hz to 20 kHz. The frequency deviation is kept constant at 200 kHz. For f_m at the lower limit of its range calculate the modulation index, the modulation sensitivity and the bandwidth. Any expression you use must be derived or explained, whichever is most appropriate.

Now calculate the same three quantities for the situation where f_m is at the upper limit of the baseband range. In addition sketch the amplitude spectrum of the modulated signal showing the signs and relative magnitude of each component.

10.18 Give the basic definition of frequency modulation and show that this leads to a phase change proportional to the integral of the baseband signal voltage.

A single sinusoid baseband at 3 kHz is frequency modulated on to a carrier at 300 kHz with a frequency deviation of 450 Hz. Derive an expression for the waveform of the modulated carrier and sketch its spectrum showing relative magnitudes.

Using a stationary phasor diagram, show how this modulation produces phase changes in the carrier. Compare this with a full AM signal at low modulation factor.

10.19 An FM mono broadcasting station is transmitting music band limited to 15 kHz. It uses a Carson bandwidth of 150 kHz. For a tone at the upper limit of the baseband calculate the modulation index. Give the full Carson spectrum of the broadcast signal modulated by that tone for a carrier at 95 MHz exactly, showing the relative amplitude of each spectral component, including sign. Derive all the theory you need to obtain the spectrum.

State the likely minimum value of modulation index for a stereo broadcast using the same baseband and broadcast bandwidth.

10.20 Define the modulation factor for an amplitude-modulated signal and the modulation index for a frequency-modulated signal.

A carrier voltage given by $v_c = 10 \cos(2\pi \times 10^6)t$ is modulated by a signal voltage given by $v_m = 2.0 \cos(2\pi \times 10^4)t$. For amplitude modulation obtain an expression for the modulated output and sketch its spectrum giving all relevant values. Give also a value for the modulation factor.

If the carrier is now frequency modulated by the signal, using a modulation index equal to the modulation factor you have just worked out, give an expression for the output waveform and sketch the new spectrum giving all relevant values. Compare the two spectra. Explain what you would expect to happen to this spectrum if the modulation index were gradually increased to a value 25 times yours.

10.21 A carrier and a baseband signal are given respectively by

$$v_c = E_c \cos \omega_c t$$

$$v_m = E_m \cos \omega_m t$$

v_m is frequency modulated on to v_c with a modulation sensitivity K such that an NBFM signal results with modulation index 0.1. Derive from first principles the spectrum of the modulated signal and sketch it showing relative phase. Now represent this spectrum on a suitable phasor diagram relating the phase deviations which occur to the constants of the NBFM. Indicate briefly the way in which this phasor diagram would differ if the modulation had been NBPM.

• 11 •
Frequency modulators

11.1 • Introduction

All frequency modulation is produced by circuits whose output frequency is linearly proportional to an input modulating voltage. This linearity is required over a small bandwidth (i.e. $\Delta f \ll f_c$) but needs to be fairly sensitive in that range (i.e. $\Delta f / v_m$ in kHz mV^{-1}) and linear (i.e. $\delta\omega \propto v_m$). Many different circuits and components can be used to do this. We shall look at three of them as representative of the main classes of method. These are: varactor tuned circuits; voltage-controlled oscillators; the Armstrong method.

11.2 • Varactor tuned circuits

The simplest carrier generation circuit is a tuned circuit oscillator as in Figure 11.1.

The output frequency is

$$f_c = \frac{1}{2\pi(LC)^{1/2}}$$

If either L or C changes in value then f_c also changes. If these component value changes are *small* then the frequency change is proportional to them. This can be shown as follows. Suppose C increases by δC. Then f_c will also change (by δf_c) such that

$$\begin{aligned} f_c + \delta f_c &= \frac{1}{2\pi[L(C + \delta C)]^{1/2}} \\ &= \frac{1}{2\pi(LC)^{1/2}(1 + \delta C/C)^{1/2}} \\ &= f_c(1 + \delta C/C)^{-1/2} \end{aligned}$$

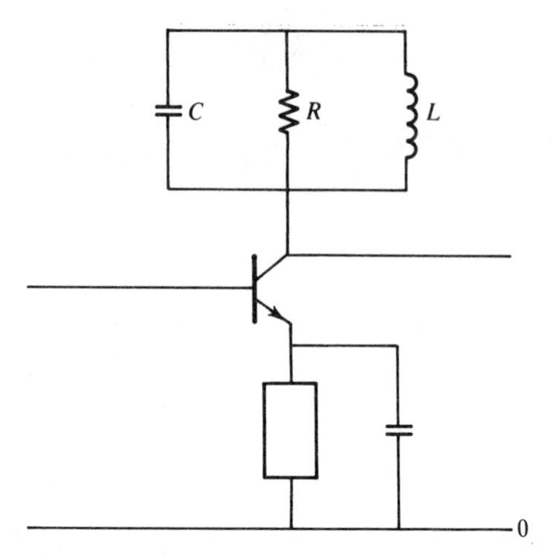

Figure 11.1 Tuned circuit carrier source

If $\delta C \ll C$ (the condition for a small change) then

$$f_c + \delta f_c = f_c(1 - \delta C/2C)$$

$$\delta f_c = \frac{-f_c}{2C}\delta C$$

That is

$$\boxed{|\delta f_c| \propto |\delta C|}$$

[A similar analysis will show that $|\delta f_c| \propto |\delta L|$ for $\delta L \ll L$.]

The question then is how to use the modulating signal, v_m, to vary C linearly over a small range. The easiest, and most usual, way to do this is to use **varactor diodes** which have lightly doped p and n regions allowing a considerable variation of the junction capacitance (C_v) with reverse bias voltage (V_b). Figure 11.2 shows a representative curve of this effect.

The term "varactor" is a contraction of "variable reactance". Varactors are designed for a wide range of maximum C values from a few picofarads to several hundred to cover the many radio frequency ranges up to microwave. For frequency modulation their important feature is that, over a limited range of bias voltages, C_v varies linearly with V_b – with a negative slope of course. Thus we can write

$$C_v = C_{v0} + \delta C = C_{v0} - kV_b \qquad (C_{v0} \gg \delta C)$$

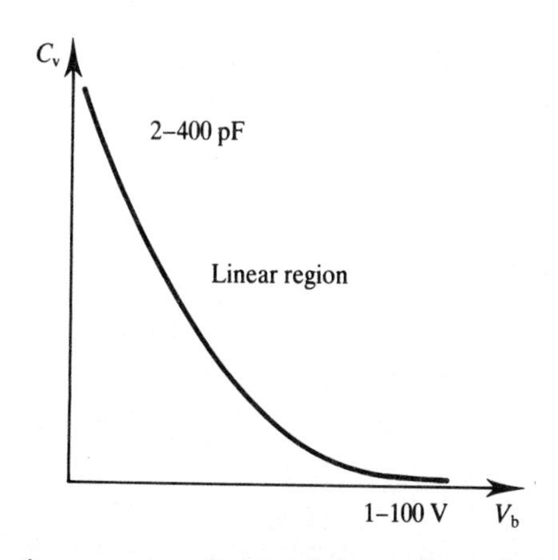

Figure 11.2 Representative varactor characteristics: values of C_v and V_b depend on type and frequency

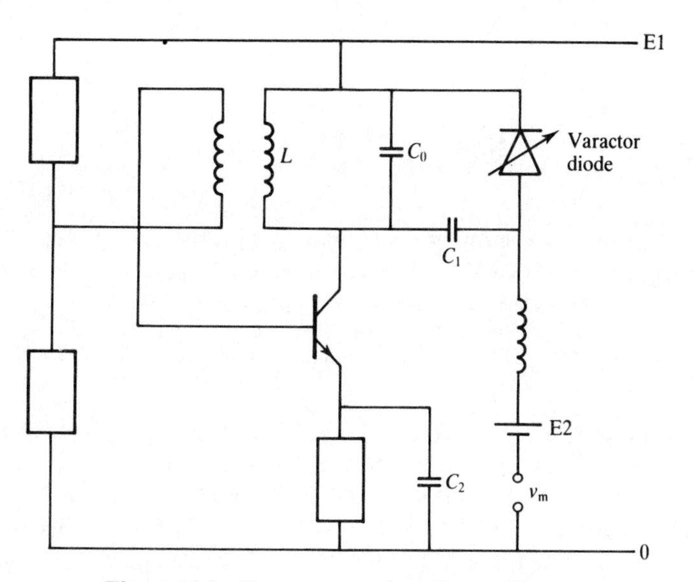

Figure 11.3 Varactor tuned carrier oscillator

This property can be utilized to vary f_c by placing the varactor in parallel with the much larger C_0 of the oscillator tank circuit as shown in Figure 11.3.

As you can see, the varactor is biased by a d.c. supply, V_{b0}, fed in via an r.f. choke. This fixes the zero value, C_{v0}, of the varactor capacitance so that the undeviated carrier frequency is given by

$$f_c = \frac{1}{2\pi[L(C_0 + C_{v0})]^{1/2}}$$

The modulating signal, v_m, is in series with the d.c. bias and provides a change in V_b, δv_b, to give the change in C_v, δC, for modulation. That is

$$C_v = C_{v0} + \delta C$$
$$= C_{v0} - kv_m$$

But

$$\delta f = \frac{-f_c}{2C}\delta C$$

so

$$\delta f = \frac{f_c}{2C}kv_m$$

or

$$\delta f = \frac{kf_c}{2(C_0 + C_{v0})}v_m = Kv_m$$

where K is the **modulation sensitivity** of the circuit with the units of Hz V^{-1}, although more often quoted in kHz mV^{-1}. For example, an MV1405 diode with a bias of 5.0 V has $C_0 = 75$ pF with $K = 25$ pF V^{-1}. That is

$$C_v = 75 - 25 . v_m \text{ pF}$$

This gives $K = 0.17$ kHz mV^{-1} at $f_c = 1.0$ MHz if used by itself (i.e. $C_0 = 0$). What L would be needed? What would be the sensitivity with a 55 pF tank capacitor?

Other voltage-sensitive components could be used to obtain a linear variation of frequency with applied signal voltage – an inductor with bias flux controlled by v_m or an FET operating at low v_D producing variable reactance in a suitable circuit.

11.3 • Voltage-controlled oscillators

Strictly speaking, any circuit whose output frequency can be controlled by an external voltage is a **voltage-controlled oscillator** (VCO). But, in practice, the term is reserved for relaxation-type oscillators which work by charging and discharging a capacitor at a constant rate between fixed voltage levels. The result is triangular or square output voltage waveforms (see Figure 11.4) from which it is easy to remove the harmonics to get a sine wave output.

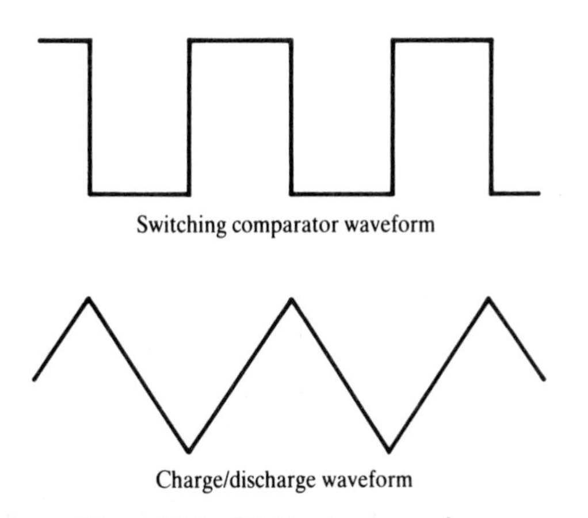

Figure 11.4 VCO output waveform

The output frequency can then be controlled in two ways, both of which essentially alter the period:

1. The charge/discharge rates can be changed by altering component values or charging current magnitude. These are used to set frequency range.
2. The upper and lower capacitor voltage limits can be changed thus altering T in proportion. These are used to set the centre frequency and to modulate.

11.4 • Multivibrators

Multivibrators are a common discrete (or i.c.) component way of doing frequency modulation although their action is complicated to explain in detail. The simplest view is that for half the cycle one transistor is off ($i_C = 0$, $V_C = V_{CC}$) and the other saturated ($i_C = V_{CC}/R_L$, $V_C \simeq 0$) and then suddenly they both switch and the situation is reversed. The effect is to make the two collector voltages form square waves of opposite phase decided by the time constant of the RC circuits attached to the bases.

To see this in more detail take the simplified circuit in Figure 11.5.

Suppose initially T1 has just been turned off and T2 just turned on. Then

1. $i_{C_1} = 0 \quad V_{C_1} = V_{CC} \quad V_{B_1} \simeq -V_{CC}$

   $i_{C_2} = V_{CC}/R_L \quad V_{C_2} \simeq 0 \quad V_{B_2} \simeq 0.7\text{ V}$
2. Now V_{B_1} starts to rise because the voltage across C_2 (about V_{CC}) starts to be discharged by current through R_2.
3. When V_{B_1} gets to around $+0.6$ V T1 starts to turn on. This drives V_{C_1} down and V_{B_2} with it via C_1.
4. This turns T2 off which causes V_{C_2} to rise and V_{B_1} with it via C_2.
5. This turns T1 *hard* on driving V_{C_1} down to near 0 V.
6. This drives V_{B_2} down to nearly $-V_{CC}$ (it was 0.7 V) and turns T2 fully off.

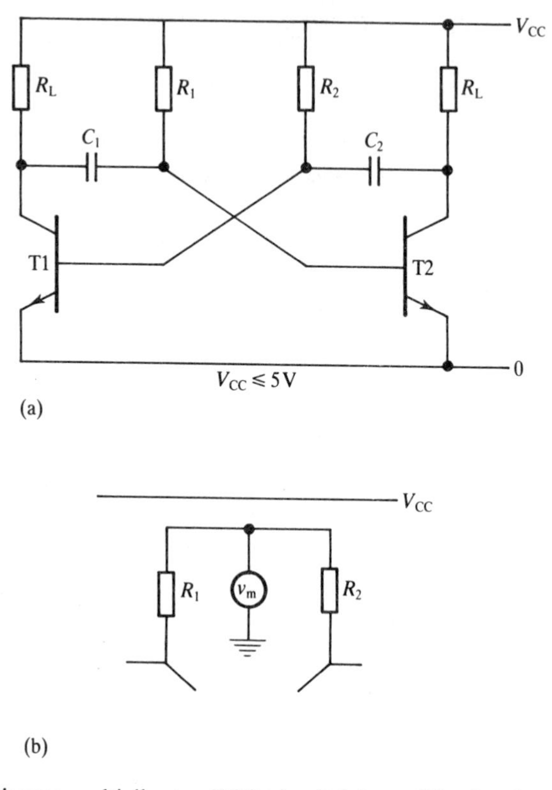

Figure 11.5 Discrete multivibrator VCO circuit (a); modification for modulation (b)

Thus the switch has been completed – very rapidly. Then the whole cycle is repeated for opposite transistors and we are back to the initial stage above. All this is more clearly seen from the voltage diagrams in Figure 11.6.

This figure shows how the period is fixed by the time taken for the base voltages to rise to turn on. They are actually rising towards V_{CC} but only reach 0.6 V at which switching occurs. Modulation can then be achieved by altering this voltage towards which these base voltages are rising and one way of doing this is shown in Figure 11.5. The modulating signal, v_m, is fed to the bases via 1 kΩ resistors and then the base voltages rise towards a level set by the potential dividers between v_m and V_{CC}.

Thus the output square waves on the collectors are modulated in period, and hence in frequency, and frequency modulation has been achieved. The C and R values set the frequency range whilst v_m can be used to set the centre frequency (with a d.c. level) and to modulate (with a variable input).

Being discrete components, these circuits can be used to quite high r.f.

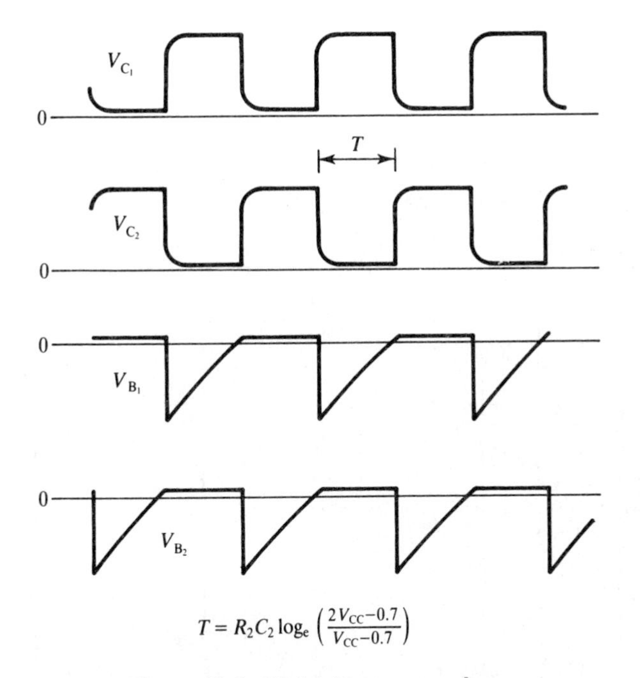

Figure 11.6 Multivibrator waveforms

11.5 • VCO integrated circuits

A **VCO chip**, such as the 566, will operate satisfactorily up to about 1 MHz and forms a convenient, easily modulated source. The **circuit schematic** of a Signetics 566 is shown in Figure 11.7.

The entire circuit is on chip except the external capacitor (C) and resistor (R) which are used to set the operating frequency range.

The 566 provides square and triangular output waveforms, as in Figure 11.8, at the same repetition frequency, f_o, given by

$$f_o = \frac{2(V_{CC} - V_{CON})}{RCV_{CC}}$$

Its operation depends on the charge and discharge of the external timing capacitor, C, at the same constant current, I, given by

$$I = \frac{V_{CC} - V_{CON}}{R}$$

The capacitor voltage then provides the triangular output waveform (at pin 4, via

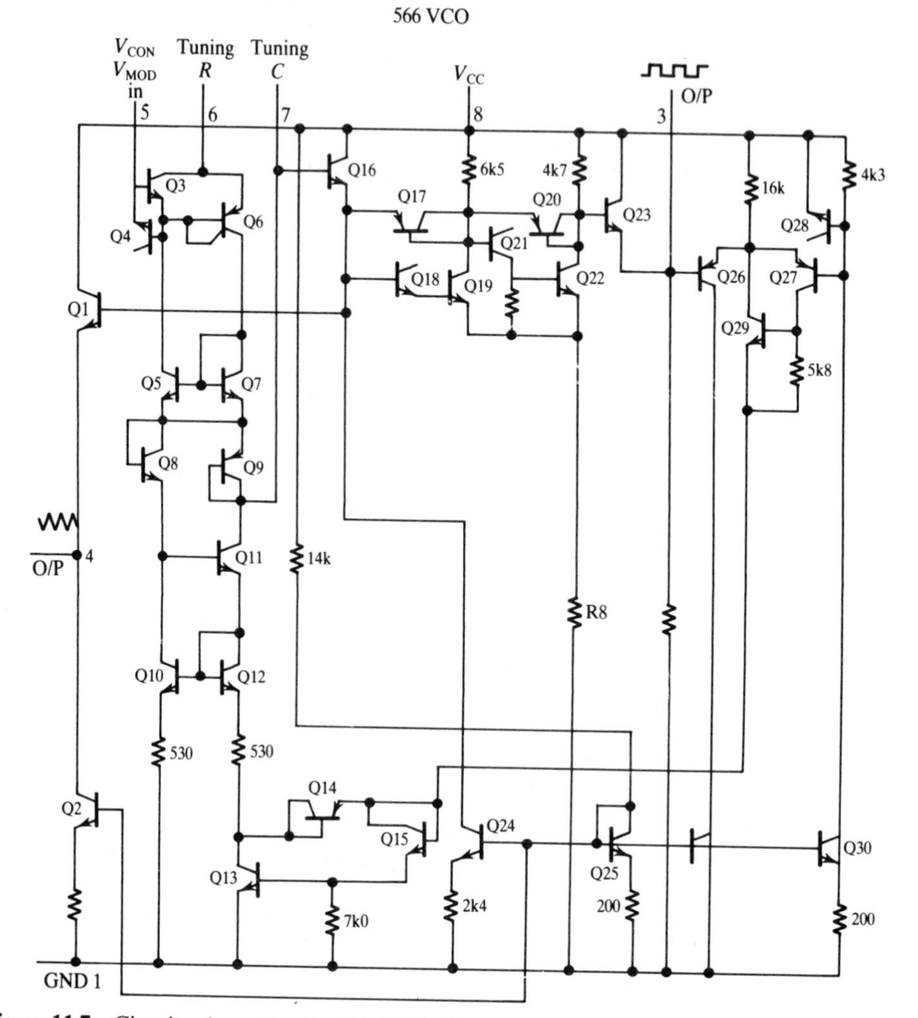

Figure 11.7 Circuit schematic of a 566 VCO chip (from National Semiconductor data sheet)

Q16 and Q1) whilst the square wave output (at pin 3) comes from the circuit used to switch the capacitor from charge to discharge and vice versa.

The quiescent value of f_o is altered by changing R (or C) so changing the operating range of the VCO. Then f_o can be varied continuously within that range by varying the fixed value of V_{CON} with a variable V_{MOD} input, so producing the frequency modulation required. A suitable circuit for doing this is shown in Figure 11.9 taken from a 566 data sheet.

The operation of this circuit depends upon the controllable constant current source provided by R and the voltages at its two ends as shown in Figure 11.10 and the equation for I above.

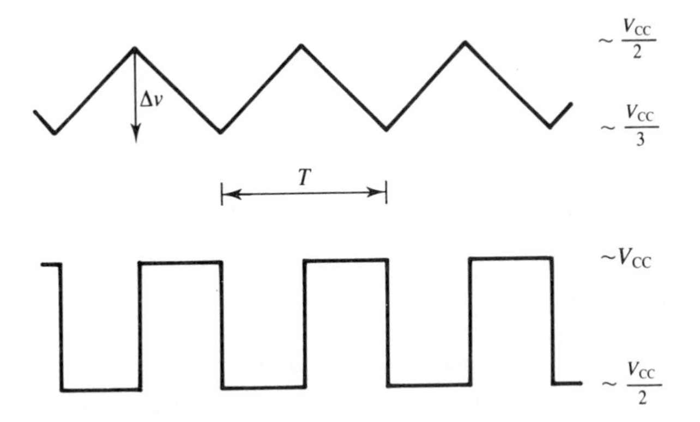

Figure 11.8 VCO output waveforms (of 566)

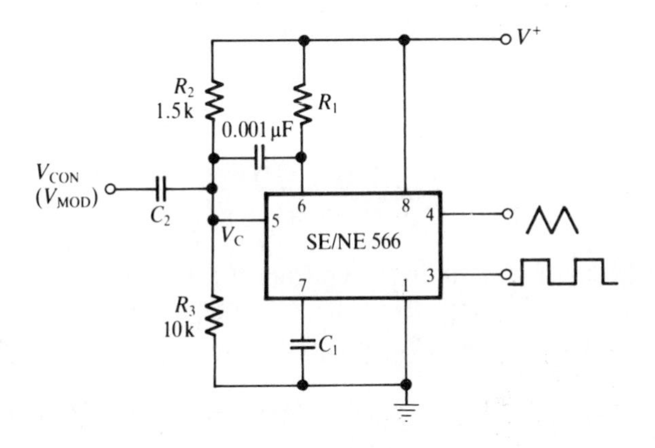

Figure 11.9 Frequency modulation using 556 VCO (from Phillips Components data sheet)

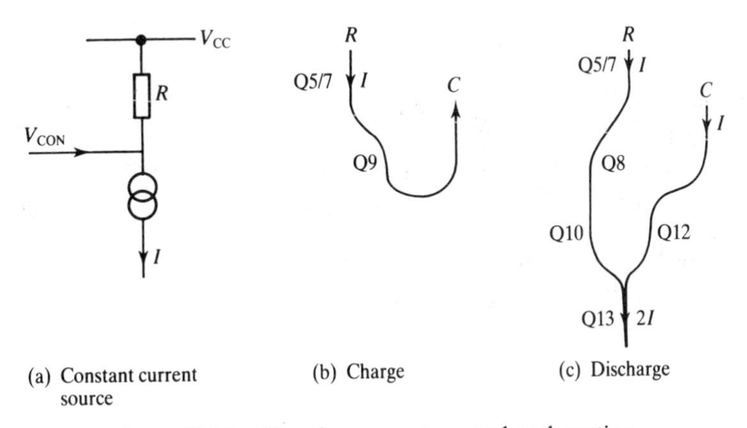

Figure 11.10 Charging current control and routing

Note that V_{CON} is connected to pin 5 but the bottom of R (pin 6) has the same potential because they are separated merely by two opposing base–emitter voltages (Q3 and Q7).

The path followed by I is then decided by whether that tail transistor (Q13) is hard on or hard off. When hard off it passes no current, so neither do Q10 or Q12 (linked by a current mirror), forcing I to flow through Q9 and then out through pin 7 into C. If hard on $2I$ flows, being I from the constant current source (via Q8 and Q10) and another I drawn from the capacitor via Q12 (whose current must equal that through Q10).

The switch from charge to discharge is done by the Schmitt trigger circuit below pin 8 at voltage levels fixed by the potential at the top of R8. This is switched between two levels ΔV apart where $\Delta V = V_{CC}/4$. Then, if the period of the output waveform is T,

$$\frac{\Delta V}{T/2} = \frac{\mathrm{d}V_{CAP}}{\mathrm{d}t} = \frac{I}{C} = \frac{V_{CC} - V_{CON}}{CR}$$

Therefore

$$f_o = \frac{1}{T} = \frac{V_{CC} - V_{CON}}{2CR\Delta V} = \frac{2(V_{CC} - V_{CON})}{CRV_{CC}}$$

which agrees with the expression above taken from the data sheet.

For example, to get $f_c = 100\,\text{kHz}$ with $R = 500\,\Omega$, then $C = 1/2Rf_c = 1/(2 \times 10^5 \times 500) = 1.0 \times 10^{-8} = 10\,\text{nF}$. This assumes that V_{CON} is set at $\frac{3}{4}V_{CC}$ by the potential divider in Figure 11.9.

A 555 timer i.c. operates in a very similar way – see Problem 11.2.

11.6 • The Armstrong method

This is a common method used where WBFM with high linearity and good frequency stability is required – as in broadcasting. The basis is to generate an accurate NBFM signal, at a low r.f. carrier, and then increase both the deviation and the carrier frequency by a combination of multiplication and down conversion, until those required are reached.

The original NBFM is done by a synthesis method, rather than any of those above, so as to retain carrier frequency stability.

Thus three processes are required:

1. NBFM by synthesis.
2. Frequency multiplication.
3. Down conversion.

The **synthesis** is based on the following equation for an NBFM signal:

$$v_{\mathrm{NBFM}} = E_c \cos \omega_c t - \beta E_c \sin \omega_c t \,.\, \sin \omega_m t$$

This equation is then created by phase shifting, multiplying, etc., as shown in Figure 11.11.

K and *M* are the amplitude factors caused by the integrator and multiplier respectively. They are adjusted to give the required β value ($\ll 0.1$) according to the relationship

$$\beta = \frac{MKE_m}{\omega_m}$$

making the deviation, Δf, small according to the equation

$$\Delta f = \frac{MKE_m}{2\pi}$$

Some representative numerical values are given with the system outlined in Figure 11.11. As you can see, this modulation is carried out at f_c and Δf values much smaller than

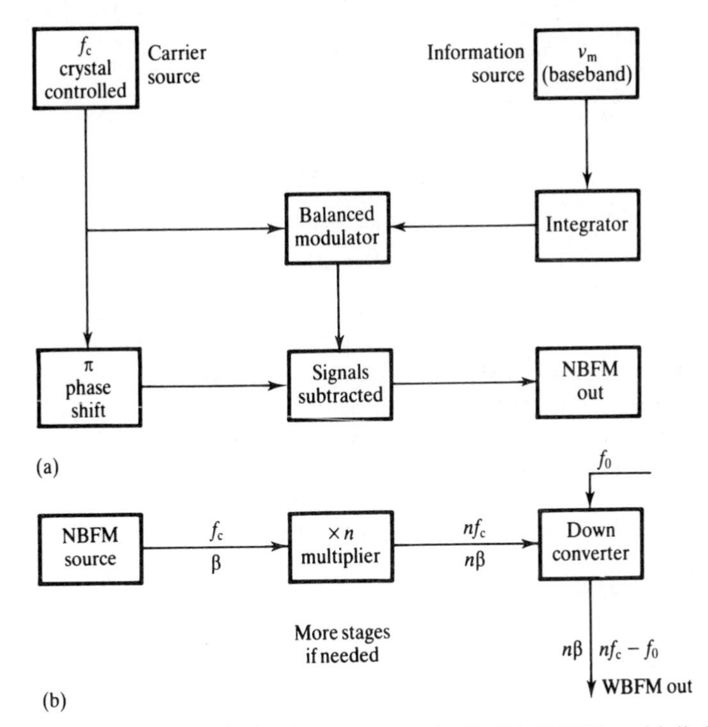

Figure 11.11 (a) NBFM by synthesis (Armstrong method); (b) NBFM multiplied and down converted to WBFM; (c) Armstrong modulator for VHF

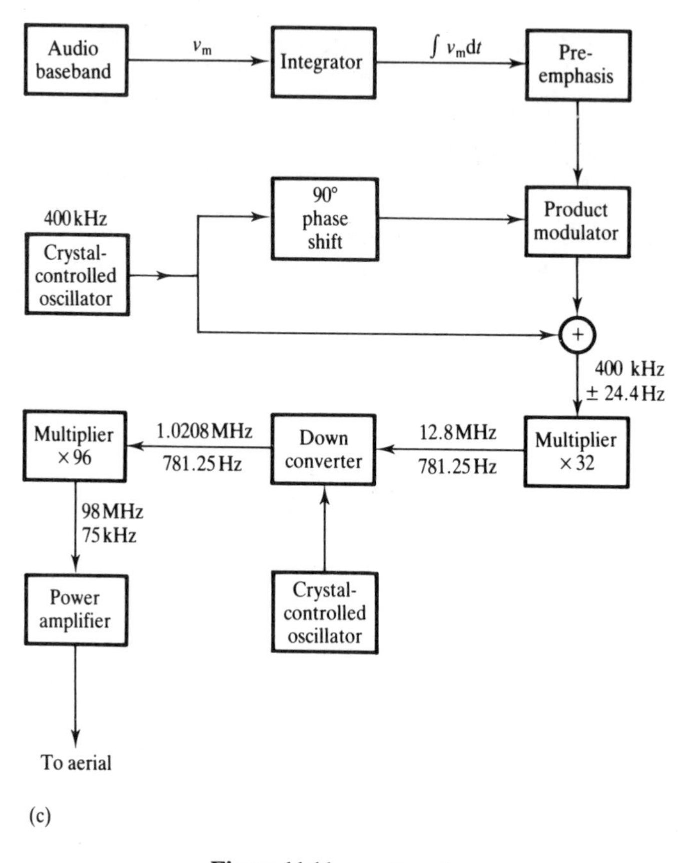

Figure 11.11 *continued*

those required for use (e.g. 100 MHz and ±75 kHz). To get there the modulated signal must now be operated on by the other two processes listed above. Note that they are both required in a carefully arranged system because multiplication alone would mean either the carrier or the deviation was of the wrong value. These two processes will not be described in detail here.

11.7 • Summary

Tuned circuit modulators use small capacitance changes in the C of a tuned circuit to produce

$$|\delta f| \propto |\delta C|$$

δC is then itself made to vary proportionately to the modulating voltage using a varactor diode with fixed d.c. bias. That is

$$\delta f = K v_m$$

where $K = -(kf_c)/2C$ and k is the diode constant, $\delta C = kv_b$.

Voltage-controlled oscillators are relaxation oscillators whose output frequencies are proportional to a control voltage. Frequency range is set by external C and R values and then frequency is varied continuously within that range by the modulating signal applied to the control voltage input. Both discrete multivibrators and i.c. VCOs are used. The 566 VCO gives an output (modulated carrier) signal of frequency

$$f_o = \frac{2}{CR} - \frac{2v_m}{CRV_{CC}} = f_c - \delta f$$

The **Armstrong method** uses very-narrow-band NBFM by synthesis, frequency multiplied and down converted to change it to the required WBFM signal. This makes everything very stable and linear.

See also Section 13.5.

11.8 • Conclusion

As can be seen, there are several methods of frequency modulation to choose from depending on application. There is even more variety in demodulation methods as will be shown in the next chapter.

11.9 • Problems

11.1 Calculate or estimate values for the modulation sensitivity under the following conditions (use $v_{OUT} = v_m$ in all):

(i) NBFM: $f_m = 1.0$ kHz, $v_m = 1.0$ V

(ii) NBFM: $f_m = 10$ kHz, $v_m = 1.0$ V

(iii) $\beta = 5$, $f_m = 600$ Hz, $v_m = 100$ mV

(iv) $\beta = 5$, $f_m = 6.0$ kHz, $v_m = 100$ mV

(v) $\Delta f = 75$ kHz, $f_m = 15$ kHz, $v_m = 75$ mV

(vi) $\Delta f = 75$ kHz, $f_m = 300$ Hz, $v_m = 75$ mV.

11.2 Figure 11.12 shows the schematic circuit of a 555 timer chip with the external timing components (R and C) added. Compare this circuit to the block diagram

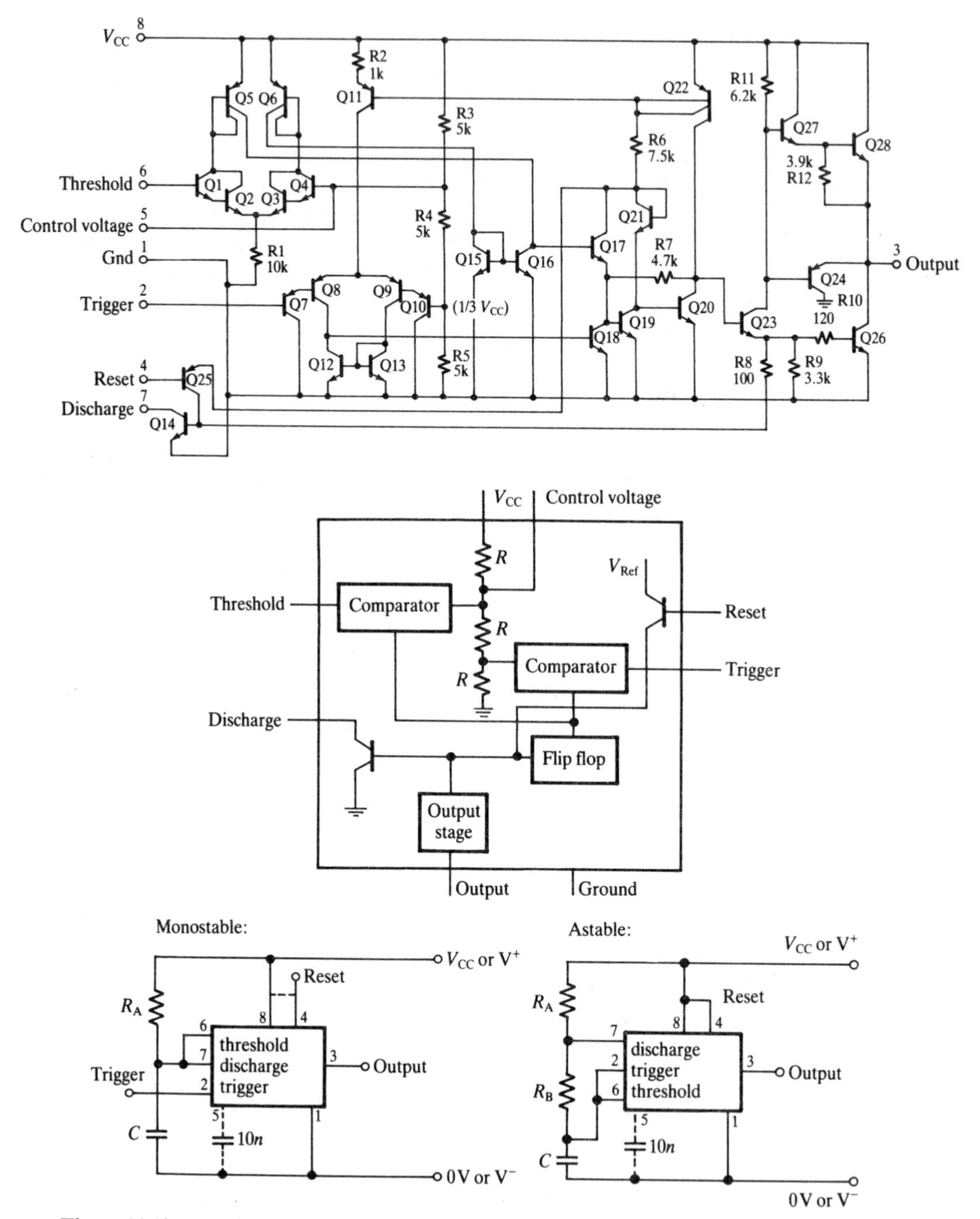

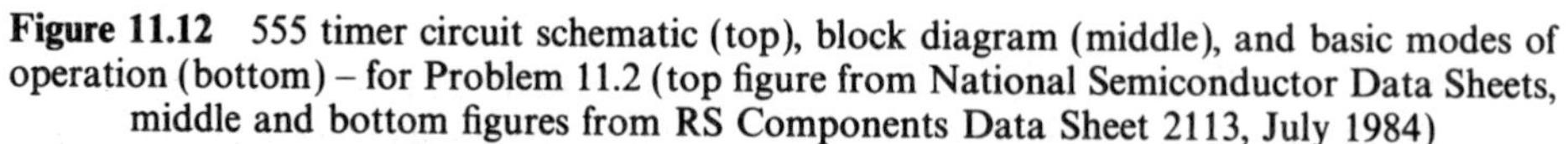

Figure 11.12 555 timer circuit schematic (top), block diagram (middle), and basic modes of operation (bottom) – for Problem 11.2 (top figure from National Semiconductor Data Sheets, middle and bottom figures from RS Components Data Sheet 2113, July 1984)

of the 555 also given, and explain as much of its operation as you can. A complete answer will only come after a period of time and several attempts.

11.3 A varactor tuned circuit modulator, with a circuit as in Figure 11.3, uses a varactor diode with characteristic as in Figure 11.2 and has capacitance given by

$$C_v = \frac{100}{(1 + 2V_b)^{1/2}}\ \text{pF}$$

With a d.c. bias of 4 V on the diode, the circuit resonates at 5.0 MHz exactly. The tank capacitor has a value of 200 pF. The diode capacitance is now modulated by an audio signal given by

$$v_m = 0.045\cos(2\pi \times 10^3)t\ \text{V}$$

Calculate:

(i) The total unbiased tuned circuit capacitance, C_0', and the value of L.
(ii) The frequency deviation, Δf, and the modulation index, β.
(iii) The amplitudes of the modulated signal spectral components within the Carson bandwidth. Sketch this spectrum.

11.4 A baseband signal in the range 20 Hz to 20 kHz is frequency modulated on to a subcarrier at 200 kHz to give a maximum phase shift of 0.2 rad. Calculate the value of B and Δf.

This signal is now converted to one with $\Delta f = 80$ kHz at $f_c = 108$ MHz. Draw the block diagram of a circuit which will do this, giving all relevant numerical values.

11.5 A frequency modulator using a 566 VCO (see Section 11.5) has an undeviated carrier frequency of 100 kHz. Making reasonable assumptions calculate suitable values for the external C and R used. Calculate also the largest signal voltage which can be used to modulate the VCO at a 1 kHz baseband frequency at the NBFM threshold.

11.6 (a) Draw a block diagram of the Armstrong narrow-band phase modulator and analyse its operation when the modulating signal is a single tone.
(b) A narrow-band frequency-modulated signal has a carrier frequency of 100kHz, modulation index 0.005, $\Delta f = 25$ kHz, and bandwidth 5.0 kHz. A wide-band FM signal with a modulation index of 10 and carrier frequency 100 MHz is to be generated from the narrow-band signal in part (a). Draw the block schematic of a system utilizing a frequency multiplier, a down converter, and a bandpass filter which allows this to be done. Give the required value of frequency

multiplication. Also, define the mixer fully by giving two permissible frequencies for the local oscillator and define the centre frequency and bandwidth of the bandpass filter.

11.7 An oscillator is required which will produce a carrier frequency of 100 MHz and can be frequency modulated to a frequency deviation of 100 kHz. The largest modulating frequency is 10 kHz. Sketch a suitable circuit for this purpose, giving likely values of all components controlling the centre frequency and its deviation. Explain the operation of the circuit and analyse it to show the factors controlling the linear relationship between modulating voltage and frequency shift.

Name two other methods which could be used to achieve the same result.

• 12 •
Frequency demodulation

12.1 • Introduction

Frequency modulation causes a carrier frequency to deviate from its rest value by an amount linearly proportional to the voltage of a baseband signal. That is

$$\delta\omega \propto v_m$$

Here we will look at the reverse process whereby the original baseband is recovered from the modulated carrier in a process called **frequency demodulation** or **detection**. This requires the reverse operation of obtaining a voltage proportional to the frequency deviation of a signal. That is

$$v_o \propto \delta\omega$$

This general situation is shown in simple diagrammatic form in Figure 12.1.

The main requirement for a circuit which will perform this operation adequately is that its output voltage (v_o) is *linearly* proportional to the deviation ($\delta\omega$) of the input carrier signal as shown in Figure 12.2.

In general an FM demodulator needs to meet three performance requirements.

1. **Linearity** – as above. v_o is linearly proportional to $\delta\omega$ over the whole modulation range ($\pm\Delta\omega$) of the input signal (e.g. ± 225 kHz at VHF). This requirement is expressed as a constant, the **demodulation sensitivity** (k), where

$$v_o = k\delta\omega$$

2. **Range** – adequate linearity must extend over the whole deviation range of the input signal as stated above, that is over $\pm\Delta\omega$.
3. **Sensitivity** – the demodulator must produce a large enough output voltage to reproduce adequately the dynamic range of the original baseband. This requires the constant, k, (defined above) to be large enough (e.g. mV kHz^{-1} at VHF).

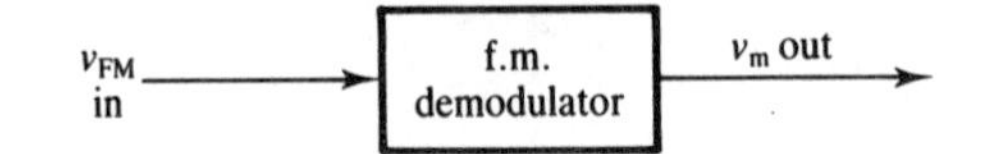

Figure 12.1 General idea of FM demodulation

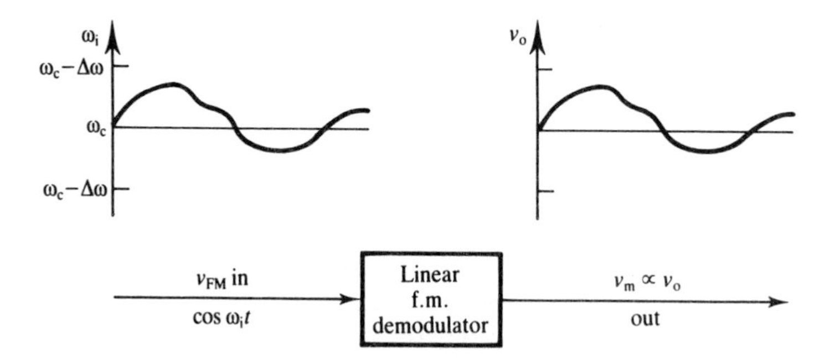

Figure 12.2 Linear FM demodulation

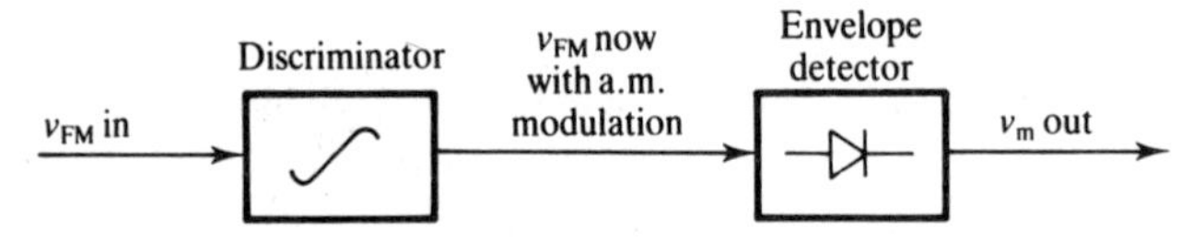

Figure 12.3 Stages of demodulation by FM–AM conversion

A lot of circuits will meet some of these requirements and many will meet them all. Here we first describe a few simpler older circuits and then look at representative circuits, in varying detail, of the four main classes of method used in current practice. These are:

1. The ratio detector.
2. The quadrature detector.
3. The zero crossing detector.
4. The phase-locked loop.

But first a short digression.

12.2 • The discriminator

Many demodulation circuits are of a general type in which demodulation is done in two stages, first converting the f.m. to a.m. and then envelope-detecting the a.m. demodulation. This is illustrated in Figure 12.3.

The first f.m. to a.m. stage is given the general name of **discriminator**, a term which is often used loosely to cover the whole of the demodulation process. The

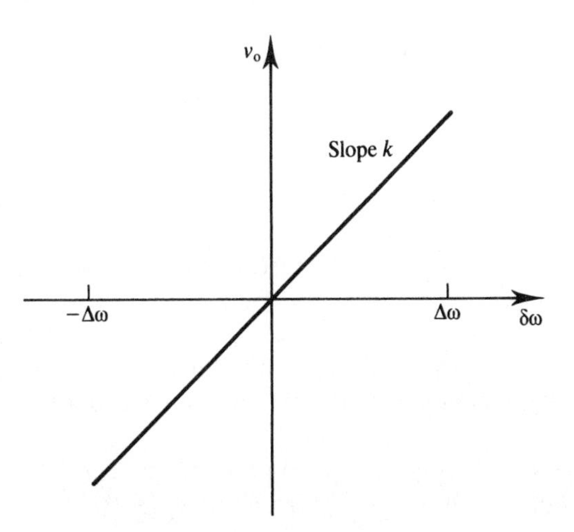

Figure 12.4 Discriminator action

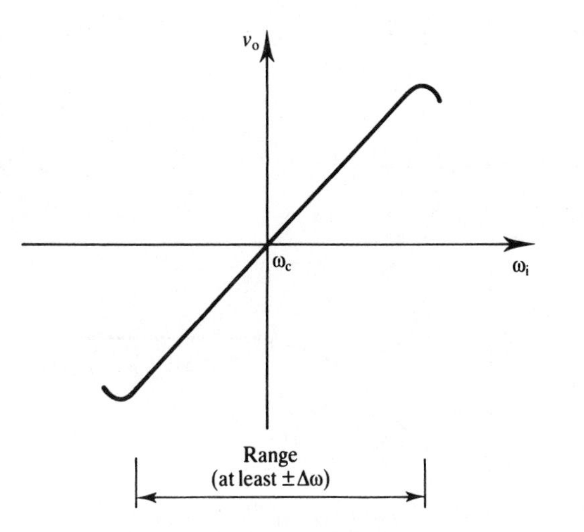

Figure 12.5 Discriminator S-characteristic

word discriminator is used here to mean a circuit which will treat signals of different frequencies differently, that is it discriminates between them. It is specifically applied in f.m. demodulation to a circuit which will convert linear FM to linear AM. This is illustrated in Figure 12.4.

The constant amplitude of the input to the discriminator is usually created deliberately by having a limiter amplifier beforehand. This prevents any unwanted AM on the input (such as noise) being detected. Because of the linearity requirement the purpose of a discriminator is best illustrated by means of its characteristic as given in Figure 12.5. This S-shaped curve is very typical of all types of discriminator circuit.

12.3 • Simple demodulators

There are two simple principles which will give useful discriminator action. These are differentiating circuits and tuned circuits.

12.3.1 • *Differentiating circuits*

These are based on the properties of inductance and capacitance whereby V and I are related by a factor ω resulting from the differentiation of one or the other of them. If one quantity is kept constant the other varies proportionally to ω as shown in Figure 12.6.

These properties can easily be used to produce f.m. demodulation using a simple filter such as the CR one shown in Figure 12.7.

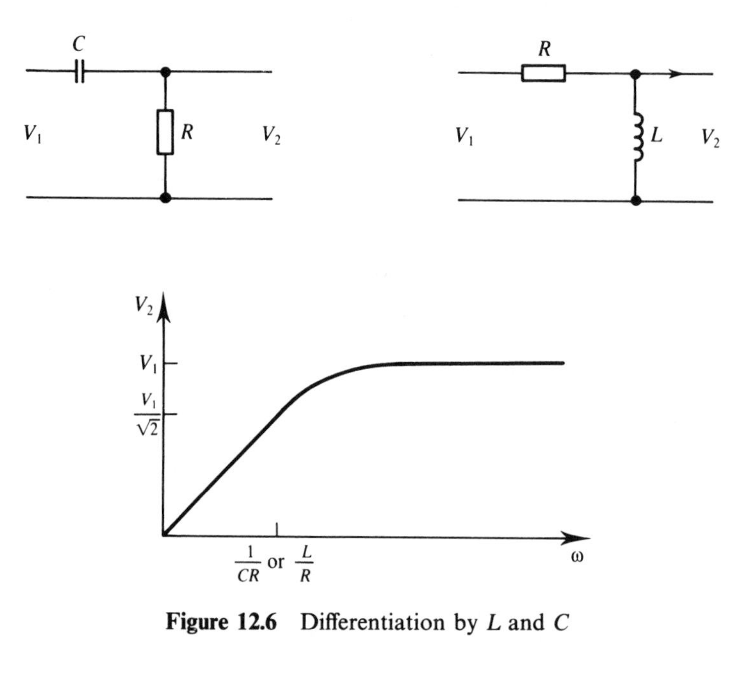

Figure 12.6 Differentiation by L and C

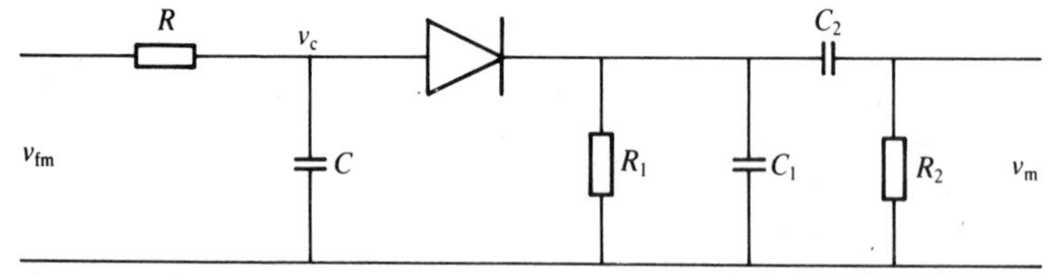

Figure 12.7 Simple low-pass CR filter as FM demodulator

The author has used a discriminator of this form very successfully as part of a laboratory experiment to illustrate frequency modulation and demodulation.

These discriminators are linear with a wide range but are not very sensitive. For example, the circuit of Figure 12.7 has (on its linear portion)

$$
\begin{aligned}
v_C &= v_{fm}/\omega CR \\
&= v_{fm}/[(\omega_c + \delta\omega)CR]^{-1} \simeq v_{fm}(\omega_c - \delta\omega)/CR
\end{aligned}
$$

therefore

$$
\begin{aligned}
\delta v_C &= (2\pi/CR)v_{fm}\delta f \\
&\simeq (2\pi v_0/CR)kv_m
\end{aligned}
$$

thus

$$\delta v_C \propto v_m$$

and demodulation has occurred. δv_C forms an a.m. envelope on top of the constant voltage input (v_{fm}) and it is then envelope detected by the diode etc.

12.3.2 • *Tuned circuits*

These have the useful property that their response changes rapidly with frequency either side of the resonance peak and is linear over a small range. This property can be used as a simple sensitive discriminator and, again, the author has used this in a laboratory experiment. Figure 12.8 illustrates the action.

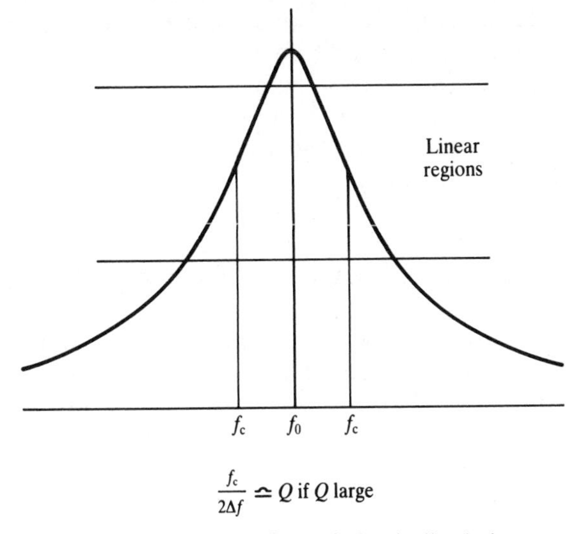

Figure 12.8 Action of tuned circuit discriminator

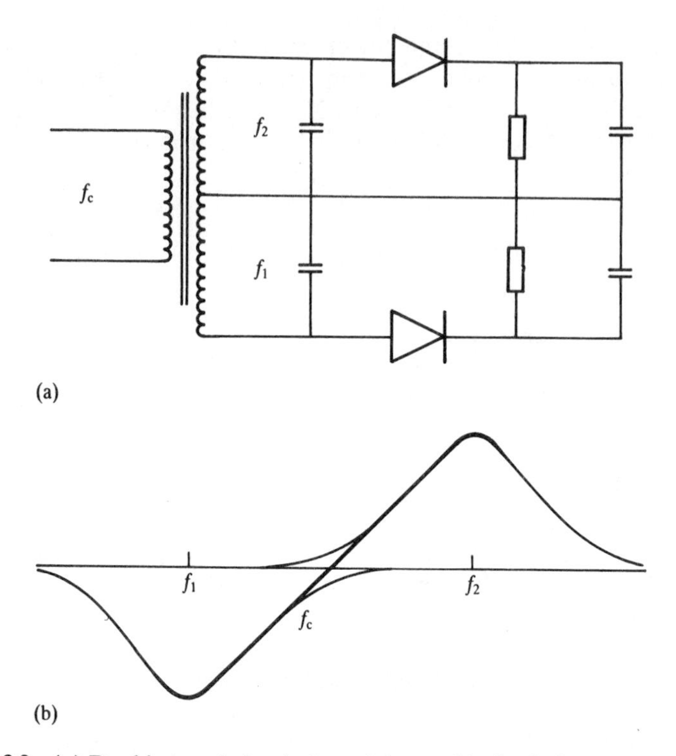

Figure 12.9 (a) Double tuned circuit demodulator; (b) discriminator characteristic

The drawback of this simple method is that its range is limited because of the limited extent of the linear slopes. This limitation can be overcome by using two tuned circuits back-to-back at slightly different resonant frequencies spaced equally either side of the carrier (or intermediate) frequency. This produces a much longer linear region between them as shown in Figure 12.9, which also shows a circuit used in radio sets years ago (Round-Travis). The back-to-back action is produced by the diodes of the envelope detectors.

12.4 • Tuned circuit phase shifts

Before considering the four main methods we need to look at another linear property of a tuned circuit used in both the ratio and quadrature detectors. This is that the current (I) through and voltage (V) across a tuned circuit are in phase at the resonance peak and change linearly in phase difference by a small amount ($\delta\phi$) for a small range of frequency ($\delta\omega$) either side of the peak. That is, at $\omega_c + \delta\omega$ when $\delta\omega \ll \omega_c$, I and V (see Figure 12.10) are $\delta\phi$ different in phase where $\delta\phi \propto \delta\omega$.

This can be shown as follows. The tuned circuit in Figure 12.10 has an admittance,

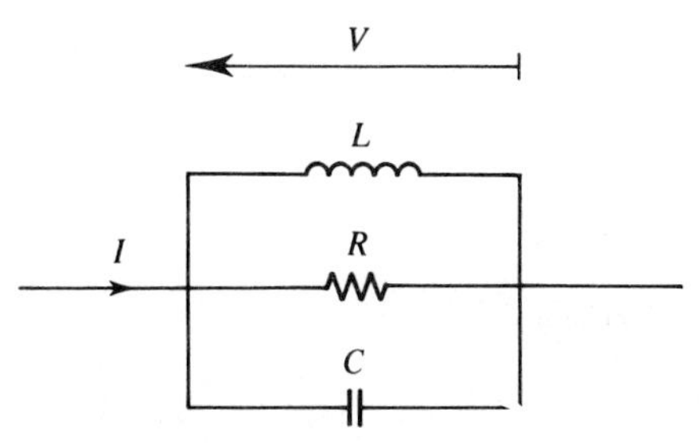

Figure 12.10 Equivalent circuit of parallel tuned circuit

Y, given by

$$
\begin{aligned}
Y &= \frac{I}{V} = \frac{|I|}{|V|}\angle\delta\phi = |Y|\angle\delta\phi \\
&= \frac{1}{R} + \frac{1}{\mathrm{j}\omega L} + \mathrm{j}\omega C \\
&= \frac{R(1-\omega^2 LC) + \mathrm{j}\omega L}{\mathrm{j}\omega LR} \\
&= \frac{\omega L + \mathrm{j}R(\omega^2 LC - 1)}{\omega LR} \\
&= \frac{[\omega^2 L^2 + R^2(\omega^2 LC-1)^2]^{1/2}}{\omega LR}\left(\frac{\omega L}{[\omega^2 L^2 + R^2(\omega^2 LC-1)^2]^{1/2}}\right. \\
&\quad \left. + \frac{\mathrm{j}R(\omega^2 LC-1)}{[\omega^2 L^2 + R^2(\omega^2 LC-1)^2]^{1/2}}\right) \\
&= |Y_{\mathrm{TC}}|(\cos\delta\phi + \mathrm{j}\sin\delta\phi) \\
&= |Y_{\mathrm{TC}}|\exp(\mathrm{j}\delta\phi)
\end{aligned}
$$

where

$$\tan\delta\phi = \frac{R(\omega^2 LC - 1)}{\omega L} = R(\omega C - 1/\omega L)$$

Now write $\omega = \omega_0 + \delta\omega$ and use $\omega_0 C = 1/\omega_0 L$. Also assume $\delta\phi$ is small so that $\tan\delta\phi = \delta\phi$. Then

$$\delta\phi = R\left((\omega_0 + \delta\omega)C - \frac{1}{(\omega_0 + \delta\omega)L}\right)$$

$$= R[(\omega_0 + \delta\omega)C] - \frac{R}{\omega_0 L}\left[1 + \left(\frac{\delta\omega}{\omega_0}\right)\right]^{-1}$$

$$= R\delta\omega\left(C + \frac{1}{\omega_0^2 L}\right)$$

$$= 2RC\delta\omega$$

That is

$$\boxed{\delta\phi = 2RC\delta\omega}$$

Of course, $\delta\omega$ is also proportional to the original baseband voltage, v_m, for an f.m. signal, so $\delta\phi$ can obviously be used for demodulation. In practice it is used to convert frequency deviation to amplitude changes (i.e. FM to AM) which are then envelope detected. Actually, doing this gets rather complicated as will be seen in the next two sections.

12.5 • The ratio detector

This is the first of the detectors to use the principle of linear phase shift with frequency near reasonance. A circuit for it is shown in Figure 12.11.

There is a superficial resemblance between this and the Round-Travis detector, although the hardware differences are obvious – only one tuned circuit (and this at f_c) and envelope detectors of opposite polarity. But the really clever difference is less obvious – the circuit arrangement allows unwanted high-frequency noise modulated

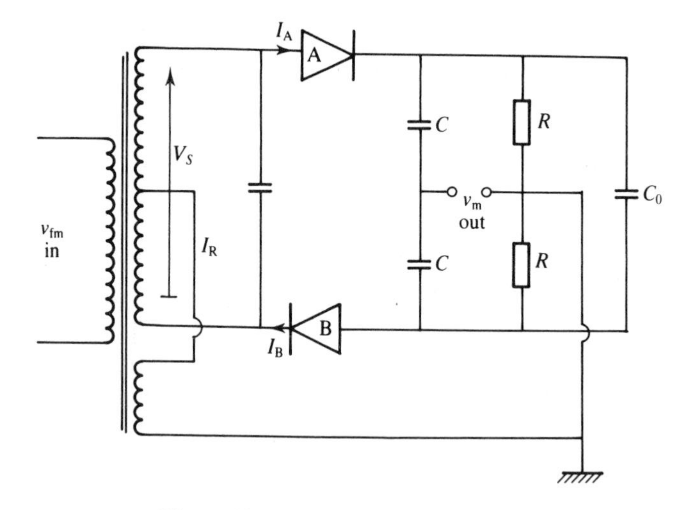

Figure 12.11 Ratio detector circuit

on to the f.m. signal to be smoothed away by the large tank capacitor, C_0. This removes the need for a limiter beforehand.

The action of the ratio detector is quite hard to understand fully. Demodulation takes place at the i.f. frequency (10.7 MHz at VHF) and the frequency-modulated signal (with 10.7 MHz as its carrier frequency) is fed from the final i.f. amplifier stage (usually the third) into the detector via a transformer with two secondary coils.

The lower coil (see Figure 12.11) provides a fairly large reference current (I_R in Figure 12.11) which is centre tapped into the upper secondary coil and divides equally to give currents of the same amplitude ($I_R/2 = I_{REF}$) and phase into each diode. These currents are not affected by the tuned circuit and act as the same phase reference into each envelope detector independent of the instantaneous frequency of the frequency-modulated carrier.

The upper secondary coil is part of a tuned circuit resonant at f_c which also produces equal amplitude currents (I_{TC}) into each diode but now of *opposite* polarity (i.e. π different in phase). These currents are *in phase* with the secondary voltage (V_S at f_i) whereas I_R is $\pi/2$ different in phase from it (because its secondary coil is an inductor). This means that the two currents into each diode are $\pi/2$ out of phase with each other (at f_c) in opposite directions. The effect of all this is shown in Figure 12.12. Note that I_{REF} is arranged to be much larger in amplitude than I_{TC}.

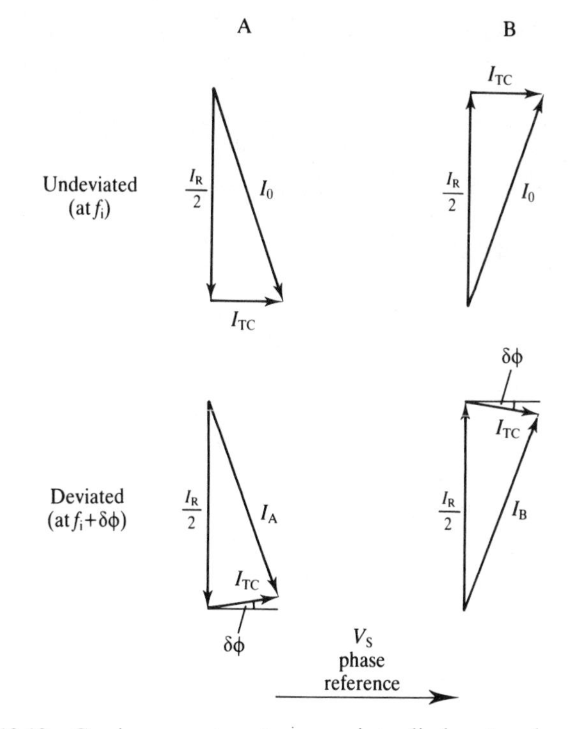

Figure 12.12 Carrier current vector sums into diodes at and near f_c

At f_i (i.e. the unmodulated intermediate carrier) the total current amplitudes into each diode are equal (i.e. $|I_{A0}| = |I_{B0}| = |I_0|$). Their phase difference is irrelevant because of the envelope detectors which produce equal but opposite d.c. currents ($|I_A| = -|I_B| = \pm|I_0|$) into the resistors, R. These produce equal d.c. voltages ($V_A = V_B = V_0$) in opposite directions across each resistor whose common point is earthed. This also puts their sum voltage ($V_A - V_B$) across C_0 and across the two C in series. The result is that the common point between the two C is also at earth potential (although not earthed) and the output voltage (v_0) is zero – as it should be for an unmodulated signal.

If the input signal is now modulated, some frequency deviation occurs and the tuned circuit operates just off resonance. This means that its current is now slightly out of phase with its voltage (by $\delta\phi$ where $\delta\phi \propto \delta f$) and, consequently (see Figure 12.12),

$$I_A = I_0 - \delta\phi \, . \, I_{TC} \quad \text{and} \quad I_B = I_0 + \delta\phi \, . \, I_{TC}$$

that is

$$I_A \neq I_B$$

whence

$$|I_A| \neq |I_B|$$

and

$$V_A \neq V_B$$

We can also write that

$$V_A = V_0 + \delta V$$

and

$$V_B = V_0 - \delta V$$

Therefore the voltage at the centre point of the capacitors, C, is now $\delta V/2$. The centre point of the resistors, R, is still at earth potential so that the output voltage, v_0, is

$$v_0 = \delta V/2$$

but

$$\delta V \propto \delta\phi$$

and

$$\delta\phi \propto \delta f$$

with

$$\delta f \propto v_{\mathrm{m}}$$

Therefore

$$\boxed{v_0 \propto v_m}$$

and the baseband is recovered! Because V_A and V_B change by equal but opposite amounts ($\pm\delta V$), their sum remains fixed at $2V_0$. This means that the voltage across the tank capacitor, C_0, is unchanged and its a.m. suppression function does not interfere with f.m. demodulation.

The name **ratio detector** arises somewhat indirectly because, whilst the sum of the detected voltages stays fixed, as above, their **ratio** changes with deviation. That is

$$\begin{aligned}\frac{|V_A|}{|V_B|} &= \frac{|V_0 + \delta V|}{|V_0 - \delta V|} \\ &\simeq 1 + 2\delta V/|V_0| \\ &= 1 + (\text{constant})\delta f\end{aligned}$$

Another common way to extract v_{m} from a ratio detector is to put a resistor in the common earth line (from the junction of the two R). Its earth current is the difference of the two detected currents which is proportional to δf. That is

$$|I_A| - |I_B| = 2\delta V \propto \delta f \propto v_{\mathrm{m}}$$

The volume control of the audio stage is often used as this earth-line resistor.

12.6 • The quadrature detector

This is another kind of **phase shift discriminator**. It uses the same principle of linear phase shift with frequency near resonance in a tuned circuit, as in the ratio detector, to obtain FM to AM conversion, but by quite a different method. The big advantage is that no transformer is needed; this allows most of the circuit to be made on one i.c. chip (e.g. 3089) which is, of course, very convenient for receiver design.

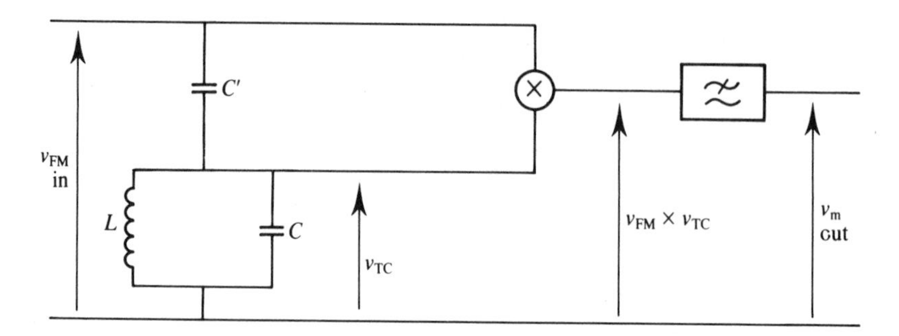

Figure 12.13 Principle of quadrature detector: C' gives fixed 90° phase shift to v_{TC}; tuned circuit adds $\delta\phi$ ($\propto \delta f$) to this

The outline of the method is shown in Figure 12.13.

Everything except the 90° phase shifter and the tuned circuit are on the chip including the i.f. amplifier/limiter (usually three differential stages). The fixed 90° phase shifter is either a C or L of impedance (at the i.f.) much larger than that of the tuned circuit. This means that the amplitude-limited frequency-modulated incoming signal is divided into two currents exactly 90° out of phase at f_i (i.e. when undeviated). These two are compared in a phase comparator producing an output of a train of pulses at repetition frequency $2f_i$ and of equal mark/space ratio.

When the incoming signal has deviation on it (i.e. it is frequency modulated), the tuned circuit introduces additional small phase shifts, just as for the ratio detector, causing the two inputs to the comparator to be of the same sign for slightly more or less than half each carrier period. This, in turn, makes the output pulses slightly wider or narrower depending on the sign of the deviation. These changes in pulse width will be linearly proportional to the deviation and hence to the original baseband voltage.

The pulse train is then smoothed (i.e. integrated) so that its amplitude averages to zero for the undeviated input and goes above or below zero for positive and negative deviations. This smoothed output is then a reproduction of the original baseband waveform and demodulation has taken place!

The action of the phase comparator is shown by the waveforms in Figure 12.14.

The diagrams in Figure 12.15 show the internal design of a simple quadrature detector chip. It is put in because its operation can be seen fairly easily.

The name **quadrature detector** comes, of course, from the fixed 90° phase shift which puts the two inputs to the phase comparator in phase quadrature.

12.7 • The zero crossing detector

This method is of rather more specialized application where exceptional linearity (0.1%) is required over a wide deviation range. This makes it especially suitable for

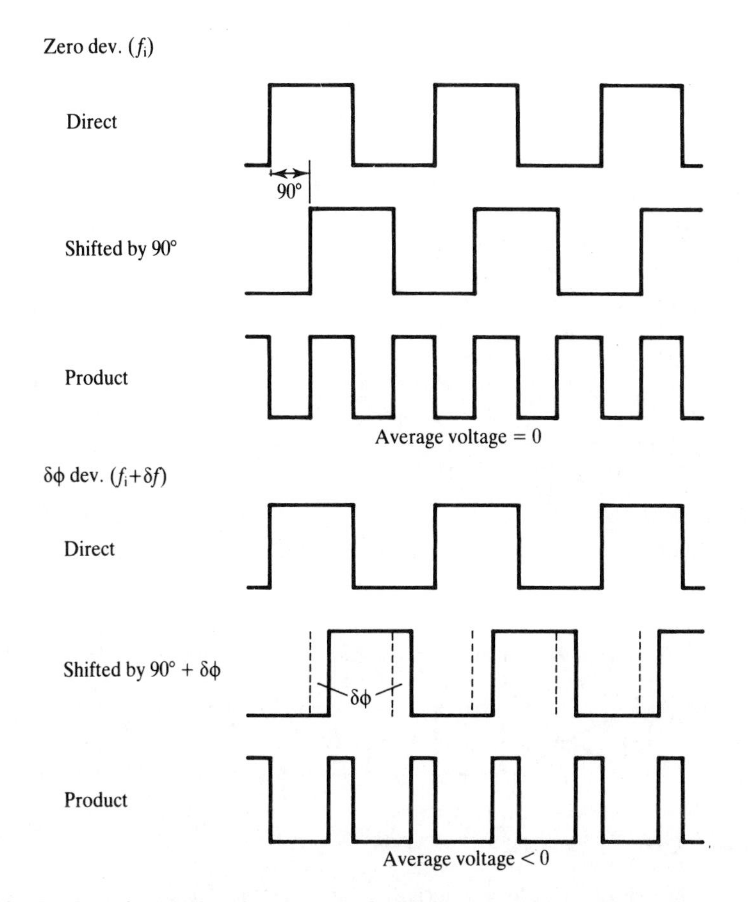

Figure 12.14 Waveforms showing action of phase comparator: d.c. average of product is proportional to $\delta\phi$ and so is proportional to v_m

such crucial uses as in line repeaters. The principle is much simpler but the practice more complex than for other detectors.

The principle is that the higher the signal frequency, the more zero crossings its waveform makes in a given time. Thus if these crossings are "counted" over a fixed time interval, the "count" is proportional to frequency and hence changes linearly with frequency deviation and, consequently, with the original baseband voltage. Thus the "count" can be used for demodulation. A block diagram of the stages needed to achieve this is shown in Figure 12.16.

The incoming modulated carrier (usually at an intermediate frequency, e.g. 10.7 MHz) is limited, as usual, to remove a.m. noise. It is then differentiated to convert zero crossings to spiky pulses and the negative-going ones removed by rectification. The positive spikes are converted to narrow rectangular pulses by a Schmitt trigger and the consequent pulse train smoothed (the integrator) to give a d.c. average of

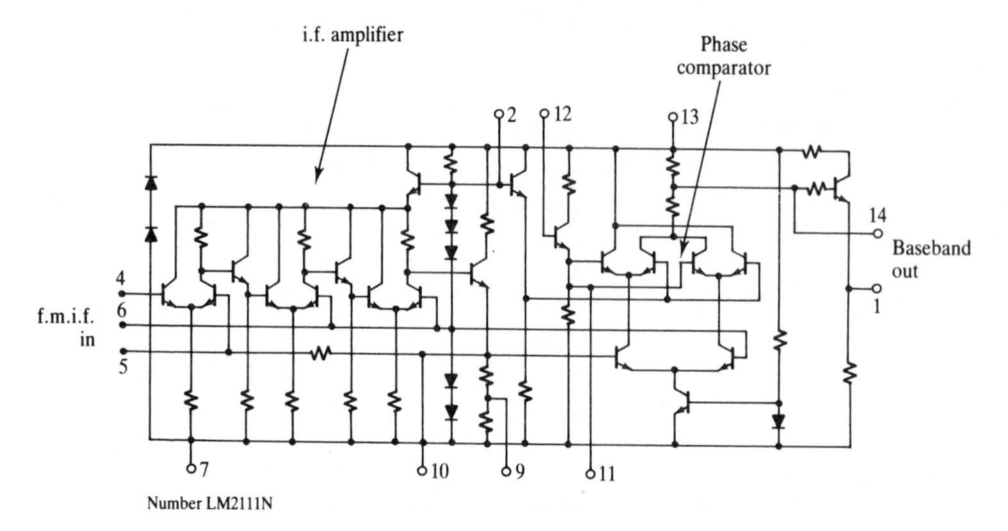

Figure 12.15 Internal details of a simple quadrature detector chip (from National Semiconductor Data Sheets)

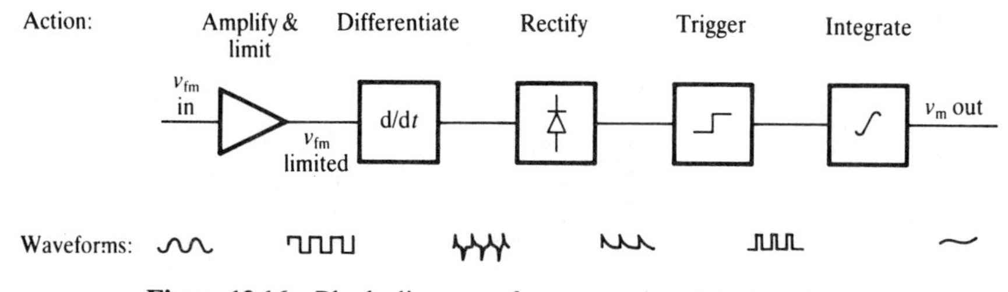

Figure 12.16 Block diagram of zero crossing detector stages

the pulse voltages. The faster the pulse rate, the larger the output voltage which means that, because of the linearities involved, the original baseband voltage waveform is reproduced faithfully.

12.8 • The phase-locked loop

This method is an example of a **feedback detector**. It is especially useful for noisy signals and where accuracy of phase and frequency is essential, as in the recovery of the 19 kHz subcarrier in a stereo broadcast signal at VHF. It has good linearity but may lack range for general-purpose f.m. demodulation. Phase-locked loops are made in i.c. form and do not need external inductors, which is an advantage over the quadrature detector.

The principle is to lock a voltage-controlled oscillator (VCO) to the signal carrier frequency (usually the i.f.) using a feedback loop. Lock is controlled by the phase

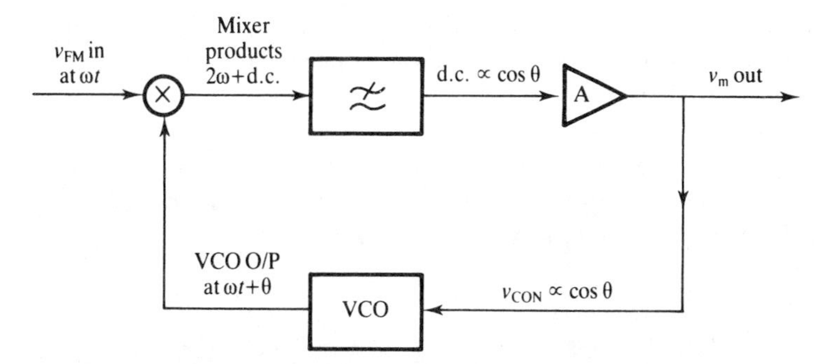

Figure 12.17 Block diagram of phase-locked loop operation

difference (θ) between the two signals. Hence the name **phase-locked loop** (PLL). The basic arrangement is shown in Figure 12.17.

The heart of the method is the VCO. This gives out a fixed amplitude square wave signal at a frequency varying linearly with the value of a d.c. control voltage (v_{CON}). When $v_{CON} = 0$ the VCO output is at its free-running or centre frequency which is set to be the carrier (or intermediate) frequency ω_i. That is

$$\omega = \omega_i + k \, . \, v_{CON}$$

where k is a constant of the system. The VCO works on the Schmitt trigger principle. A capacitor is charged and discharged linearly at a constant rate between a low reference voltage (used to fix the centre frequency) and v_{CON}. A comparator switches at these voltages, both switching the capacitor from charge to discharge (and vice versa) and providing the square wave voltage output. Strictly it is the period of the capacitor triangular voltage waveform which is varied by v_{CON}. The operation of this type of circuit has already been described in Chapter 11 (see Figure 11.10).

The feedback loop is used to arrange that v_{CON} is proportional to the frequency deviation, making v_{CON} also the demodulated signal. This equality is provided in the following way. We have that

$$\omega = \omega_i + K \, . \, v_m$$

and fix that

$$\omega = \omega_i + (\text{constant}) \cos \theta$$

By making

$$v_{CON} \propto \cos \theta$$

we get

$$v_{CON} \propto v_m$$

The key step in this series is the third one, that of making v_{CON} proportional to $\cos\theta$. This is done by means of the **phase comparator** which produces an output proportional to $\cos\theta$ – which is then fed back as v_{CON}.

The comparator is merely an analogue mixer with its two inputs at the same frequency but with a phase difference between them. Its output contains two signals, one at twice the carrier frequency ($2\omega_i$) and the other a d.c. voltage proportional to $\cos\theta$. The following analysis shows how this occurs.

Writing

$$v_{FM} = V_i \cos \omega_i t$$

and

$$v_{VCO} = V_0 \cos(\omega_i t + \theta)$$

multiplying

$$\begin{aligned} v_{FM} \cdot v_{VCO} &= V_i V_0 \cos \omega_i t \cos(\omega_i t + \theta) \\ &= \tfrac{1}{2} V_i V_0 [\cos(\omega_i t + \theta + \omega_i t) + \cos(\omega_i t + \theta - \omega_i t)] \\ &= \tfrac{1}{2} V_i V_0 [\cos(2\omega_i t + \theta) + \cos\theta] \end{aligned}$$

The $2\omega_i$ component is removed by the low-pass filter leaving only a d.c. voltage proportional to $\cos\theta$. This is amplified up to the level required for v_{CON} thus completing the feedback loop with

$$\boxed{v_{CON} \propto \cos\theta \propto v_m}$$

Note that $\theta = \pi/2$ at zero deviation. Some PLL circuits will also operate with $\theta = 0$ at zero deviation if required.

Note also that only the fundamental of the square wave VCO output has been considered. This is because any multiplication products with the higher harmonics would also be removed by the low-pass filter. However, it does mean that ω_i could itself be a higher harmonic with the VCO fundamental much lower than the carrier.

PLLs are altogether very versatile circuits with many applications, of which this is only one. Also, there is a great deal more to their detailed theory than is given here, where only a basic view of their operation is considered.

12.9 • Summary

Early f.m. demodulators were all some form of **discriminator** converting FM to AM and then envelope detecting. Simple and early examples of these are described. The operation of the common present-day discriminator, the **ratio detector**, is described in detail and is quite hard to understand fully. It depends for its linearity on the linear phase change with frequency near resonance in a tuned circuit and this is analysed for a parallel tuned circuit, getting

$$\delta\phi = 2CR\delta\omega$$

The same principle is used in another common demodulator, the **quadrature detector**, whose operation is also described in detail. Its convenience in use arises because it can largely be made in one i.c. with a minimum of readily available external components.

Two other methods using different principles are described in less detail. The **zero crossing detector** "counts" the frequency continuously by making a running average of the rate at which the carrier waveform crosses the zero voltage level. It has especial use for accurate demodulation of signals of small bandwidth. The **phase-locked loop** locks a VCO on to the instantaneous frequency of the incoming signal by phase difference in a feedback loop. This is also an accurate method and is conveniently in i.c. form.

12.10 • Conclusion

This is the end of the description of frequency modulation. There is still one more analogue modulation method to look at and we do so in the next chapter. This is phase modulation which is rapidly becoming as important, if not more so, than the other two because of its multilevel digit capability. Read on.

12.11 • Problems

12.1 Carry out the following calculations:

(i) $\Delta f = 75$ kHz, $K = 40$ mV kHz^{-1}. Find v_{out}.
(ii) $\beta = 5$, Carson B/W = 110 kHz, $v_{out} = 2.0$ mV. Find K.
(iii) $v_{out} = 2.5$ V p–p, $K = 10$ mV kHz^{-1}. Find maximum deviation.
(iv) $f_c = 100$ MHz, $\Delta f = 75$ kHz. Find K to give v_{out} of 1.0 V.
(v) $f_c = 100$ kHz, $\Delta f = 3$ kHz. Find K to give $v_{out} = 1.0$ V.

12.2 In Figure 12.18 $|v_{FM}| = 5$ V, $L = 0.3\ \mu$H, $R = 100\ \Omega$, $f_c = 100$ MHz, and $\Delta f = 10$ MHz.

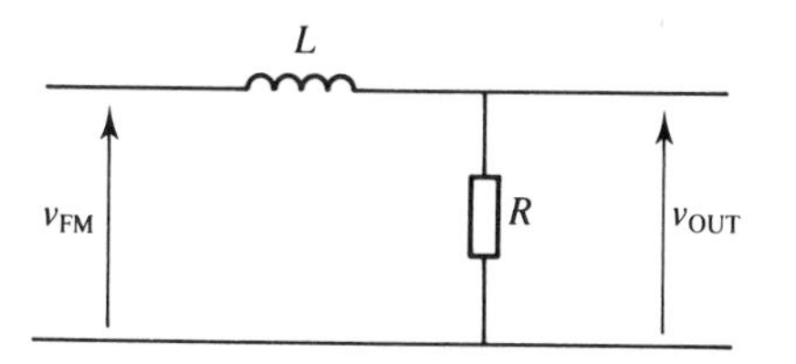

Figure 12.18 Circuit for Problem 12.2

Show that $|v_{out}| \simeq 2.5$ V maximum and that it has a maximum amplitude modulation index on it of about 0.08.

Obtain a value for the demodulation sensitivity.

12.3 For a quadrature detector explain the need for a fixed phase shifter before the tuned circuit.

If the input signal is standard FM of ± 75 kHz deviation on an i.f. carrier of 10.7 MHz, estimate:

(i) The Q of the tuned circuit to keep adequate linearity.
(ii) A value for a capacitor to form the fixed phase shifter. Make sure you understand any assumptions you need to make.

12.4 For a phase-locked loop sketch the form of the variation of the control voltage into the VCO against (i) frequency deviation, (ii) phase difference at the comparator. Explain why it is useful to set the phase difference at $\pi/2$ for an undeviated input signal. Consider the merits and demerits of setting it at zero instead.

12.5 Name three principles which can be used to demodulate f.m. signals. Describe in detail the operation of a practical system using *one* of these principles and assess its merits relative to methods using the other two principles.

12.6 Explain what a discriminator is when used for demodulation of a frequency-modulated signal. Give the expected characteristic. Explain the operation of one type of f.m. demodulator which is available in integrated circuit form.

12.7 The input to the circuit shown in Figure 12.19 is a sinusoid of frequency $f_0 + \delta f$, where f_0 is the resonant frequency of the LCR circuit and $\delta f \ll f_0$. C_1 is such that $1/\omega C_1 \gg R$.

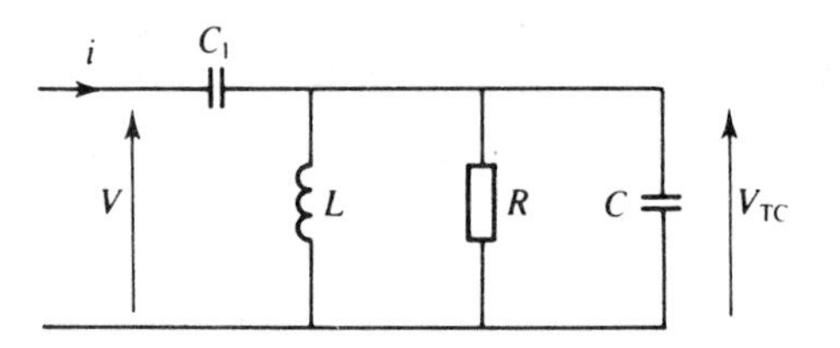

Figure 12.19 Circuit for Problem 12.7

Show that V_{TC} is related to V by

$$V_{TC} = \frac{\omega C_1}{R} V \angle(\pi/2 + \delta\phi)$$

where $\delta\phi = -4\pi RC\delta f$. Show also that if V and V_{TC} are multiplied together, their product is given by

$$V \times V_{TC} = V_0^2 CC_1 2\pi\delta f$$

where $V = V_0 \cos \omega t$.

Explain briefly how this product relationship above can be used to make an f.m. demodulator.

12.8 Discuss the desirable properties of an f.m. demodulator to be used with high-quality audio broadcast reception.

Sketch a circuit for one type of phase shift discriminator and explain its method of operation.

• 13 •
Phase modulation

13.1 • Introduction

Phase modulation is the third and last of the analogue modulation methods. It is also an angle modulation like FM, and is very similar to it in lots of ways. It is obtained by altering the phase constant, ϕ_c, of the carrier in proportion to a baseband voltage shown, as before, on the carrier equation as

$$v_c = E_c \cos(\omega_c t + \underbrace{\phi_c}_{\text{PM}})$$

Modulation changes ϕ_c by $\delta\phi$ where

$$\delta\phi = K_P v_m$$

Definition of phase modulation

The analysis is very similar to that for FM and so will not be given in full detail here.

13.2 • General analysis

Assume $\phi_c = 0$ when $v_m = 0$. Then

$$v_{PM} = E_c \cos(\omega_c t + K_P v_m)$$

When v_m is a sinusoid written as

$$v_m = E_m \cos \omega_m t$$

the modulated signal becomes

$$v_{PM} = E_c \cos(\omega_c t + K_P E_m \cos \omega_m t)$$

which is more usually written as

$$v_{PM} = E_c \cos(\omega_c t + \beta_P \cos \omega_m t)$$

where β_P is the **phase modulation index**.

Then, by analogy to FM, there will be a maximum **phase deviation**, $\Delta\phi$, where

$$\Delta\phi = \beta_P = K_P E_m$$

But, because there is a 2π ambiguity, that is

$$\Delta\phi + 2\pi \equiv \Delta\phi$$

useful phase changes must lie between $+\pi$ and $-\pi$.

Expansion of the equation in β_P above gives

$$v_{PM} = E_c[\cos \omega_c t \cos(\beta_P \cos \omega_m t) - \sin \omega_c t \sin(\beta_P \cos \omega_m t)]$$

which is very similar to the corresponding equation for FM and leads to narrow-band and wideband modulations – NBPM and WBPM.

13.3 • Narrow-band phase modulation

The general expression for NBPM is

$$\begin{aligned} v_{NBPM} &= E_c \cos \omega_c t - \beta_P E_c \sin \omega_c t \cos \omega_m t \\ &= E_c \cos \omega_c t - \tfrac{1}{2}\beta_P E_c[\sin(\omega_c - \omega_m)t + \sin(\omega_c + \omega_m)t] \end{aligned}$$

which is superficially similar to the expression for NBFM. However, further analysis shows the difference because

$$\begin{aligned} v_{NBPM} &= E_c \cos \omega_c t - \tfrac{1}{2}\beta_P E_c\{\cos[\pi/2 - (\omega_c - \omega_m)t] + \cos[\pi/2 - (\omega_c + \omega_m)t]\} \\ &= E_c \cos \omega_c t - \tfrac{1}{2}\beta_P E_c\{\cos[(\omega_c - \omega_m)t - \pi/2] + \cos[(\omega_c + \omega_m)t - \pi/2]\} \\ &= E_c \cos \omega_c t + \tfrac{1}{2}\beta_P E_c\{\cos[(\omega_c - \omega_m)t - \pi/2 + \pi] \\ &\quad + \cos[(\omega_c + \omega_m)t - \pi/2 + \pi]\} \\ &= E_c \cos \omega_c t + \tfrac{1}{2}\beta_P E_c\{\cos[(\omega_c - \omega_m)t + \pi/2] + \cos[(\omega_c + \omega_m)t + \pi/2]\} \end{aligned}$$

This is, of course, just a carrier and a single sideband pair whose amplitude spectrum is exactly the same as for AM and NBFM with bandwidth $2F_M$.

Obviously the phase spectrum *is* different and this is where the uniqueness of

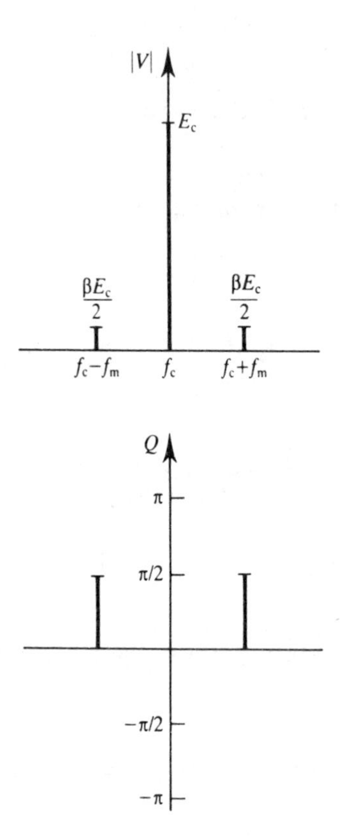

Figure 13.1 Amplitude and phase spectra for NBPM

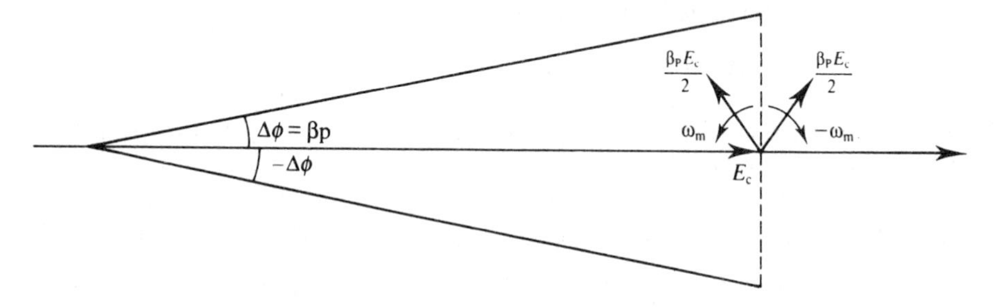

Figure 13.2 NBPM in stationary phasors

NBPM lies. The $\pi/2$ phase shifts cause the side frequencies to add vectorially at right angles to the carrier phasor, like FM, and not in line with it, like AM. This is shown on the phasor diagram in Figure 13.2.

The criterion for the signal to be narrow band, as for FM, lies in the value of β_P, that is NBPM needs $\beta_P \leqslant 0.2$. This means phase deviation is less than or equal to $\pm 11°$ (0.2 rad).

13.4 • Wideband phase modulation

The general expression for PM above is

$$v_{\text{PM}} = E_c[\cos \omega_c t \cos(\beta_P \cos \omega_m t) - \sin \omega_c t \sin(\beta_P \cos \omega_m t)]$$

As for FM, Bessel function expressions are needed to expand this further. These can easily be obtained from those for FM (Section 10.6) by writing $\cos \omega_m t = \sin(\pi/2 - \omega_m t)$ etc., and are

$$\cos(\beta_P \cos \omega_m t) = J_0(\beta_P) - 2J_2(\beta_P) \cos 2\omega_m t + 2J_4(\beta_P) - \ldots$$
$$\sin(\beta_P \cos \omega_m t) = 2J_1(\beta_P) \cos \omega_m t - 2J_3(\beta_P \cos 3\omega_m t + \ldots$$

leading to the full expression for WBPM showing the signs of all side frequencies:

$$\begin{aligned} v_{\text{WBPM}} = & \ldots + J_4(\beta_P) \cos(\omega_c - 4\omega_m)t \\ & + J_3(\beta_P) \cos[(\omega_c - 3\omega_m)t - \pi/2] - J_2(\beta_P) \cos(\omega_c - 2\omega_m)t \\ & - J_1(\beta_P) \cos[(\omega_c - \omega_m)t - \pi/2] + J_0(\beta_P) \cos \omega_c t \\ & - J_1(\beta_P) \cos[(\omega_c + \omega_m)t - \pi/2] - J_2(\beta_P) \cos(\omega_c + 2\omega_m)t \\ & + J_3(\beta_P) \cos[(\omega_c + 3\omega_m)t - \pi/2] + J_4(\beta_P) \cos(\omega_c + 4\omega_m)t \\ & - \ldots \end{aligned}$$

The amplitudes are, of course, exactly the same as for WBFM (for the same β value) with the same bandwidths and again the difference lies only in the phases which are not the same.

13.5 • Comparison of PM and FM

These two methods are closely related and, in some respects, very similar. They both have the effect of changing carrier phase as a function of baseband voltage. This can be seen clearly from their general expressions

$$v_{\text{FM}} = E_c \cos\left(\omega_c t + \beta \int v_m \, dt \right)$$
$$v_{\text{PM}} = E_c \cos[\omega_c t + \beta_P v_m]$$

Thus it is not surprising that the fuller expressions for NBPM and WBPM are very similar to those for NBFM and WBFM. These show that for both narrow and wide modulations the amplitude spectra are the same, the difference between PM and FM lying in the phases of the harmonic sidebands.

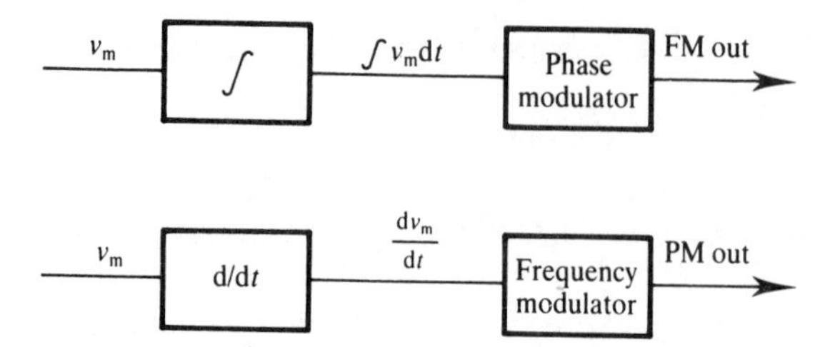

Figure 13.3 FM and PM obtained from each other

The similarity gives simple ways of obtaining one modulation type from the other by integrating or differentiating the baseband first. Figure 13.3 illustrates this.

The relationships can easily be shown analytically: if

$$v_m = E_m \cos \omega_m t$$

then

$$\int v_m \, dt = E_m/\omega_m \sin \omega_m t$$

and

$$\begin{aligned} v_{PM} &= E_c[\cos \omega_c t \cos(\beta_P/\omega_m \sin \omega_m t) - \sin \omega_c t \sin(\beta_P/\omega_m \sin \omega_m t)] \\ &= E_c[\cos \omega_c t \cos(\beta \sin \omega_c t) - \sin \omega_c t \sin(\beta \sin \omega_m t)] \\ &= v_{FM} \text{ (writing } \beta_P/\omega_m \equiv \beta) \end{aligned}$$

with a similar analysis for differentiation.

As we shall see, this method is a convenient way to get NBFM using an NBPM modulator because the circuitry is simpler and does not need tuned circuits.

For simple baseband waveforms the difference between FM and PM can be seen clearly (see Figure 10.1) but it is not at all easy to see with a continuously varying baseband. Try sketching them for a sinusoidal baseband. This does bring home the great similarity between the two methods.

13.6 • Use of PM

There are two common uses:

1. For NBFM generation in such applications as CB and mobile radio.
2. As WBPM for digital signals in BPSK as described in the next chapter.

The first needs simpler circuitry (i.e. no tuned circuits) than the direct f.m. method for applications where only narrow-band modulation is required. The second can also be done simply – using a multiplier – and has very wide applications.

13.7 • Generation and demodulation of PM

Narrow-band phase modulation can be obtained from a simple *RC* low-pass filter as in Figure 13.4, with *C* consisting of a fixed capacitor in series with a varactor (providing $\delta C \propto v_m$).

The unmodulated carrier goes in and a phase-modulated one comes out. The analysis is as follows:

$$\frac{v_{\text{NBPM}}}{v_c} = \frac{1/j\omega C}{R + 1/j\omega C} \qquad c = c_0 + c_r$$

$$= \frac{1}{1 + j\omega CR}$$

$$= \frac{1}{(1 + \omega^2 C^2 R^2)^{1/2}} \angle\theta \qquad (\tan\theta = -\omega CR)$$

If ωCR is kept small (<0.2) then $\theta \simeq -\omega CR$. So if C changes by δC, then θ changes by $\delta\theta$ where

$$\theta + \delta\theta = -\omega(C + \delta C)R$$

Therefore

$$\delta\theta = -\omega_c RKv_m$$

that is

$$\delta\theta \propto -v_m$$

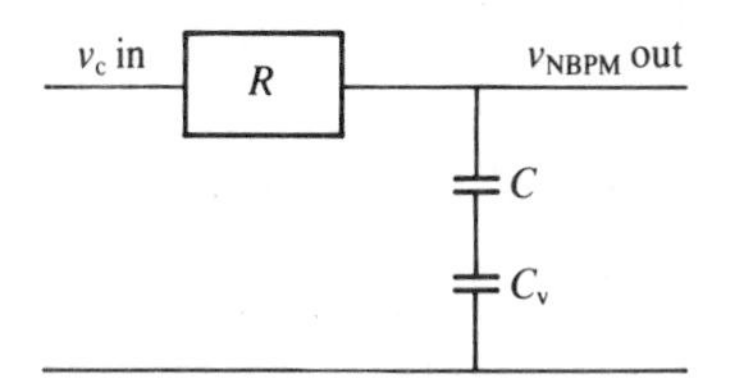

Figure 13.4 *RC* NBPM modulator

And thus linear phase modulation has been achieved. Of course the assumptions made mean that the fractional change in $\delta\theta$ must be small ($<11°$) so that only NBPM can be achieved. But, because the modulation is linear, it can be used for analogue basebands and in particular, with a pre-integrator, to give NBFM as already mentioned above.

By contrast WBPM is really only used for discrete basebands with fixed voltage levels between which it is switched. The simplest of these is the binary PSK method described in the next chapter and its detailed description will be left to then. There the phase is either 0° (representing a one) or 180° (representing a zero).

Demodulation of continuous PM is rarely needed as it is not used as a mode of transmission. And for the description of the demodulation of binary PSK see the next chapter.

13.8 • Summary

Phase modulation occurs when θ_c changes linearly proportional to v_m, that is

$$v_{PM} = E_c \cos(\omega_c t + K_P v_m)$$

where K_P is the **phase modulation sensitivity**.

For single sinusoid basebands

$$v_{PM} = E_c \cos(\omega_c t + \beta_P \cos \omega_m t)$$

where β_P is the **phase modulation index** and $\beta_P = K_P E_m$.

NBPM occurs when $\beta_P \leqslant 0.2$ and gives one sideband pair, both negative, as

$$v_{NBPM} = E_c \cos \omega_c t - \tfrac{1}{2}\beta_P E_c \sin(\omega_c - \omega_m)t - \tfrac{1}{2}\beta_P E_c \sin(\omega_c + \omega_m)t$$

Its **stationary phasor** diagram shows clearly how it differs from FM and AM.

WBPM gives $\beta + 1$ sideband pairs with the same amplitude spectrum as WBFM but with sign and phase differences (of $\pi/2$). The full expression for WBPM involves Bessel functions and is given in Section 13.4.

PM and FM are very similar and one can be obtained from the other by integrating or differentiating the baseband before modulating.

PM is a convenient way of getting NBFM (e.g. for mobile radio). It is also widely used for the modulation by frequency translation of digital signals where its multilevel capability is of great importance.

13.9 • Conclusion

This is as good a lead-in as any to the next chapter which describes a very important class of usage for these analogue modulation methods – the simple, but widely used, shift keying methods. Read on.

13.10 • Problems

13.1 For $f_m = 3$ kHz, $E_m = 3.0$ V, $\beta = 4.0$, $E_c = 2.0$ V, and $f_c = 200$ kHz, calculate:

(i) The phase deviation ($\Delta\phi$).
(ii) The phase modulation sensitivity (K_P).
(iii) The maximum and minimum instantaneous phase deviations.
(iv) The bandwidth.
(v) The amplitudes and signs of the carrier and all side frequencies within the bandwidth.

[Compare with Problem 10.1.]

13.2 A square wave baseband at 2.5 kHz repetition frequency and ± 6 V magnitude is frequency modulated on to a carrier at 25 kHz with phase modulation sensitivity of 0.25 rad V^{-1}:

(i) What is the modulation index?
(ii) What is the phase deviation?
(iii) What is the bandwidth?
(iv) Sketch the modulated waveform.
(v) Draw the instantaneous amplitude spectrum at the extremities of deviation.
(vi) Draw the p.m. spectrum showing amplitude and sign.

[Compare with Problem 10.5.]

13.3 Draw the phasor diagram for $\beta_P = 0.2$ showing especially:

(i) maximum and minimum phase deviations
(ii) the phasor positions at $\pm|v_m|$.
(iii) estimate how much AM occurs.

[Compare with Problem 10.6.]

13.4 See Problem 10.8.

13.5 Discuss the relative advantages of using FM and PM for

(i) speech basebands
(ii) digital basebands
(iii) keying modulations
(iv) linearity of modulation and demodulation
(v) obtaining one from the other
(vi) wideband modulating.

13.6 Sketch the stationary phasor diagram for NBPM at $\beta = 0.1$. Attempt also a sketch at $\beta = 1.0$.

• 14 •
Binary (keying) modulation

14.1 • Introduction

Here we look at a rather special area of application of the analogue modulation methods already described. The methods themselves are not new but the particular applications make them appear very different. This application is to digital basebands where the baseband voltage changes only between fixed levels. The most common such baseband is a binary digit (bit) stream giving the application its usual name of **binary modulation**. It is concerned with the modulation, by frequency translation, of digital basebands so that they can be transmitted readily by line or radio links.

14.2 • The binary baseband

Each analogue modulation method leads to its corresponding binary method. The basebands involved are all digital and the simplest of these is the binary digit which will be used here to develop the basic theory of the methods. In particular a steady 101010 bit stream will be used because that is the one with most frequent voltage changes and hence of greatest baseband bandwidth. It is the "worst case" situation.

Such a bit stream can either by **unipolar** ($1 \equiv V$; $0 \equiv 0$) or **bipolar** ($1 \equiv V$; $0 \equiv -V$), where V is a fixed voltage level set by the system used. Both these are shown in Figure 14.1.

To analyse the modulation methods we need to know the spectrum of each type. This is best expressed as its Fourier series (derive them for yourself if you are uncertain). For the waveforms in Figure 14.1:

Unipolar bit stream: $$s_1(t) = \tfrac{1}{2} + \frac{2}{\pi}[\cos \omega_0 t - \tfrac{1}{3}\cos 3\omega_0 t + \tfrac{1}{5}\cos 5\omega_0 t - \ldots]$$

Bipolar bit stream: $$s_3(t) = \frac{4}{\pi}[\cos \omega_0 t - \tfrac{1}{3}\cos 3\omega_0 t + \tfrac{1}{5}\cos 5\omega_0 t - \ldots]$$

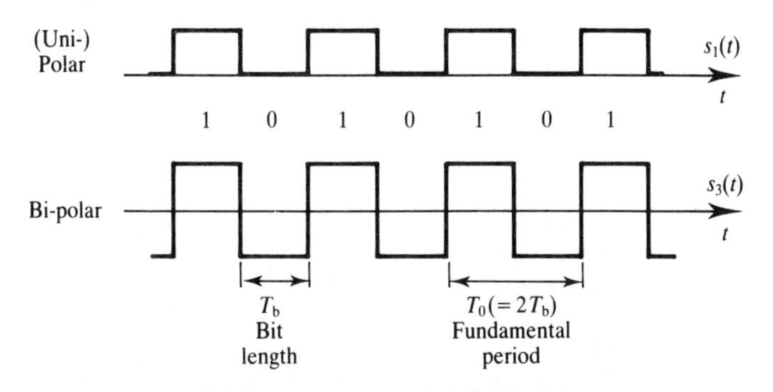

Figure 14.1 "Worst case" bit streams

where $\omega_0 = 2\pi/2T = \pi/T$ is the fundamental frequency of the bit stream [$s_2(t)$ will appear later in Section 14.7].

14.3 • Summary of methods

There are three basic methods, each corresponding to one of the three analogue modulation methods. They go by the somewhat curious names of

Amplitude shift keying (ASK)	using AM
Frequency shift keying (FSK)	using FM
Phase shift keying (PSK)	using PM

The term "shift keying" arises historically from the early days of signalling by telegraphy in the nineteenth century. Then symbols were transmitted by sending current pulses down a line by making and breaking a circuit with a *key* (i.e. a switch) which *shifted* (i.e. changed) the value of the current (from 0 to 1) to represent a digit. The Morse "key" and the teleprinter "key" still use the term. Then, as the technology developed from these simple unmodulated pulse origins to more sophisticated methods, the term "keying" was kept because of the similarity of usage. The technology was indissolubly wedded to it and so we are stuck with it today. At least it is a distinctive term. Figure 14.2 illustrates the differences between the three methods by showing a "worst case" bit stream modulated by each.

14.4 • An historical note

Apart from their present-day importance, methods of sending digital data date from the earliest days of electrical communication. Nineteenth-century wired systems (such as used on the railways) merely sent information as pulses of unmodulated d.c. current. When wireless (i.e. radio) communication became possible in the first decade

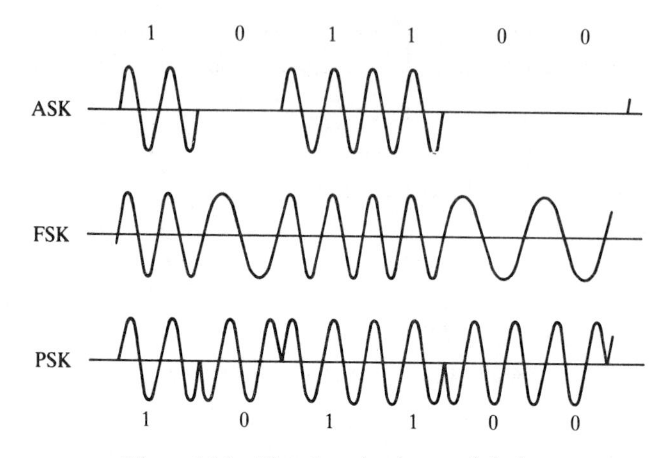

Figure 14.2 The three keying modulations

of the twentieth century and grew rapidly under the influence of Marconi to provide a commercial ship–shore service, it was done by on–off keyed Morse – a type of ASK. This very straightforward method is still in widespread use, its main disadvantage being noise degradation which distorts the modulation envelope and can cause errors. Similar methods were used in the early days of teleprinters.

When the need to send teleprinter messages down telephone lines grew in importance in the 1930s the digital signals had to be modulated at audio frequencies. ASK was tried but it soon became apparent that low-deviation FSK had advantages at the digit rates used. As a result the nominal voice frequency telephone channel (0–4 kHz) can be divided formally into elaborate sets of adjacent teletype channels separated by a small frequency difference depending in size on the baud rate of the digital baseband. These are still very much in use today.

Then, after the Second World War, there was a rapid increase in radio communication, especially at microwave frequencies on short- and medium-distance links and then to satellites. At first they all used FM for its noise immunity properties, but the "digital revolution" starting in the 1970s forced a changeover to digital modulation techniques. Higher-level PSK systems are now standard.

And now the latest techniques of the 1980s, optical fibres, have gone back to simple ASK – because, as yet, the technology cannot do anything more advanced. The light is just switched on and off by the data. But coherent modulation by microwave basebands will come soon.

14.5 • Areas of use

Nowadays keying methods are used in three different areas and the theoretical requirements differ in some respects between them. The three areas are as follows:

1. Audio frequency multiplexing on to voice frequency channels. For teletype or data. Mainly FSK.
2. Radio teletype (RTTY) transmission at HF and VHF. Teletype mainly – FSK.
3. Digital signals on microwave radio – terrestrial and satellite. Mainly PCM using PSK and PAM.

From this list it is clear that there are two main areas of technology from which these basebands come:

1. Teletype codes – digital but not binary with up to $7\frac{1}{2}$ digits per character depending on the system.
2. Binary digit streams – either direct as data or from PCM-coded analogue signals.

Some attempt is made in the descriptions which follow to show the differences in emphasis caused by particular usages, but the whole subject is very detailed and it will be necessary to consult reference books to find out more.

14.6 • Amplitude shift keying

This is the simplest of the three methods and also the oldest by far. It is not much used nowadays at a.f. and r.f., because it is so easily corrupted by unwanted noise, but it is worth looking at for its simplicity of analysis. It is the main method for optical fibre transmission.

The amplitude of a carrier sinusoid is shifted from 0 volts for a logic 0 to some fixed value, V, for a logic 1 as in Figure 14.3 (or vice versa).

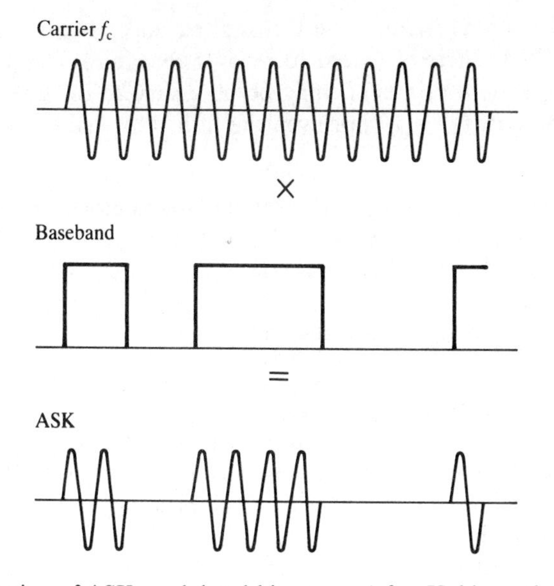

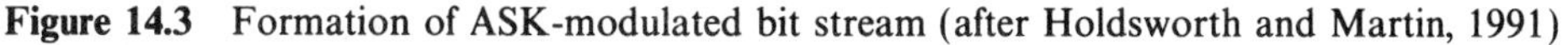

Figure 14.3 Formation of ASK-modulated bit stream (after Holdsworth and Martin, 1991)

This method gives 100% amplitude modulation and is, therefore, often called on–off keying (OOK).

Modulation is done merely by multiplying the carrier by the baseband. Writing the carrier with unity amplitude gives

$$v_c = \cos \omega_c t$$

Then

$$\begin{aligned} v_{ASK} &= v_c s_1(t) \\ &= \cos \omega_c t \left(\tfrac{1}{2} + \frac{2}{\pi}(\cos \omega_0 t - \tfrac{1}{3}\cos 3\omega_0 t + \tfrac{1}{5}\cos 5\omega_0 t + \ldots) \right) \\ &= \tfrac{1}{2}\cos \omega_c t + \frac{2}{\pi}\cos \omega_c t \cos \omega_0 t - \frac{2}{3\pi}\cos \omega_c t \cos 3\omega_0 t + \ldots \\ &= \tfrac{1}{2}\cos \omega_c t + \frac{1}{\pi}\cos(\omega_c - \omega_0)t + \frac{1}{\pi}\cos(\omega_c + \omega_0)t \\ &\quad - \frac{1}{3\pi}\cos(\omega_c - 3\omega_0)t - \frac{1}{3\pi}\cos(\omega_c + 3\omega_0)t + \frac{1}{5\pi}\ldots \end{aligned}$$

which is merely the standard spectrum of a square pulse train frequency translated up to f_c. This is shown in Figure 14.4 for a bit rate $1/T_b$ ($= 2f_0$) modulated on to a carrier f_c.

The bandwidth which needs to be transmitted depends on the requirements of the application. Where there is a need to recover recognizably square pulses at the receiver (e.g. to operate a printer) it may be necessary to include at least the third harmonic sideband pair ($f_c \pm 3f_0$) or even the fifth. This leads to the criterion

$$B_{ASK} = 6 \,.\, f_0 = 3 \text{ (bit rate maximum)}$$

that is

$$B_{ASK} = 3 \text{ (symbol rate)}$$

This bandwidth is marked on Figure 14.4.

Take as an example the standard symbol speed of 66 words per minute (wpm) used by many automatic teletype machines. This corresponds to 50 baud (symbols) per second which needs an f_0 of 25 Hz. Thus its third harmonic bandwidth is 75 Hz (or 120 Hz with guard bands).

But where all that is needed is to be able to decide whether a one or a zero was received and an exact bit shape is not required, then the first harmonic pair only

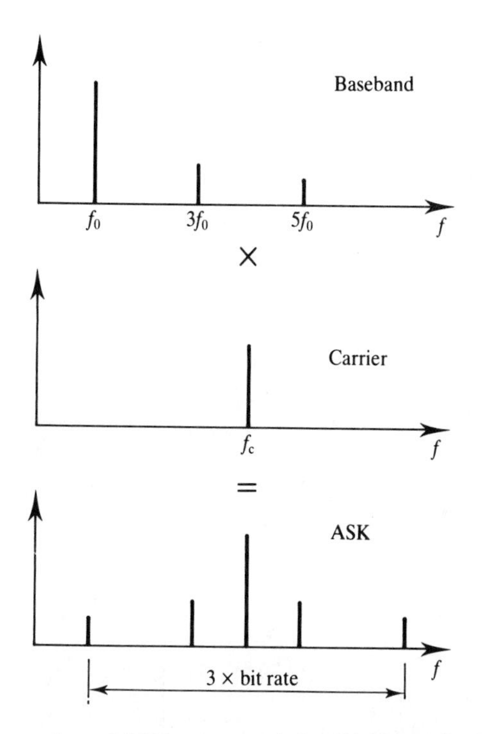

Figure 14.4 Formation of ASK spectrum (after Holdsworth and Martin, 1991)

may be adequate. This leads to the criterion

$$B_{\text{ASK}} = \text{bit rate}$$

The recommended maximum rates for audio FSK modems are set this way (see Figure 14.7). It is also used for PSK in microwave digital radio but for rather different reasons (see Section 14.9).

14.7 • Frequency shift keying

ASK is very vulnerable to noise corruption, especially from impulses which upset the start/stop arrangements of a teleprinter. To overcome this the simple solution is to send ones and zeros at different frequencies, shifting from one to the other as required. Hence we get **frequency shift keying** which can be regarded as a form of FM using a fixed deviation and only two modulated frequencies. The effect has already been shown in Figure 14.2. This is how it is obtained in audio frequency teleprinter modems (see Section 14.8).

However, the simplest way to *analyse* this modulation (i.e. to see what its spectrum and bandwidth are) is to treat it as two separate interleaved digit streams – one for

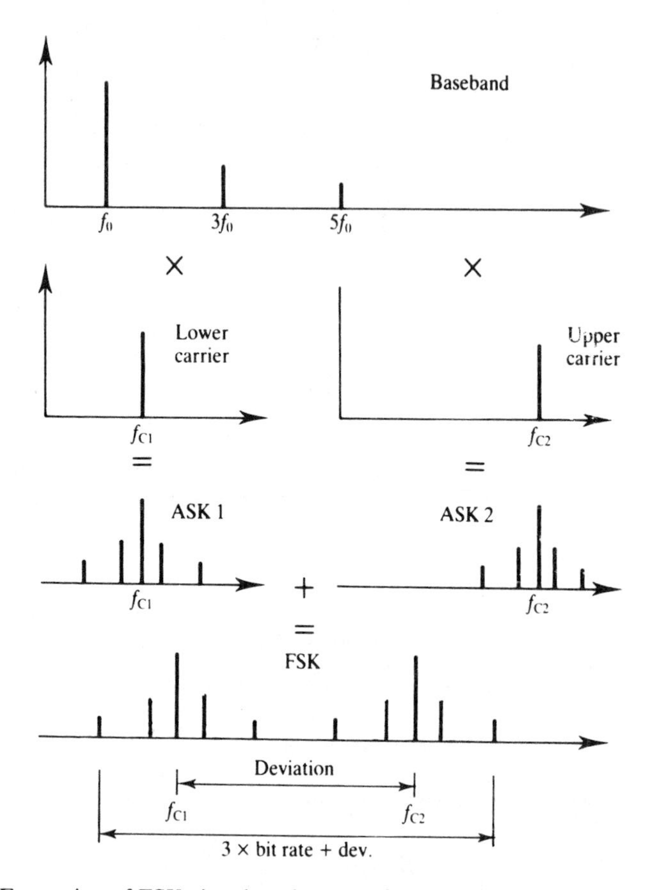

Figure 14.5 Formation of FSK signal as the sum of two ASK signals (after Holdsworth and Martin, 1991)

the ones and one for the zeros. Then each is ASK modulated at a different frequency and the two added to give the complete FSK waveform. This process is shown in Figure 14.5.

There are now two "carrier" frequencies and so

$$v_{\text{FSK}} = \cos \omega_1 t \, . \, s_1(t) + \cos \omega_2 t \, . \, s_2(t)$$

where $s_2(t)$ is the complement of $s_1(t)$ so that $s_2(t) = 1 - s_1(t)$. Therefore

$$v_{\text{FSK}} = \cos \omega_1 t \left(\tfrac{1}{2} + \frac{2}{\pi}(\cos \omega_0 t - \tfrac{1}{3} \cos 3\omega_0 t + \ldots) \right)$$

$$+ \cos \omega_2 t \left(\tfrac{1}{2} - \frac{2}{\pi}(\cos \omega_0 t - \tfrac{1}{3} \cos 3\omega_0 t + \ldots) \right)$$

that is

$$v_{\mathrm{FSK}} = \tfrac{1}{2}\cos\omega_1 t + \frac{1}{\pi}\cos(\omega_1 - \omega_0)t + \frac{1}{\pi}\cos(\omega_1 + \omega_0)t$$

$$- \frac{1}{3\pi}\cos(\omega_1 - 3\omega_0)t - \frac{1}{3\pi}\cos(\omega_1 + 3\omega_0)t + \ldots$$

$$+ \tfrac{1}{2}\cos\omega_2 t - \frac{1}{\pi}\cos(\omega_2 - \omega_0)t - \frac{1}{\pi}\cos(\omega_2 + \omega_0)t$$

$$+ \frac{1}{3\pi}\cos(\omega_2 - 3\omega_0)t + \frac{1}{3\pi}\cos(\omega_2 + 3\omega_0)t + \ldots$$

which is, as expected, merely two ASK spectra centred around ω_1 and ω_2 and separated by $2\Delta f$ $(= f_2 - f_1)$, the **frequency shift**. Figure 14.6 shows this spectrum for a bit rate $1/T_b$ of fundamental frequency f_0 $(= 1/T = 1/2T_b)$.

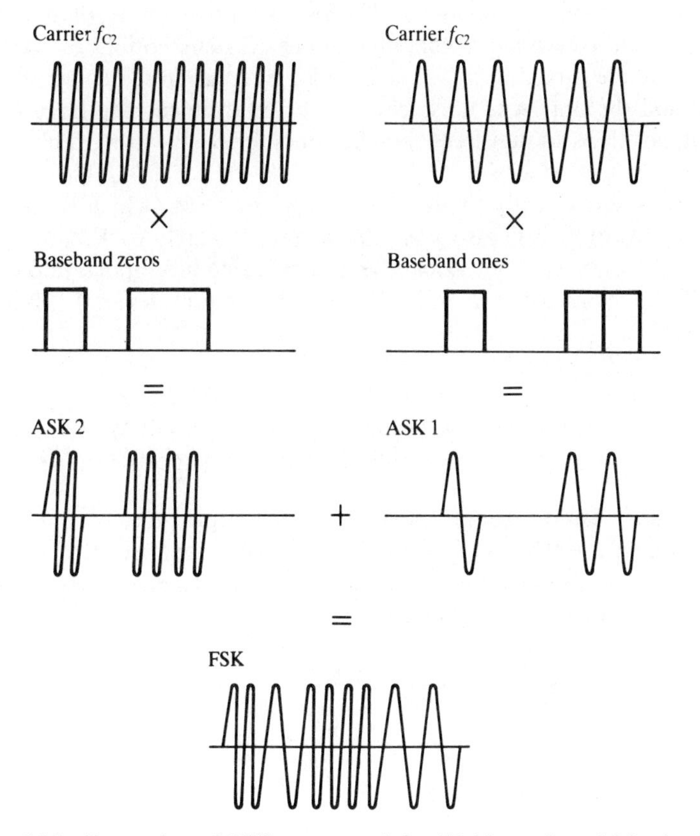

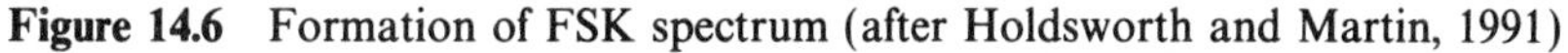
Figure 14.6 Formation of FSK spectrum (after Holdsworth and Martin, 1991)

For its **bandwidth** let us first take the third harmonic sideband pair criterion. Then

$$\begin{aligned} \mathrm{B/W} &= 3f_0 + 2\Delta f + 3f_0 \\ &= 6f_0 + 2\Delta f \\ &= 6(1/2T_b) + \text{shift} \\ &= 3(1/T_b) + \text{shift} \end{aligned}$$

that is

$$B_{\mathrm{FSK}} = 3 \text{ (bit rate)} + \text{frequency shift}$$

If the first harmonic pair only had been taken the bandwidth would, of course, be just

$$\text{bit rate} + \text{shift}$$

Note that these are inevitably larger than B_{ASK} by the amount of the frequency shift, which is an unavoidable disadvantage of FSK. But this is more than compensated by its main advantage which is that it is much less susceptible to corruption by unwanted amplitude modulation – due to noise or transients. The size of the signal is irrelevant and the only way noise can give error is by causing large shifts in the zero crossing positions so that frequency becomes seriously wrong. This is much less of a problem.

The analysis above really treats FSK as AM, not FM (which is actually closer to the truth). However, it is also possible to treat it purely as FM by taking each harmonic of the "worst case" bit stream separately using bits bipolar about the centre frequency. If the levels are $+1$ and -1 then the amplitudes and "shifts" for the harmonics are

$$\begin{aligned} &f_0 \rightarrow \text{amplitude } 4/\pi \rightarrow \text{shift } \pm\Delta\omega/\pi \rightarrow \mathrm{B/W}\ 2(\Delta f + f_0) \\ &3f_0 \rightarrow \text{amplitude } 1.3/\pi \rightarrow \text{shift } \pm\Delta\omega/3\pi \rightarrow \mathrm{B/W}\ 2(\Delta f + 3f_0) \\ &5f_0 \rightarrow \text{amplitude } 0.8/\pi \rightarrow \text{shift } \pm\Delta\omega/5\pi \rightarrow \mathrm{B/W}\ 2(\Delta f + 5f_0) \end{aligned}$$

B/W is the Carson bandwidth of 2(shift + f_m) in Chapter 12.

As an example, a 100 baud ($f_0 = 50$ Hz) signal on a channel with shift of ± 60 Hz (a standard usage) gives the following bandwidths by the two methods. They compare well. Try working them out for yourself.

Harmonic	AM (2ASK) method	FM method
f_0	220	250
$3f_0$	420	350
$5f_0$	620	530

14.8 • Usage of FSK

FSK is much used in practice. There are two main areas of use:

1. At audio frequencies for multiplexing on to 4 kHz telephone channels for teletype or data.
2. At HF and VHF for RTTY (radio teletype) transmissions for teletype.

For the **voice frequency** multiplexing there is an elaborate hierarchy of mark/space frequencies and channel separations depending on the baud rate (and hence bandwidth) required. These are laid down by CCITT recommendations and are used internationally by agreement. Channel bandwidths range from 120 Hz (± 30 shift) to 960 Hz (± 240 shift) fitting respectively 24 and 3 teletype channels into one telephone channel. Some examples of the frequencies used are given in Figure 14.7 for standard voice frequency carrier telegraphy (VFCT) channels.

By convention the lower frequency is the "mark" (1) and the upper the "space" (0) with a nominal centre frequency which is not itself used directly. The deviations quoted are the differences from this centre frequency rather than that from mark to space.

Two examples of recommended usage are: up to 50 baud in 120 Hz channels with 60 Hz shift; up to 100 baud in 240 Hz channels with 120 Hz shift.

For **radio teletype** use at HF and VHF two types of FSK are used – f.s.k. and a.f.s.k. For f.s.k. the carrier itself is shifted directly by the baseband. For a.f.s.k. the shifting is done at audio frequencies first and then the result is SSB modulated on to the radio frequency carrier. Hence the "a" in a.f.s.k. means "audio". On the whole f.s.k. is used at HF whilst a.f.s.k. is more common at VHF because of carrier frequency stability problems. For amateur use shifts of 170 and 850 are used with, for example, the mark at 1275 Hz and the space at 1445 or 2125 Hz.

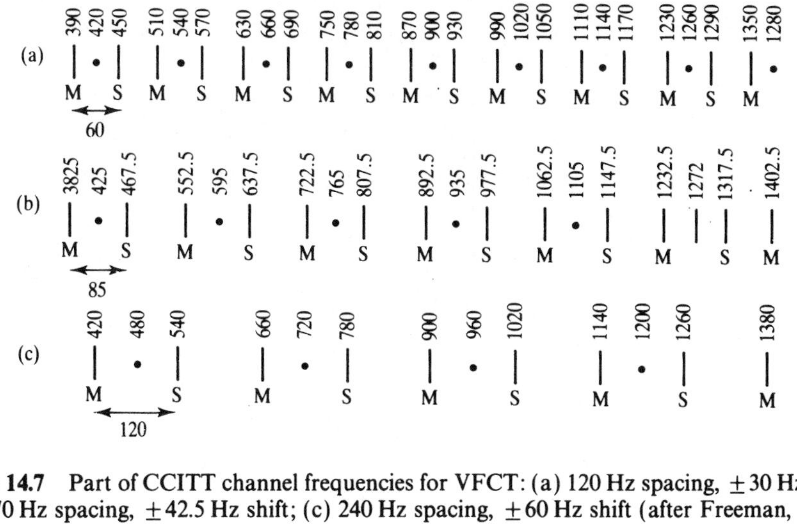

Figure 14.7 Part of CCITT channel frequencies for VFCT: (a) 120 Hz spacing, ± 30 Hz shift; (b) 170 Hz spacing, ± 42.5 Hz shift; (c) 240 Hz spacing, ± 60 Hz shift (after Freeman, 1981)

14.9 • Phase shift keying

PSK has advantages over both ASK and FSK. It has the same bandwidth as ASK which is, of course, less than that of FSK by the amount of shift. It is less susceptible to noise corruption than FSK and, of course, very much better than ASK.

PSK distinguishes types of digit by changing the phase of the carrier. For binary signals (BPSK) there are only two types of digit (1 and 0) and consequently only two phases (0° and 180°) are needed as shown in Figure 14.8.

These can also be shown in an informative manner on a phase diagram of the "constellation" type as in Figure 14.9.

Note that we now need a bipolar baseband because one phase is merely minus the other (180° phase change ≡ multiplying by −1). Also note how the phases can be shown on a phase diagram. This is a very useful pictorial representation especially for multilevel digit systems where their spangled appearance attracts the name "constellation".

The analysis uses the "worst case" bipolar bit stream ($s_3(t)$) and multiplies the carrier with it to get BPSK as follows:

$$v_{\mathrm{BPSK}} = v_c \, . \, s_3(t)$$

$$= \cos \omega_c t \, . \frac{4}{\pi}\left(\cos \omega_0 t - \tfrac{1}{3}\cos 3\omega_0 t + \tfrac{1}{5}\cos 5\omega_0 t - \ldots\right)$$

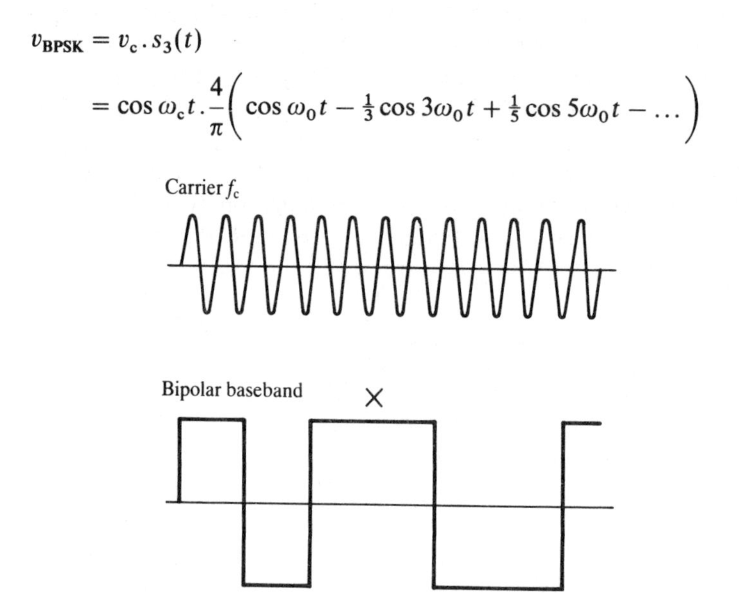

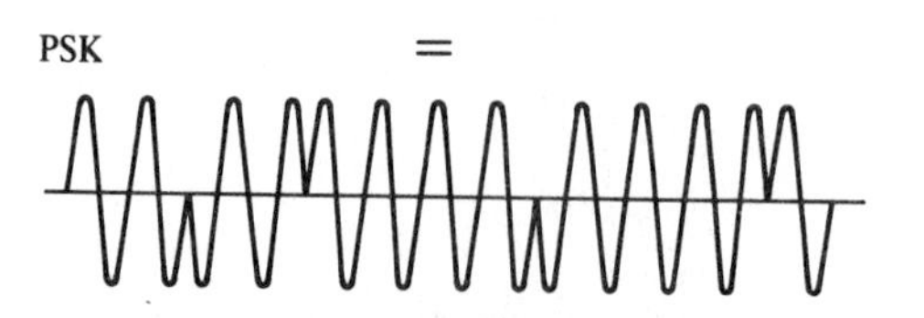

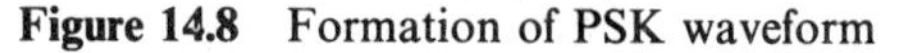

Figure 14.8 Formation of PSK waveform

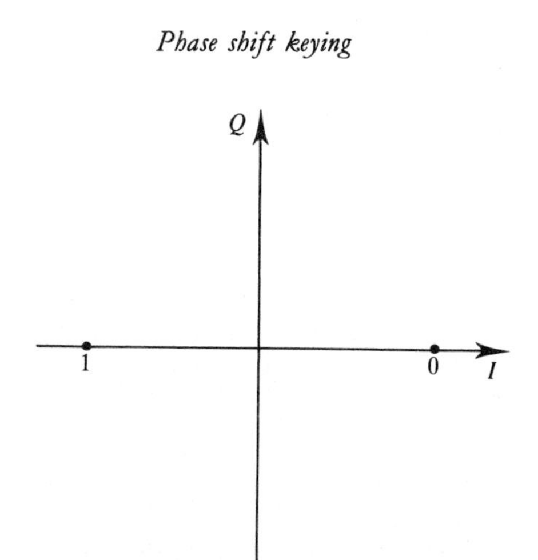

Figure 14.9 PSK "constellation" phase diagram

$$= \frac{4}{\pi}\left(\cos \omega_c t \,.\, \cos \omega_0 t - \tfrac{1}{3} \cos \omega_c t \,.\, \cos 3\omega_0 t + \ldots \right)$$

$$= \frac{2}{\pi}[\cos(\omega_c - \omega_0)t + \cos(\omega_c + \omega_0)t - \tfrac{1}{3}\cos(\omega_c - 3\omega_0)t$$

$$- \tfrac{1}{3}\cos(\omega_c + 3\omega_0)t + \ldots]$$

which is a series of diminishing sidebands as in Figure 14.10.

Obviously this is the same spectrum as ASK but without the carrier (the DSBSC version?) and its bandwidth may be calculated in the same way, so, if the third harmonic pair are included, we get:

$$B_{\mathrm{BPSK}} = 3 \text{ (bit rate)}$$

(but see below).

Phase shift keying has three advantages over the other two keying methods. Two of these have been mentioned before whilst the third is new:

1. Less susceptible to noise degradation.
2. Less bandwidth than FSK (same as ASK).
3. Bandwidth reduced by multilevel schemes.

All these make PSK a very desirable method for some applications – high bit rates and high carrier frequencies, for example. As such it has become the only method used for digital transmission on microwave radio, although it is used at other frequencies too.

For example, 8-level PSK is used at **audio frequencies** to get 4800 bits per second into a 4 kHz voice frequency channel with a carrier of 1800 Hz occupying most of the useful channel (300 to 3400).

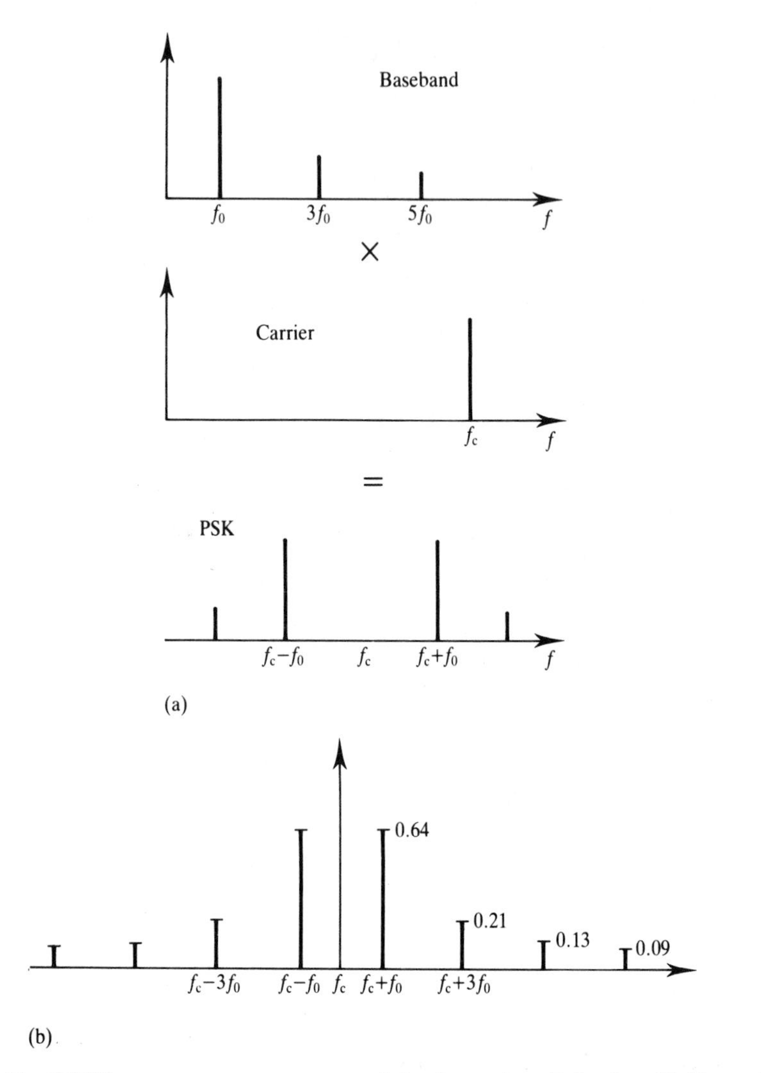

Figure 14.10 BPSK worst case spectrum and its formation ((a) after Holdsworth and Martin, 1991)

At **gigahertz radio frequencies** the requirement is for bit rates in the megahertz range for both terrestrial (16-QAM standard) and satellite (QPSK standard) links. Those abbreviations are explained in Section 14.10 and mean, respectively, 4 and 2 bits per symbol, so increasing information rates for a given bandwidth.

At these high frequencies, **bandwidth** calculations are based on quite different considerations from those we have looked at already. The main one is the avoidance of intersymbol interference (ISI), described in Section 18.4, which leads to the rather simple rule that **bandwidth equals symbol rate**.

14.10 • Multilevel PSK

Multilevel digits have been mentioned several times. This occurred mostly in connection with PSK but it is possible to have them in ASK and FSK too. Figure 14.11 shows these multilevel digits. However, it is in PSK where they have proved to be useful and it is that which is discussed here.

First of all: why are multilevel digits a good idea? It is because each of them can be made to convey more than one binary digit, so increasing the bit rate for a given symbol rate. For example, if 1 symbol is to convey 2 bits, then 4 distinct symbols are needed to cover all possible combinations of 2 bits (00,01,10,11) and 4-PSK (QPSK) is obtained. The need to provide for all combinations means that useful multilevel systems must have a number of symbols which is a power of 2. That is, 3 bits per symbol requires 8 levels; 4 bits needs 16 levels; and so on, with the general need being that n bits requires M levels where

$$M = 2^n$$

[Note: the term "level" arises from ASK use but is used generally to mean "distinguishable symbol".]

In PSK multilevel symbols arise first of all by making smaller changes in the carrier phase. Any phase from 0° to 360° can be used, not just 0° and 180°, and the most common of these uses 90° differences at 45°, 135°, 225°, and 315° to get **quadrature phase shift keying** or QPSK (see Figure 14.12).

Phase can be subdivided further (e.g. in 8-PSK) but there is a limit to how fine

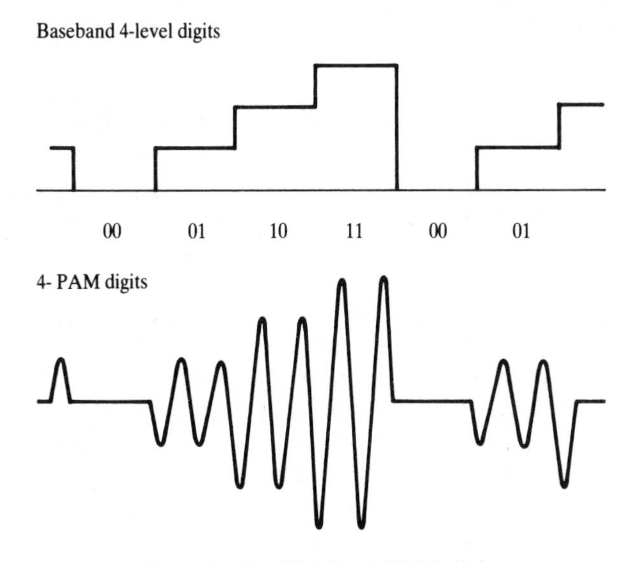

Figure 14.11 Multilevel PAM digits

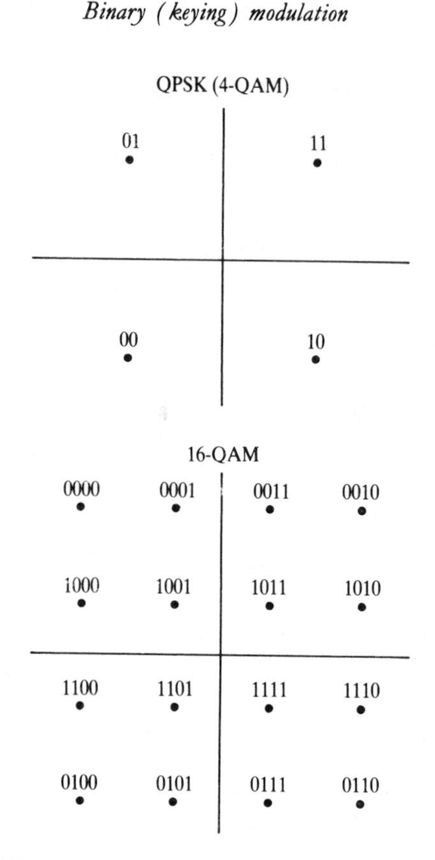

Figure 14.12 QPSK and 16-QAM constellations (binary numbers Gray coded)

the division can be because they have to be separated accurately at the receiver. To overcome this restriction, changes of amplitude are used as well in systems and are called **phase amplitude modulation** (PAM) or **quadrature amplitude modulation** (QAM). A common one of these is 16-QAM also shown in Figure 14.12. Note how the symbols are all separated by the same distance ($2d$) on both the in-phase (x) and quadrature (y) axes and that, for a given amplitude, the phase differences are quite large. This square grid constellation can be extended further by more $2d$ distances to get larger squares with 64, 256, etc. symbols – the mind boggles.

Of course there are snags in all this delightful gain in bit rate. The system is now amplitude dependent again and so is susceptible to noise and interference, which means that higher transmitter powers are needed. Also the levels get proportionately closer together the more levels there are and so the whole system has to be very linear, which is a particular difficulty for efficient use of the r.f. power amplifier in the transmitter (especially at the microwave frequencies normally used). Readers should look at more specialized books to learn details of all this.

14.11 • Summary

Binary, or **keying, modulations** result from the analogue modulation of digital signals, binary or multilevel. Because of their origins and discrete levels they are treated separately from other uses of analogue modulation. There are three distinct classes:

1. Amplitude shift keying (ASK) a form of AM
2. Frequency shift keying (FSK) a form of FM
3. Phase shift keying (PSK) a form of PM

ASK is the product of a carrier (v_c) and a unipolar bit stream ($s_1(t)$)

$$v_{ASK} = v_c \cdot s_1(t)$$

FSK is best regarded as the sum of two ASK modulations on two carrier frequencies separated by a **frequency deviation,** Δf:

$$\Delta f = f_2 - f_1$$

Each carrier is multiplied by a unipolar bit stream, one for the ones ($s_1(t)$ as for ASK) and its inverse ($s_2(t)$) for zeros:

$$v_{FSK} = v_{c1} \cdot s_1(t) + v_{c2} \cdot s_2(t)$$

PSK is the product of a carrier and a bipolar bit stream ($s_3(t)$)

$$v_{PSK} = v_c \cdot s_3(t)$$

Each produces spectra containing side frequencies around the carrier frequency at harmonics of frequencies of half the bit rate ($f_0 = f_b/2$).

Bandwidth required depends upon application. For most audio uses (e.g. modems) side frequencies up to f_0 or $3f_0$ are regarded as necessary. Then

$$B_{ASK} = f_b \quad \text{or} \quad 3f_b$$

$$B_{FSK} = f_b + \Delta f \quad \text{or} \quad 3f_b + \Delta f$$

$$B_{PSK} = f_b \quad \text{or} \quad 3f_b$$

For radio use the bandwidth for PSK is usually taken as f_b only. For multilevel PSK the symbol rate (f_s) is used.

The main applications are

ASK	some audio modems; optical fibres
FSK	most audio modems; HF and VHF radio
PSK	bandwidth-efficient audio modems; microwave radio

PSK makes especial use of multilevel digits in the form of QPSK and 16-QAM.

14.12 • Conclusion

From this chapter we now see how to modulate binary (and other) digital basebands. But where do these basebands come from in the first place? Some obviously come directly from digital sources (e.g. computer data) but a very important majority come originally from analogue signal sources (e.g. telephone conversations) and are converted into digital form close to source. What methods are used to do this? To find out read the next chapter on sampling.

14.13 • Problems

14.1 A 600 baud data stream is modulated as follows:

(i) ASK at $f_c = 1500$ Hz
(ii) BPSK at $f_c = 1500$ Hz
(iii) FSK at $f_c = 1500$ Hz and $\Delta f = 200$ Hz.

For each case sketch the signal waveform and its spectrum. Also state the type of bit stream you use and assign bandwidths.

14.2 An FSK system uses 600 Hz and 1200 Hz for 1 and 0 respectively. Obtain a value for the maximum digit rate which can be transmitted using two different criteria – three times bit rate and bit rate. Both signals are to be transmitted down a normal telephone channel of band limits of 300 Hz and 3400 Hz.

14.3 Carry out the following calculations on audio FSK modems:

(i) 120 Hz spacing; ± 30 Hz shift; find highest baud rate.
(ii) 100 baud; ± 60 Hz shift; find likely channel spacing.
(iii) 1200 baud; channel separation 2000 Hz; find likely deviation.
(iv) frequencies 500 ± 200 Hz; 600 baud; find channel spacing.

14.4 Using the criterion of bandwidth equals symbol rate, assign bandwidths for:

(i) ASK: 1.0 kbps on 100 kHz
(ii) PSK: 1.0 kbps on 100 kHz
(iii) FSK: 1.0 kbps using 90 and 100 kHz
(iv) PSK: 10 Mbps on 5.0 GHz
(v) ASK: 4 kbps on 64 kHz
(vi) QPSK: 10 Mbps on 5.0 GHz.

14.5 Suggest C and R values to envelope-detect a 100 baud 101010 ASK bit stream using a 1.0 kHz carrier. Sketch.

14.6 Look up and sketch the signal space diagrams (constellations) for the following modulations:

(i) QPSK
(ii) 8-PSK
(iii) 16-QAM.

Use Gray-coded versions.

14.7 Leased lines can transmit a data rate of 9600 bits per second using only the standard voice frequency bandwidth. Explain how this is possible and why "leased lines" are needed.

14.8 Give the main relative advantages and disadvantages of ASK, PSK, and FSK modulation methods.

14.9 Listen to data being loaded from a tape recorder into your personal computer (if it will do it that way) and decide which type of binary modulation is being used. Assign values if you can.

14.10 Explain the types of modulation used to impress a binary digital signal on to a continuous high-frequency carrier. Obtain an expression for the bandwidth required when a frequency shift keying method is used, relating your answer to the bit rate.

Several teleprinter channels are to be transmitted simultaneously over a single telephone channel. Explain a simple multiplexing scheme which will do this and calculate the number of channels which can be sent.

14.11 A binary data stream of bit length 1.0 μs is to be transmitted on a radio link at 5.0 MHz carrier frequency using BPSK modulation. Calculate the channel bandwidth required explaining the origin of any expressions you use. In the light of your result, discuss the suitability of the carrier frequency chosen and, if necessary, recommend a different one.

Sketch the waveform of the modulated signal, at the original carrier frequency, when sending a continuous 101010 bit stream.

Other binary modulation methods could be used. For each of them explain the reasons why they should, or should not, be recommended as alternatives.

Finally state briefly the main reason why QPSK modulation is an improvement over BPSK.

• 15 •
Sampling

15.1 • Introduction

Now we take the first step towards looking at the digital signals which are so very much a feature of present-day communications. The need for these preliminary ideas arises because, whilst many digital signals originate as such, a very large proportion start life as analogue signals. How does the signal change from one to the other? The first part of the answer is **sampling**. For it is an absolutely astonishing fact that, if you send an analogue signal for only very short intervals of time with long gaps in-between, the **complete** original signal can be recovered at the receiving end. The only condition is that these samples are taken fast enough to meet a minimum threshold – the **Nyquist rate** – equal to twice the highest signal frequency.

15.2 • The action of sampling

The basic process of sampling is the **gating** of an **analogue signal** by a periodic pulse train which will only allow the signal through whilst each pulse is on.

The gating signal, or **sampling function** $s(t)$, has pulses of constant height, length τ seconds, and separation T seconds as shown in Figure 15.1(b). The analogue signal, or **baseband** v_m, is level shifted so that no part of it is negative, but is otherwise unchanged. This is to make all samples positive, as shown in Figure 15.1(d). For a simple cosine signal this (assuming unity amplitude) becomes

$$v_m = 1 + \cos \omega_m t$$

The actual gating is done by **multiplying** these two signals together to give the sampled signal as shown in Figure 15.1(c) and (d). The effect is merely to take slices, or **samples**, out of the raised baseband.

Here, T is the **sampling interval**, τ is the **sampling time**, and f_s $(= 1/T)$ is the **sampling frequency**.

Samples are kept narrow ($\tau \leqslant T$) so that their flat tops accurately represent the

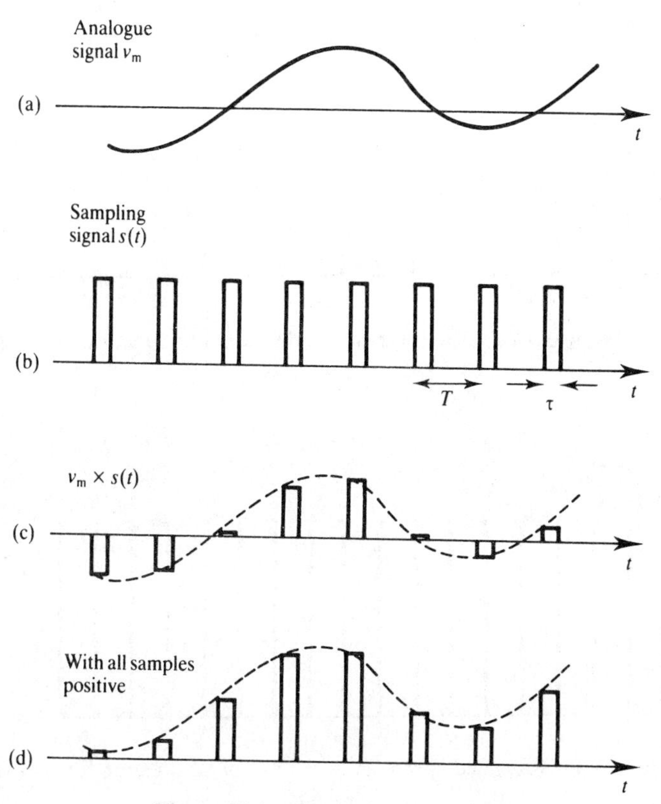

Figure 15.1 The action of sampling

baseband height at one instant. The action is represented by the equation

$$v_s = v_m \cdot s(t)$$

Before carrying out the analysis of this, we need to look at the sampling function itself in detail.

15.3 • The sampling function

The sampling function, $s(t)$, is a train of narrow pulses as already shown in Figure 15.1. By Fourier analysis its spectrum is given by

$$s(t) = \frac{\tau}{T}\{1 + 2[(\mathrm{sinc}\,\pi\tau/T)\cos\omega_s t + (\mathrm{sinc}\,2\pi\tau/T)\cos 2\omega_s t + \ldots]\}$$

This plots as in Figure 15.2 showing a series of harmonics of the pulse repetition frequency $(1/T = f_s)$ with a sinc envelope having zeros at harmonics of $1/\tau$.

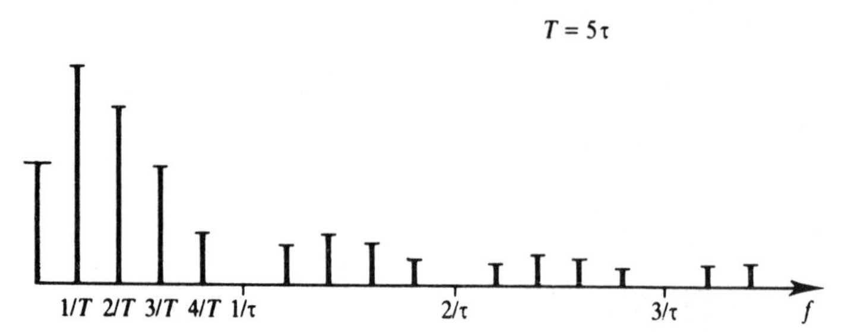

Figure 15.2 The spectrum of the sampling function

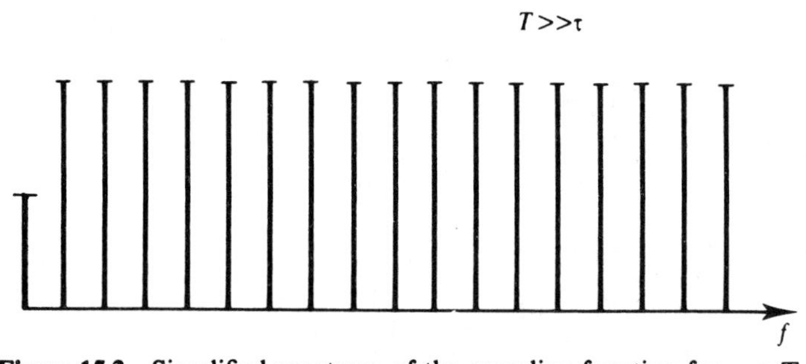

Figure 15.3 Simplified spectrum of the sampling function for $\tau \ll T$

Because we use short samples a simplification can be made based on the assumption that $\tau \ll T$, and we can write, for the first few harmonics, that

$$\operatorname{sinc}(n\pi\tau/T) \simeq 1$$

and

$$s(t) = \frac{\tau}{T}(1 + 2\cos\omega_s t + 2\cos 2\omega_s t + 2\cos 3\omega_s t + \ldots)$$

This gives a constant amplitude spectrum as shown in Figure 15.3.

The effect of this simplification is that the first zero of the sinc envelope (at $1/\tau$) has moved up in frequency so that we are looking only at the top of the first peak which is so wide that it now appears effectively flat with all harmonics of f_s at the same amplitude. Of course this requires τ/T to be sufficiently small, as it is in practice (e.g. at most $T/10$ for an LM398).

Now what happens when you use this simplified function to sample a baseband?

15.4 • Analysis of sampling

The action of sampling is the multiplication of the raised baseband by simplified (constant amplitude) sampling function, $s(t)$. That is

$$v_s = v_m \cdot s(t)$$

$$= E_m(1 + \cos \omega_m t)\frac{\tau}{T}(1 + 2\cos \omega_s t + 2\cos 2\omega_s t + \ldots)$$

$$= \frac{E_m \tau}{T}(1 + \cos \omega_m t + 2\cos \omega_s t + 2\cos \omega_s t \cos \omega_m t + 2\cos 2\omega_s t$$

$$+ 2\cos 2\omega_s t \cos \omega_m t + \ldots)$$

$$= (E_m \tau / T)[1 + \cos \omega_m t + \cos(\omega_s - \omega_m)t + 2\cos \omega_s t + \cos(\omega_s + \omega_m)t$$

$$+ \cos(2\omega_s - \omega_m)t + 2\cos 2\omega_s t + \cos(2\omega_s + \omega_m)t + \ldots]$$

which is a d.c. level plus the original baseband plus a set of 100% amplitude modulations on the sampling frequency and its harmonics. This spectrum is shown in Figure 15.4(a) and, to make the position absolutely clear, the corresponding spectrum for an actual base**band** of frequencies is also shown (in Figure 15.4(b)). Note that the somewhat strange spectral shape of the baseband spectrum has no special significance. Its distinctive skew shape is used merely to allow the upper and lower sideband mirror shapes to be displayed unambiguously.

Note particularly that the original baseband itself is there as a separate unspoiled entity and can easily be separated out by using a low-pass filter to remove all the higher frequency a.m. components. This explains the "astonishing fact" mentioned at the start of the chapter. *But* it is, as ever, not quite as simple as that because of the condition also mentioned then – the **Nyquist criterion** that the sampling rate (f_s) must be at least twice the highest baseband frequency (f_m). That is

$$f_s \geqslant 2f_m \; (= f_N) \qquad \textbf{Nyquist criterion}$$

When sampling is exactly at $2f_m$ it is said to be at the **Nyquist frequency**, f_N. This criterion was named after its discoverer (in 1928) and is illustrated diagrammatically in Figure 15.5.

This figure illustrates the practical point of the Nyquist criterion, namely that, unless it is met, it is impossible to recover the original baseband by filtering. If sampling occurs at too slow a rate the lower sideband ($f_s - f_m$) of the sampling frequency overlaps the baseband, irretrievably corrupting it (Figure 15.5(c)). Actually, sampling at the Nyquist frequency makes the two bands just touch so that whilst, theoretically, an ideal filter could recover the baseband, no such filter exists, of course.

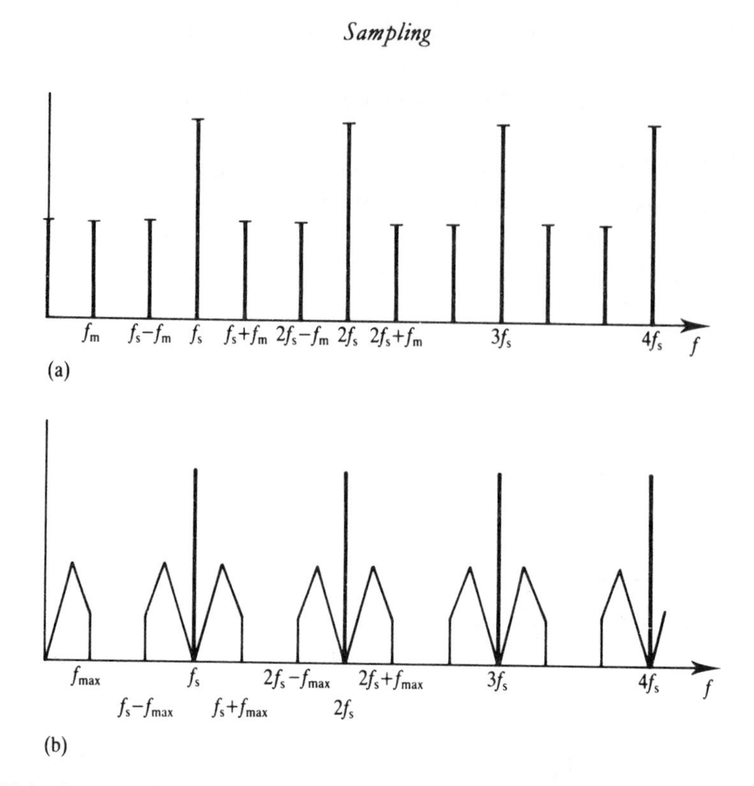

Figure 15.4 Spectra of sampled signals: (a) single frequency baseband ($f_s = 3.2f_m = 1.6f_N$); (b) band of signal frequencies (0–f_{max}) – $f_s = 3.2f_{max} = 1.6f_N$)

In reality one has to sample sufficiently above f_N to allow a gap for filtering of at least $0.2f_N$. Figure 15.5(a) shows that.

At first sight it looks as if the faster f_s is the better, because it makes the gap for filtering so much wider. But this is not an advantage because the actual samples become closer together in time and one of the major advantages of sampling is lost, namely that of being able to **time-multiplex** other sampled signals in the gaps between these samples. This technique is described in detail in Chapter 19. A simple way to look at the Nyquist criterion qualitatively is to see that, if samples are taken too slowly, the detail of the original analogue waveform is lost and so it cannot possibly be recovered intact. An even worse effect is when a recovered baseband signal turns out to be a spurious one. This is the problem of aliasing.

15.5 • Aliasing

One serious problem which is avoided by keeping to the Nyquist criterion is that of **aliasing**. A lower sideband of f_s can appear within the baseband range and be thought to be part of it. It has disguised itself as baseband and so has adopted an **alias** or

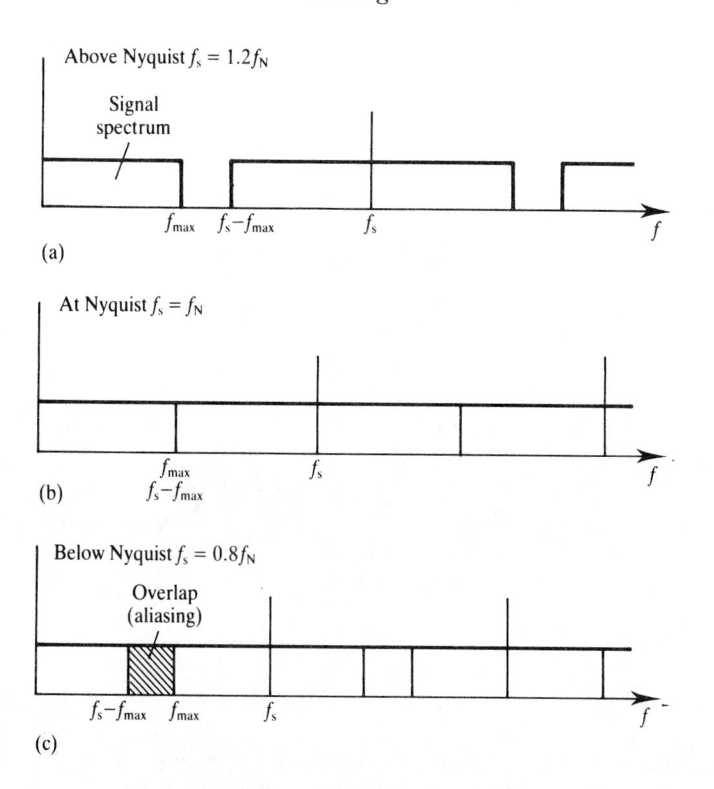

Figure 15.5 The Nyquist criterion

false name. This phenomenon is illustrated in Figure 15.6 for a baseband sampled at only $1.6f_m$ ($0.8f_N$).

Look at the spectrum first. The lower side frequency of f_s is at $0.6f_m$ (i.e. $1.6f_m - f_m$) and could easily be thought to be part of the original baseband – or even, if the filter cut-off were low enough, to be the *only* baseband. Disastrous!

Now look at the waveform and see how this wrong result can be explained on the samples themselves. The shorter wavelength sinusoid is the original f_m itself and the samples are positioned at sampling intervals of T_s starting from a sample at $v = 0$ on the left. T_s is related to the baseband period, T_m, by

$$f_s = 1/T_s = 1.6f_m = 1.6/T_m$$

therefore

$$T_s = T_m/1.6$$

$$\boxed{T_s = \tfrac{5}{8}T_m}$$

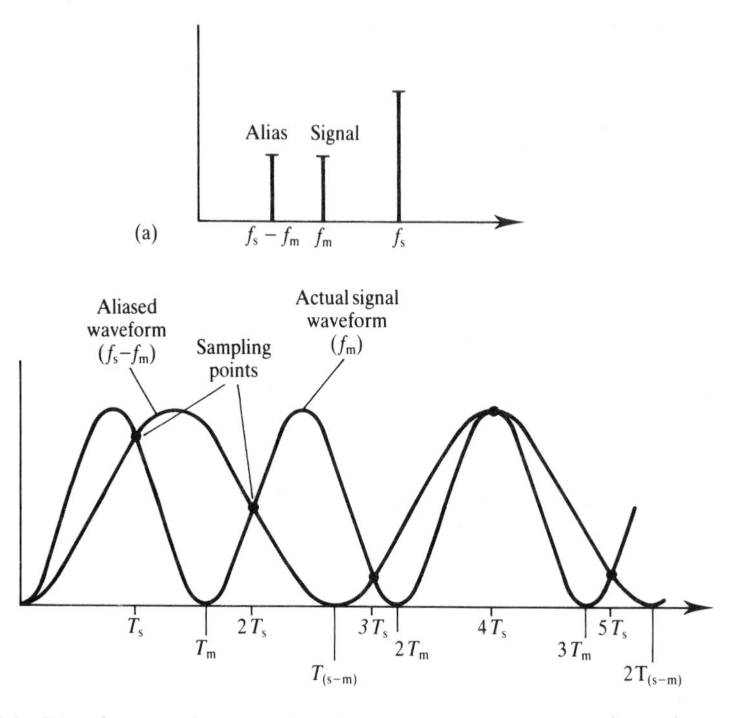

Figure 15.6 Waveforms and spectra showing aliasing: (a) spectrum of signal sampled below f_N (at $f_s = 0.8f_N$); (b) waveforms showing how samples of f_m fit f_s–f_m also ($T_s = \frac{5}{8}T_m$, $T_{(s-m)} = \frac{5}{3}T_m$)

But, because the samples occur too infrequently, it is possible to draw another longer-wavelength sinusoid through the sample tops as shown in Figure 15.6. For all the samples know this could well be the baseband – by measuring its period (say T_x) you can see that it is indeed the unwanted lower sideband of f_s (at f_s–f_m) as below:

$$1.5T_X = 2.5T_m$$

therefore

$$T_x = 5/3T_m$$

$$\boxed{f_X = 0.6f_m = f_s - f_m}$$

Thus aliasing can be a serious problem if the baseband includes any unwanted frequencies higher than the highest signal frequency (e.g. noise). This then aliases into the required baseband on recovery. To avoid this problem, in practical systems, the baseband is always band limited by a low-pass filter to include only the intended band of frequencies. This filter is called an **anti-aliasing filter**.

15.6 • Sample and hold

For the analogue pulse modulations in the next chapter the analysis so far is adequate. However, the major use for sampling is to convert analogue signals to digital ones. To do this a different type of sample is needed – one which keeps its height until the next sample occurs. This is the process known as **sample and hold**; it produces a stepped waveform as shown in Figure 15.7. The constant height steps are necessary to provide time for the digitizing to take place.

This waveform is obtained by using a **sample-and-hold circuit** whose action is to charge up a capacitor quickly in a very short time τ (the sampling time), and then to hold this sampled voltage connected to the input of the digitizing circuit until the next sample has to replace it. This cycle is repeated after each sampling interval of T_s. Figure 15.8 shows the basic parts of such an S/H circuit.

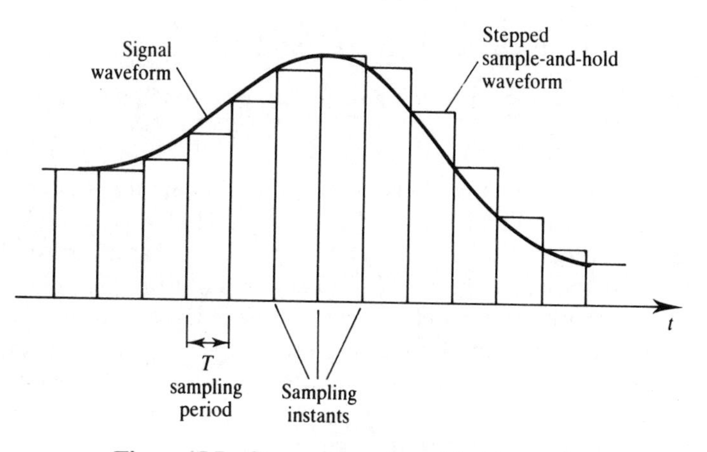

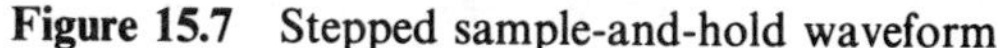

Figure 15.7 Stepped sample-and-hold waveform

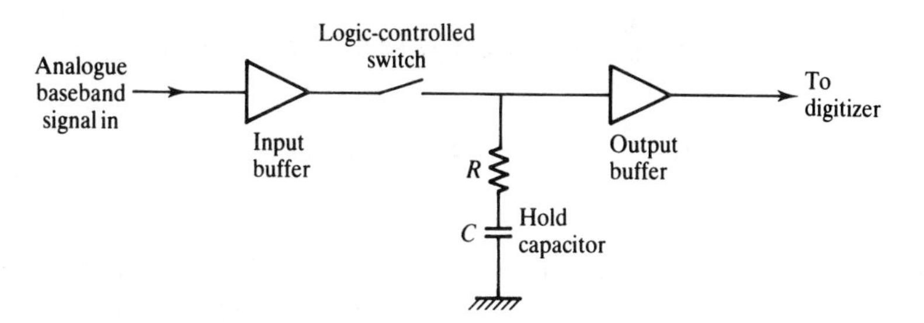

Figure 15.8 Schematic of a sample-and-hold circuit

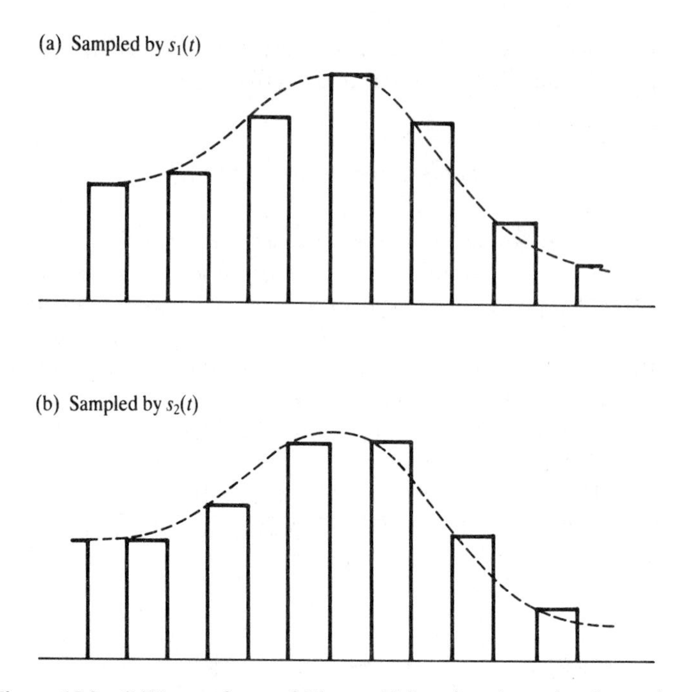

Figure 15.9 S/H waveform of Figure 15.7 as interleaved pulse trains

Note how the analogue source thinks it is being sampled for the required very short times but the digitizing circuit thinks the sample height lasts the whole sampling interval through. Neat, isn't it? This schematic is based on that of an LF398.

The stepped waveform of Figure 15.7 is quite different from the pulsed sampled waveform in Figure 15.4 but the baseband can still be recovered from it by filtering just the same. This can be proven by treating the stepped waveform as the sum of two interleaved square pulse trains whose amplitudes are those of the analogue based at the start of each pulse. This way of looking at it is shown in Figure 15.9.

If both pulse trains have unsampled height V, then

$$s_1(t) = \frac{V}{2} + \frac{V}{\pi}(\sin \omega_0 t + \tfrac{1}{3}\sin 3\omega_0 t + \dots)$$

$$s_2(t) = \frac{V}{2} - \frac{V}{\pi}(\sin \omega_0 t + \tfrac{1}{3}\sin 3\omega_0 t + \dots)$$

Now make $V = 1 + \cos(n\omega_m T)$ and $1 + \cos[(n + \frac{1}{2})\omega_m T]$ respectively, where T is the pulse train repetition period, and the required proofs follow. Readers may care to complete the analysis.

15.7 • Summary

Sampling is the process of taking instantaneous values of a continuous signal over a very short period of time (τ) at regular intervals a much longer time (T) apart. Here, τ is the sampling time, T the sampling interval, and $f_s = 1/T$ the sampling frequency.

The **sampling function**, $s(t)$, is a train of equal height pulses, τ long and T apart, which **gates** the baseband to produce the **sampled signal**, v_s, whose spectrum contains d.c., the baseband, and a series of full a.m. modulations about the harmonics of f_s. For a single sinusoid baseband this is ($\tau \ll T$)

$$v_s = (E_m\tau/T)[1 + \cos\omega_m t + \cos(\omega_s - \omega_m)t + 2\cos\omega_s t + \cos(\omega_s + \omega_m)t$$
$$+ \cos(2\omega_s - \omega_m)t + 2\cos 2\omega_s t + \cos(2\omega_s + \omega_m)t + \ldots]$$

Clearly the baseband can then be recovered by filtering provided the **Nyquist criterion** is met, that is

$$f_s \geqslant 2f_m (= f_N)$$

where f_N is the **Nyquist frequency**.

If $f_s < f_N$ **aliasing** results and $f_s - f_m$ falls below f_m and corrupts the baseband. To avoid this, systems deliberately include a low-pass **anti-aliasing filter** to restrict the baseband bandwidth to B Hz where

$$B \leqslant f_N/2$$

In practice sampling is usually a preliminary to digitizing and sample heights are held constant between sampling instants to produce a stepped **sample-and-hold** waveform of step length T. This allows time for digitizing to take place.

15.8 • Conclusion

The next problem is how to transmit these samples. One way is to do it directly in an analogue fashion as in the next chapter. But by far the most important way is to send them as binary numbers obtained from the stepped S/H waveform via an A/D converter. The operation of these converters is not discussed in this book but the resulting binary signals are. First, however, let us look at pulse analogue signals.

15.9 • Problems

15.1 A sinusoid of $f_m = 1.0$ kHz is sampled at $f_s = 4f_m$. The maximum amplitude is 1.0 V and $\tau = 0.01T$.

(i) Sketch the sampled waveform.
(ii) Sketch its spectrum showing where amplitude falls to zero.

15.2 An analogue signal is band limited to a telephone channel.

(i) What is its highest baseband frequency?
(ii) What is its Nyquist frequency? Explain your choice.
(iii) What sampling frequency is normally used? Explain why.
(iv) What decides the bandwidth of the sampled signal? Give a likely value, explaining where you got it from.

15.3 A baseband signal is a square wave at 10 kHz repetition frequency. It is to be sampled.

(i) What sampling frequency would you use? Give reasons.
(ii) If sampled at 25 ksamples per second what signal could you recover?
(iii) If sampled at 100 ksamples per second what signal could you recover?

15.4 The sampled signal in Problem 15.1 is to be transmitted by ASK.

(i) What bandwidth would you suggest if

(a) it were the only signal being used
(b) it were one of several time-multiplexed signals?

(ii) τ is increased to $\tau = 0.1T$. Sketch again

(a) the sampled waveform
(b) the spectrum showing zero point.

Then assign a bandwidth as in part (i).

15.5 A telephone channel is normally sampled at 8.0 ksamples per second.

(i) Why is this fast enough?
(ii) Why not sample at a higher frequency to be sure?
(iii) What are the problems if you try to reduce the sampling frequency?

15.6 (i) State the Nyquist sampling criterion.
(ii) Demonstrate, with the aid of suitable sketches, that if the Nyquist criterion is upheld, an original continuous signal may be regenerated from its samples taken at uniformly spaced time intervals. State any assumptions you make and say where the practical case deviates from the ideal. Explain carefully what happens if the sampling frequency is too low.
(iii) The BBC has been transmitting digitially encoded audio signals around the UK for many years. Assume you have been asked to design this system with the following constraints:

dynamic range	$\leqslant$96 dB
message bandwidth	20 kHz

Select a suitable sampling frequency, explaining your reasons for the choice, and hence calculate the minimum transmission bandwidth, assuming the use of baseband digital transmission and even-order parity for error detection. [Hint: use an approximate formula to find the number of binary digits required for each sample.]

Briefly discuss the advantages of this distribution system compared with baseband analogue transmission.

15.7 A square wave, $v(t)$, of 10 kHz repetition frequency and amplitude ± 1, is sampled at 1.2 times the Nyquist frequency of its third harmonic. Its spectrum is given by

$$s(t) = \frac{\tau}{T} + \frac{2}{\pi}\left[\sin\left(\frac{\pi\tau}{T}\right)\cos\omega_s t + \tfrac{1}{2}\sin\left(\frac{2\pi\tau}{T}\right)\cos 2\omega_s t + \ldots\right]$$

where $\omega_s = 2\pi/T$.

(i) Obtain a complete expression for the sampled signal up to $2\omega_s$ and its side frequencies.
(ii) Show how your answer to (i) simplifies if τ is very much smaller than T.
(iii) Sketch the spectrum of (ii). Say how it differs from the expected spectrum of a sampled signal.
(iv) Suggest a sensible value for τ, explaining your choice.
(v) Explain how the original signal can be recovered from the sampled signal.

• 16 •
Pulse analogue modulation

16.1 • Introduction

A sampled signal can be transmitted as it is or in modified forms, all of which are essentially analogue in nature. These are the **pulse analogue modulations** which were developed after the Second World War for telemetry purposes such as on guided missile systems.

There are three classes of this type of modulation in a short chain of development driven by the need to reduce noise corruption. These are

1. Pulse amplitude modulation PAM
2. Pulse width modulation PWM
3. Pulse position modulation PPM

PWM is also called PDM or PTM for Duration or Time respectively. In each of them a continuously varying (i.e. analogue) quantity is used to represent the original sample height, as below

PAM	$v_{PAM} \propto v_s$	(i.e. samples unchanged)
PWM	$\tau \propto v_s$	
PPM	$t_d \propto v_s$	

where v_s is sample height, v_{PAM} is PAM pulse height, τ is PWM pulse width, and t_d is PPM pulse delay time (from sampling instant). The three methods are illustrated in Figure 16.1.

As explained for sampling itself, coding signals in these ways, as well as reducing noise corruption, allows transmission in TDM as discussed in Chapter 19.

Now for a brief description of each method.

16.2 • Pulse amplitude modulation (PAM)

Here the samples are sent unchanged, except for scaling. Thus the PAM spectrum is the same as that of the samples and, assuming these are narrow ($\tau \ll T$), it is as

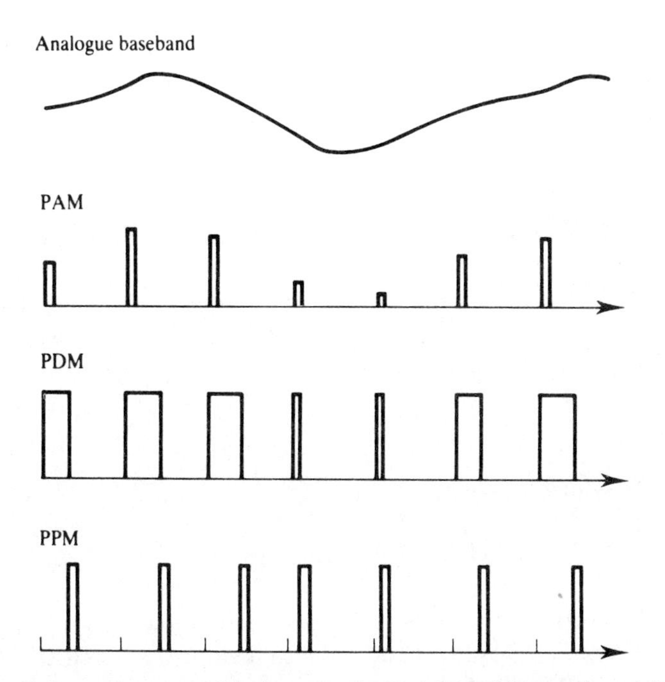

Figure 16.1 Pulse analogue modulation waveforms (after Holdsworth and Martin, 1991)

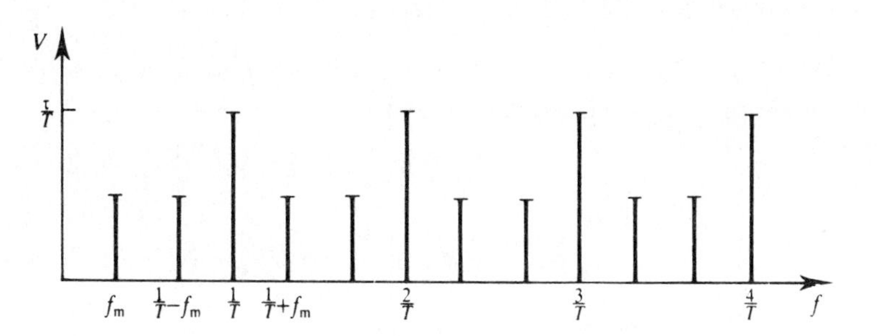

Figure 16.2 PAM spectrum ($\tau \ll T$) for single frequency baseband

given at the start of Section 15.4 and repeated here for a single sinusoid baseband:

$$\begin{aligned} v_{\text{PAM}} = (\tau/T)[1 &+ \cos\omega_{\text{m}}t + \cos(\omega_{\text{s}} - \omega_{\text{m}})t + 2\cos\omega_{\text{s}}t \\ &+ \cos(\omega_{\text{s}} + \omega_{\text{m}})t + \cos(2\omega_{\text{s}} - \omega_{\text{m}})t + \ldots] \end{aligned}$$

This spectrum is the same as that of the sampled signal in Figure 15.4 but is shown again in Figure 16.2.

This method is very straightforward as the samples are sent without further processing and, in fact, can be switched straight into an r.f. modulator by a telemetry switch. Thus the technology involved is very simple. Unfortunately this is outweighed by a serious disadvantage which is that the height of these narrow pulses is very easily altered by unwanted noise or interference, so further coding is necessary. The first method is as follows.

16.3 • Pulse width modulation (PWM)

PWM is also called pulse duration (PDM) or pulse time modulation (PTM). Here the transmitted signal pulses are all of the same height but their width is varied proportional to the original sample height. That is

$$\tau \propto v_s$$

or, for a single sinusoid baseband,

$$\tau = \tau_0(1 + \cos \omega_m t)$$

The analysis of its spectrum is very easy. The expression for a train of narrow pulses (the sampling function, $s(t)$) is taken, as given in Section 15.3, and the expression above is substituted in the overall constant multiplier. That is

$$\begin{aligned} v_{PWM} &= [\tau_0(1 + \cos \omega_m t)/T][1 + 2\cos\omega_s t + 2\cos 2\omega_s t + \ldots] \\ &= (\tau_0/T)(1 + \cos \omega_m t + 2\cos \omega_s t + 2\cos \omega_m t \cos \omega_s t + 2\cos 2\omega_s t \\ &\quad + 2\cos \omega_m t \cos 2\omega_s t + \ldots) \end{aligned}$$

or

$$\begin{aligned} v_{PWM} = {} & (\tau_0/T)[1 + \cos \omega_m t + \cos(\omega_s - \omega_m)t + 2\cos \omega_s t \\ & + \cos(\omega_s + \omega_m)t + \cos(2\omega_s - \omega_m)t + \ldots] \end{aligned}$$

This spectrum is now the same as the PAM spectrum in Figure 16.2 but only because we have used the simplified version of $s(t)$ for $\tau \leqslant T$. If the full version of $s(t)$ were used it would also be necessary to modify the sinc function coefficients because these contain τ too, giving

$$v_{PWM} = (\tau_0/T)\left(1 + \cos \omega_m t + \frac{2}{\pi}\sin[\tfrac{1}{2}\omega_s \tau_0(1 + \cos \omega_m t)] \cos \omega_s t + \ldots\right)$$

This still has the original baseband as a separate term easy to recover. The other terms would give Bessel function expressions, as for FM, apparently leading to multiple sidebands about each harmonic of the sampling frequency which could alias into the baseband. But because τ will always be much smaller than T, these sidebands die away rapidly giving effectively NBFM and so no problem arises. Thus the simpler approach is valid.

This method reduces the susceptibility to noise corruption because the exact heights of the nominally constant height pulses do not matter now, only their length. However, this does depend on the exact positions of two zero crossings, and noise can shift these but with much less consequent error than in amplitude. None the less, this was still found to be not totally adequate and so a further development occurred.

16.4 • Pulse position modulation (PPM)

In this method the pulses are all of the same height and width but occur delayed by a small time from the exact sampling repetition instant. This delay time (t_d) is proportional to sample height (v_s). That is

$$t_d = t_{d0} \cdot v_s$$

or, for the usual single sinusoid baseband,

$$t_d = t_{d0}(1 + \cos \omega_m t)$$

There is no simple way to analyse such a signal with its varying pulse positions. Some idea can be obtained by taking the simplified version of $s(t)$ and modifying both T and τ by writing

$$t \equiv t + t_d = t + t_{d0}(1 + \cos \omega_m t)$$

and

$$\begin{aligned} T &\equiv T_s + \delta T \\ &= T_s + \mathrm{d}t_d/\mathrm{d}t \,.\, \delta t \\ &= T_s - \omega_m t_{d0} \sin \omega_m t \,.\, T_s \\ &= T_s(1 - \omega_m t_{d0} \sin \omega_m t) \end{aligned}$$

thus

$$v_{PPM} = (\tau/T_s)(1 - \omega_m t_{d0} \sin \omega_m t)^{-1} \left[1 + 2\cos\left(\frac{2\pi[t + t_{d0}(1 + \cos \omega_m t)]}{T_s(1 - \omega_m t_{d0} \sin \omega_m t} \right) + \ldots \right]$$

Further simplification can be made if the signals are assumed to be sent in TDM. Then

$$\tau \leqslant T_s \quad \text{and} \quad T_m$$

This makes the factor $\omega_m t_{d0}$ small so (writing $2\pi/T_s = \omega_s$) v_{PPM} becomes

$$v_{PPM} = (\tau/T_s)(1 < \omega_m t_{d0} \sin \omega_m t)\{1 + 2\cos \omega_s[(t + t_{d0}) + t_{d0} \cos \omega_m t] + \ldots\}$$

The terms in ω_s and its harmonics will produce narrow f.m.-type spectra (as $\beta = \omega_s t_{d0}$ is small) which should not alias into the baseband region. This means a further simplification to

$$v_{PPM} = (\tau/T)(1 + \omega_m t_{d0} \sin \omega_m t + \ldots)$$

and a term in ω_m could be recovered by filtering and then converted back to the baseband itself by integrating. However it is analysed, a PPM signal does not contain the baseband as a separate term which could be filtered off, so it is usual to convert it back to PWM on reception.

The chief advantage of PPM is that it is less suceptible to noise corruption than PWM because only one of its zero crossings matters – the one defining the end of the delay time.

16.5 • Generation and recovery

PAM is merely the original samples and so is generated by the same methods, for example

$$v_{PAM} = s(t)v_m$$

and recovered by low-pass filtering. Figure 16.3 shows both.

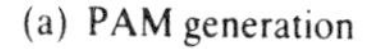

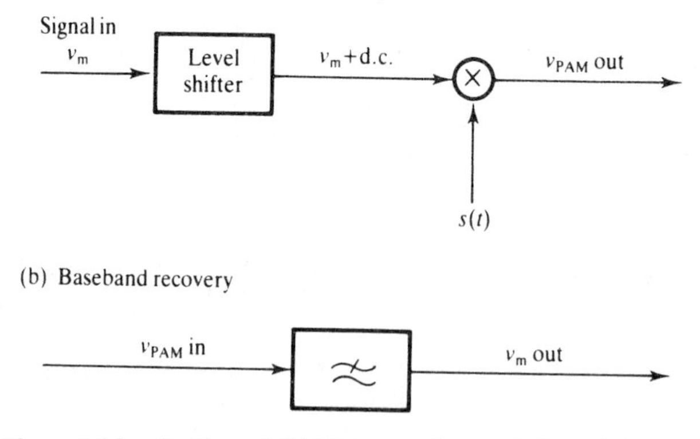

Figure 16.3 Outline of PAM generation and demodulation

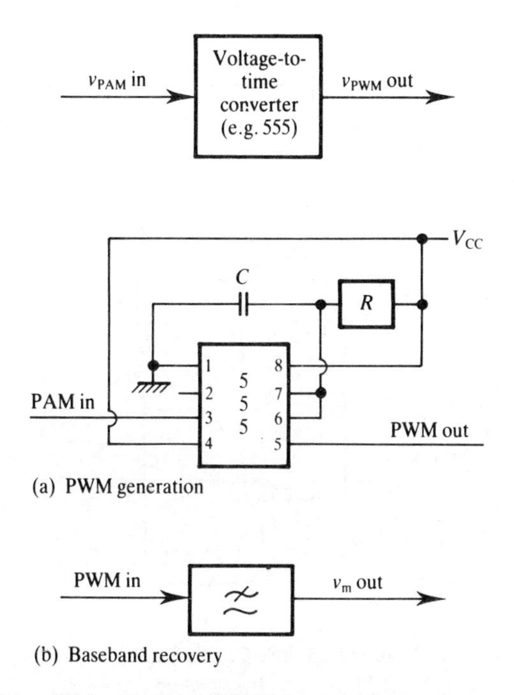

(a) PWM generation

(b) Baseband recovery

Figure 16.4 Outline of PWM generation and recovery

PWM is generated by some form of multivibrator using sample heights as a control voltage to give output pulse widths as in Figure 16.4.

PPM can be generated by any system in which pulses of fixed height and duration are produced after a time delay from the sampling instant – with the delay being proportional to sample height. One way to do this is to start from PWM as shown in Figure 16.5(a).

Recovery is either by low-pass filtering and integration or by conversion to PWM. Both PWM and PPM can be obtained directly from the unsampled baseband. One simple way to do this is with a timer i.c. such as a 555 as in Figure 16.5(b). The manufacturer's data sheet explains how. Recovery can also be demonstrated this way.

16.6 • Summary and conclusion

Sampled signals can be sent in an analogue form either directly (PAM) or in coded form (PWM and PPM):

PAM	pulse amplitude modulation	sample height $\propto v_m$
PWM	pulse width modulation	sample width $\propto v_m$
PPM	pulse position modulation	sample delay $\propto v_m$

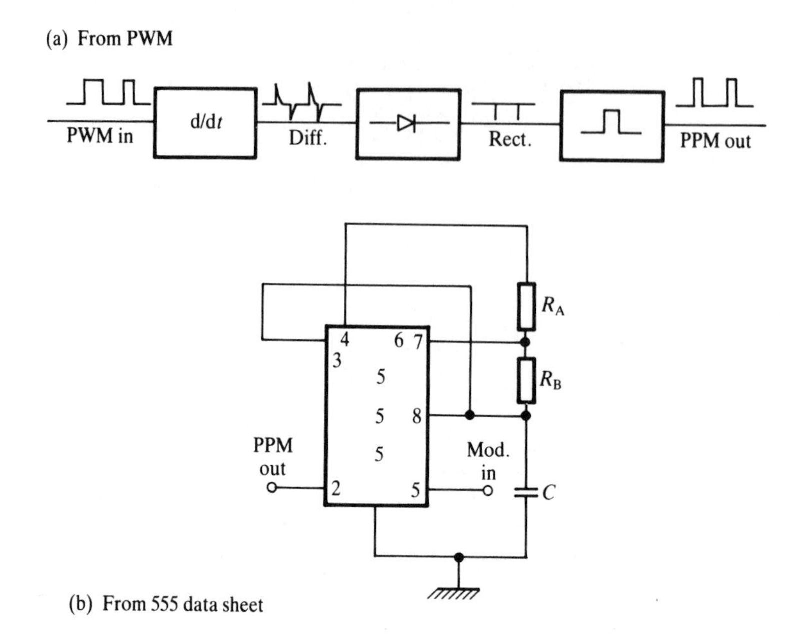

Figure 16.5 PPM generation

PAM is easily corrupted but the other types less so, which is the reason for their development. Analysis shows that PAM and PWM contain the baseband in an explicit form allowing it to be recovered easily on reception by filtering, but PPM needs to be converted back to PWM first.

All three can be generated by standard circuits available in i.c. form.

Many satisfactory working systems were designed around these methods enabling information to be sent satisfactorily in TDM. But then the digital revolution came along and a much better and easier way of representing sample height arrived – PCM.

16.7 • Problems

16.1 Sketch the following PAM signals:

(i) sine wave
(ii) square wave
(iii) triangular wave.

For each waveform do *two* sketches, one at the Nyquist frequency and the other at twice the Nyquist frequency.

16.2 What are the minimum sampling frequencies for

(i) a 1 kHz sine wave
(ii) a 1 kHz square wave
(iii) a telegraph channel
(iv) a telephone channel
(v) hi-fi radio
(vi) a TV channel?

16.3 Sketch frequency domain plots for PAM modulation of the following waveforms:

(i) a 1 kHz sine wave
(ii) a 1 kHz square wave
(iii) a 1 kHz pulse train for mark/space ratios of 0.5 and 0.1.

In each case sample at 1.5 times the Nyquist frequency.

16.4 Work out the transmission bandwidths required to receive the following PAM signals (NF = Nyquist frequency): Go up $2f_s$ and its sidebands.

(i) 1 kHz sine wave at NF
(ii) 1 kHz square wave at 1.2NF
(iii) 10 kHz saw-tooth at 2NF
(iv) 1 kHz pulse train (mark/space 0.1) at 1.4NF
(v) 1 kHz sine wave fully a.m. on to 10 kHz carrier, $m = 0.5$
(vi) BCD signal at 100 baud.

16.5 Show that a PDM spectrum is the same as a PAM one (for same f_m, T, and τ) at low harmonics of f_s. [Replace T by $(\tau_0/2)(1 + \cos \omega_m t)$.] Suggest a way of obtaining a PDM spectrum

16.6 If $\tau = 0.1T$, for how many harmonics of f_s is the Statement in Problem 16.5 true to 1%? Try also $\tau = 0.01T$ and $\tau = 0.001T$, suggest the largest practical value of τ.

16.7 For signal sampled with $\tau = 0.05T$, how many signals would you expect to be able to time-multiplex together?

16.8 Define the Nyquist frequency and explain its meaning.

A telemetry link is used to transmit data using pulse analogue modulation methods. Describe the main types of pulse analogue signal which could be used, showing why improvement in performance can be expected if the transmission channel is noisy.

The original data is obtained from 10 transducers each producing an output voltage varying continuously with time at frequencies within those of a standard telephone channel. Calculate the bandwidth required by the resultant pulse analogue modulated signal using the method you have decided to be least corruptible by noise. Use likely values for any other parameters you need, explaining your choices.

• 17 •
Pulse code modulation

17.1 • Introduction

Pulse code modulation (or PCM) literally means **modulating** a signal by converting it to **pulses** and then **coding** these. However, in practice, it is used in a much more specific sense to mean sampling (the pulses) a continuous telephone channel signal at its Nyquist rate (8 kHz) and expressing the sampled voltages as 8 bit binary numbers (the coding). Here the term modulation is being used in the second sense, as defined in Section 6.3, of altering the nature of a baseband.

Many analogue signals are digitized at different sampling rates than these producing signals very similar to PCM, but, in general use, the term itself is taken to mean digitizing under the specific conditions above. As such PCM has become a standard method worldwide for telecommunications and is probably nowadays the single most important aspect of technology in communications. Thus a clear understanding of its details is most important.

17.2 • Why digitize?

Why go to all the trouble of taking a perfectly unambiguous analogue signal of limited bandwidth (4 kHz) and converting it to an apparently formless stream of bits occupying a much larger bandwidth (64 kHz)? The answer is partly the general advantage of sampling that allows time division multiplexing of signals and hence a more efficient use of transmission links. But still – why digitize?

Again the answer is partly a general one that bit streams are not easily corrupted by interference and noise. Specifically, sample height can be transmitted far more accurately, under given conditions, than when using any of the analogue sampled methods of Chapter 16.

But now another general advantage appears, which is that digital signals are very easy to process by cheap standard techniques. This makes complicated mass communication systems, such as telephone networks, much easier to implement and allows new techniques to be used – optical fibres for example.

Thus the **advantages** are

1. TDM possible
2. Less corruptible
3. Easily processed.

These three advantages far outweigh the main general **disadvantages** which are that PCM occupies much more bandwidth than the original baseband and that, when an error does occur, the result can be catastrophic with an outage much longer than the single bit in error. That is

4. Much larger bandwidth
5. Does not degrade gracefully.

There are also several less general disadvantages often in specific applications: for example, "twisted-pair" transmission lines for telephones; quantization noise; poor reproduction of low signal levels; etc.

17.3 • Obtaining a PCM signal

The original analogue baseband is first sampled using a sample-and-hold method as described in the previous chapter. This "held" signal is applied to an analogue-to-digital conversion (A/D or ADC) circuit which converts each sample height to the nearest sequential binary number using the two processes of **quantization** and **encoding** carried out together. After transmission these coded signals are converted back to the original baseband by a decoder or digital-to-analogue converter (D/A or DAC) followed by a low-pass filter. Both these processes are shown as a block diagram circuit in Figure 17.1. The parallel in–serial out (PISO) and serial in–parallel out

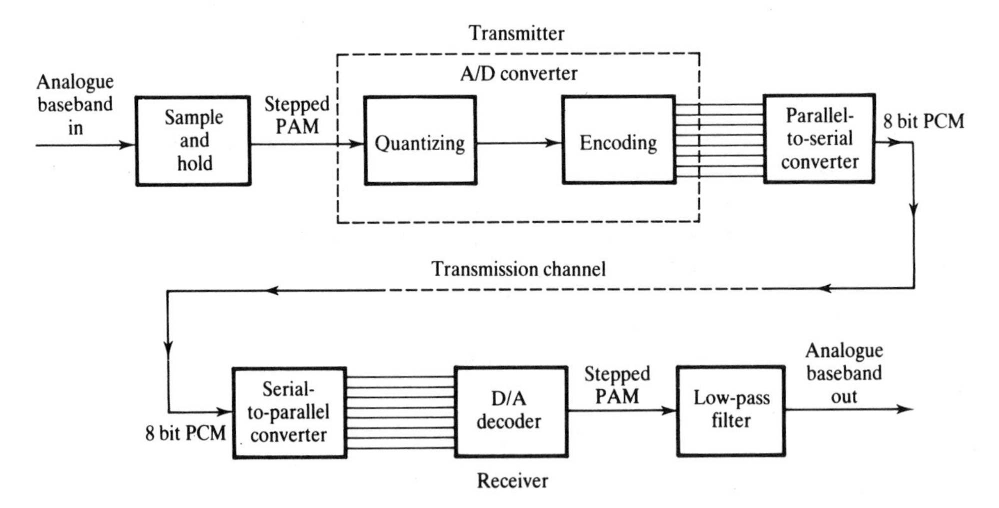

Figure 17.1 Transmission system using PCM

(SIPO) shift registers are needed because most A/D and all D/A circuits have simultaneous output (or input) of all eight binary digits in a number.

Quantization is the process of assigning each actual sample height to its nearest numbered level. The whole possible range of sample heights (0–10 V is common) is divided into a large number of small equal voltage steps (the **quantization interval**) with half steps at top and bottom. **Encoding** is done by assigning a number to each step level, starting sequentially from zero at the lowest level, and then expressing this number in binary form with n digits.

The quantization is done in such a way that there are always 2^n levels so that all possible n-digit numbers are utilized. For example, this means 256 levels for 8 bit

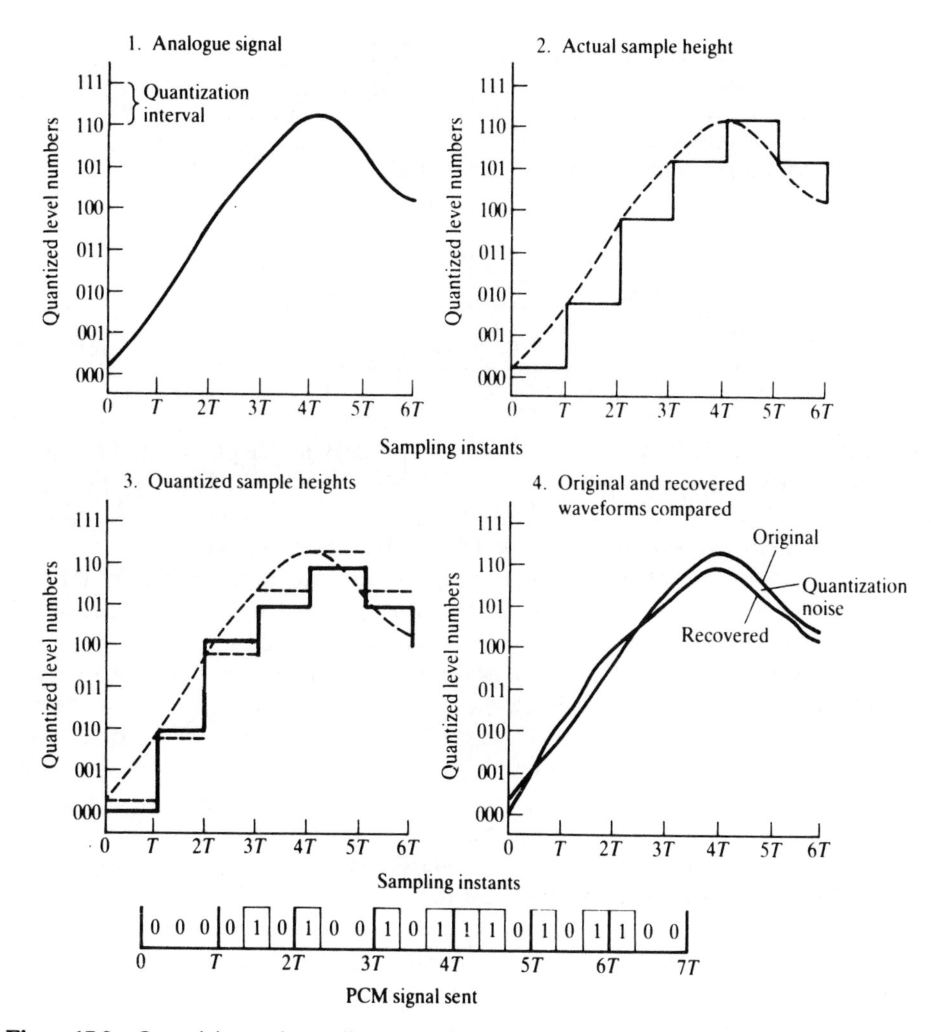

Figure 17.2 Quantizing and encoding an analogue signal (after Holdsworth and Martin, 1991)

numbers and 8 levels for the simplified 3 bit diagram in Figure 17.2. This diagram illustrates the whole process.

Both processes are done together as stated above.

17.4 • PCM bandwidth

The bandwidth of a PCM signal is much larger than that of the original baseband and also larger than the corresponding PAM signal. This can be shown as follows:

Maximum baseband frequency (i.e. bandwidth)	f_{max}
Nyquist frequency	$f_N = 2f_{max}$
Minimum sampling frequency	$f_s = f_N = 2f_{max}$
Bit rate (n bits per sample)	$f_b = nf_s = 2nf_{max}$
PCM transmission bandwidth (= bit rate)	$B = f_b = 2nf_{max}$

Therefore

$$B = 2nf_{max}$$

For example, if $f_m = 4$ kHz and $n = 8$ as is standard in telephone traffic, then $B = 2.8.4 = 64$ kHz; compare with baseband B/W = 4 kHz. The bandwidth of an individual bit is taken as equal to the bit rate which is a standard criterion for many digital signal transmissions for reasons which will be explained later.

At first sight you might consider that a 16 fold increase in bandwidth would be quite unacceptable. After all, we are always having it drummed into us that bandwidth is a precious commodity which costs money and must be used as economically as possible. So why is the use of PCM so widespread? The answer lies in the noise immunity of binary digital signals which is so much better than that of any of the analogue sampled signals in the last chapter. Bit shape does not matter as it is only necessary to know whether or not a one is present, and there has to be a great deal of noise/interference to cause that to be in error – in effect a one has to be mistaken for a zero and vice versa, which means the noise voltage has to be at least half as large as the signal voltage for it to happen. Thus PCM noise immunity is so good it is worth using much more bandwidth to get it.

The use of binary signals becomes even more advantageous when one also considers the ease of processing which comes with its digital nature. Multiplexing, addressing, regeneration, and all manner of d.s.p. actions can be carried out easily by standard methods.

17.5 • Quantization noise

One disadvantage of PCM is the degradation of signal accuracy because the samples are now quantized. This is because a recovered sample height can be at least half a quantization interval in error as can be seen in Figure 17.2.

This uncertainty in sample height appears as if it were an extra unwanted noise signal added to the recovered baseband and causes an additional signal-to-noise power ratio $(S/N)_Q$ of value 2^{2n} obtained as follows.

For a signal quantized over the range $+V$ to $-V$ volts

maximum signal voltage is V volts
maximum signal power is $S_Q = V^2$ watts

For n bits per sample the quantization interval is $2V/2^n$ volts. Therefore

maximum possible error in any recovered sample height is this at $V/2^n$ volts
so equivalent maximum noise power is $N_Q = V^2/2^{2n}$ watts

So the best additional signal-to-noise ratio $= S_Q/N_Q = V^2 \div (V^2/2^{2n}) = 2^{2n}$. That is

$$(S/N)_Q = 2^{2n}$$

For example, if $n = 8$ $(S/N)_Q = 2^{16} = 6.6 \times 10^4$ ($n = 3$ in Figure 17.2 gives 64). From this it can easily be seen that $n = 8$ is a reasonable general-purpose choice for number of bits per sample but note that this is the best possible S/N obtainable. Whilst the signal power can only be less, the noise power always stays the same, so S/N can only get worse. This is especially bad for very small signal levels, a problem which the next section tackles.

17.6 • Companding

A further source of error arising from the quantizing process is that very small signals become comparable in size with the quantization interval itself. This will cause gross errors in them on recovery and can be a serious problem. One way of reducing this effect is to increase the number of levels by increasing n, the number of bits per sample, and is one reason for using at least 8 bit coding. But even then the problem still remains.

Companding is the composite name given to a method of increasing the accuracy of transmission of these small signal levels. It is quite simple really. The analogue signal is distorted in a controlled manner before quantizing by **com**pressing its larger values and then ex**panding** them again after recovery at the receiver. This has the effect of providing many more quantization levels closer together for small signals (and correspondingly fewer for the larger signals) whilst keeping the quantization

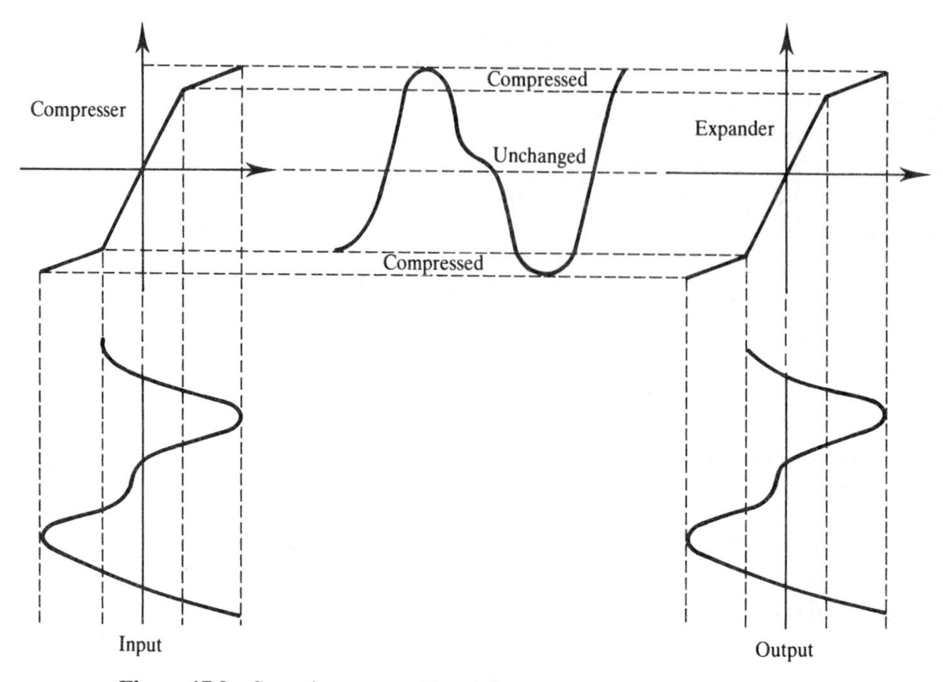

Figure 17.3 Steps in companding (after Holdsworth and Martin, 1991)

interval constant in the actual quantizer part of the A/D converter. These processes are shown in Figure 17.3 in simplified form.

The actual compression is carried out by a non-linear amplifier of known characteristic with another of exactly the inverse characteristic for the expansion at the other end. **Compander characteristics** are quite complex. The basis for design is to keep the $(S/N)_Q$ roughly the same at all signal levels, but to do this accurately requires a continuously variable logarithmic curve of which two versions are in common use, having, unfortunately, been developed separately in the USA and Europe. These are:

the A law in Europe
the μ law in USA

Both are expressed mathematically below and shown graphically in Figure 17.4:

$$y = \frac{1 + \log(Ax)}{1 + \log(A)} \qquad \text{for } x \text{ between } 1/A \text{ and } 1$$

$$= \frac{Ax}{1 + \log(A)} \qquad \text{for } x \text{ between } 0 \text{ and } 1/A$$

$$y = \frac{\log(1 + \mu x)}{\log(1 + \mu)} \qquad \begin{aligned} x &= v_{IN}/v_{IN}(\max) \\ y &= v_{OUT}/v_{OUT}(\max) \end{aligned}$$

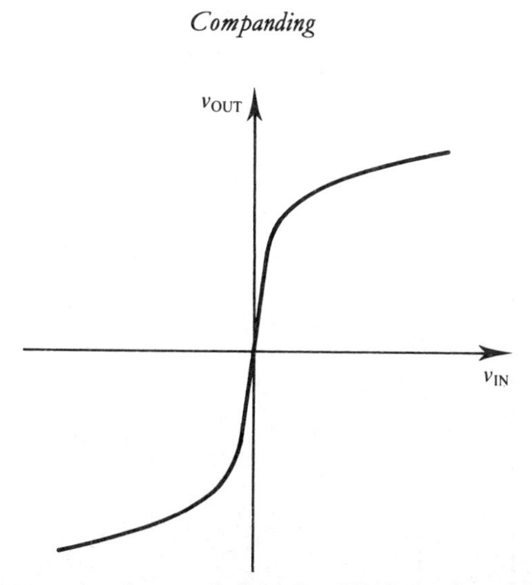

Figure 17.4 Compander characteristics (after Holdsworth and Martin, 1991)

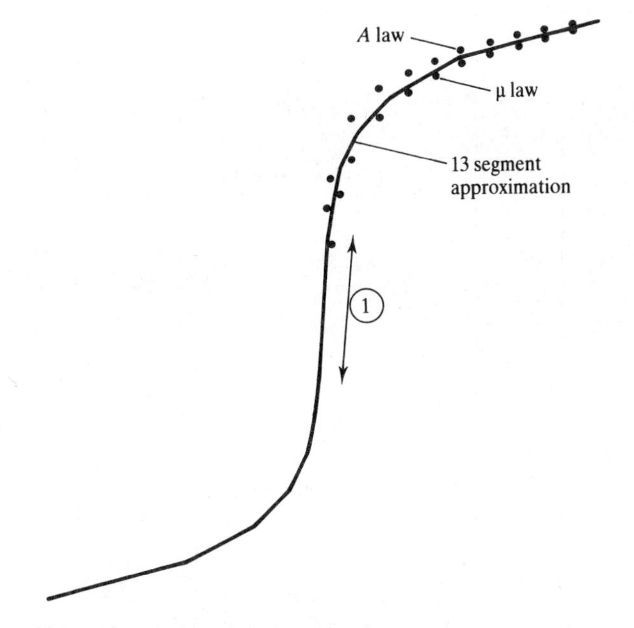

Figure 17.5 The 13 segment piecewise linear compressor characteristic

The most common values for the constants are $A = 87.6$ (CCITT) and $\mu = 100$ (USA/Japan). (See Coates' book for more detail.)

In practice these continuous curves are realized by a piecewise linear approximation usually with 13 separate segments (one through zero and then six symmetrically each side) as shown in Figure 17.5 where it lies neatly between the curves for the A and μ values above. Nowadays companding is increasingly being done after the analogue

Table 17.1 Near instantaneous companding

	Twelve-digit codeword												Eight-digit codeword							
Segment	I'_{11}	I'_{10}	I'_9	I'_8	I'_7	I'_6	I'_5	I'_4	I'_3	I'_2	I'_1	I'_0	I_7	I_6	I_5	I_4	I_3	I_2	I_1	I_0
HI	1	1	K	L	M	N	–	–	–	–	–	–	1	1	1	1	K	L	M	N
GH	1	0	1	K	L	M	N	–	–	–	–	–	1	1	1	0	K	L	M	N
FG	1	0	0	1	K	L	M	N	–	–	–	–	1	1	0	1	K	L	M	N
EF	1	0	0	0	1	K	L	M	N	–	–	–	1	1	0	0	K	L	M	N
DE	1	0	0	0	0	1	K	L	M	N	–	–	1	0	1	1	K	L	M	N
CD	1	0	0	0	0	0	1	K	L	M	N	–	1	0	1	0	K	L	M	N
BC	1	0	0	0	0	0	0	1	K	L	M	N	1	0	0	1	K	L	M	N
AB	1	0	0	0	0	0	0	0	K	L	M	N	1	0	0	0	K	L	M	N
B′A	0	0	0	0	0	0	0	0	K	L	M	N	0	0	0	0	K	L	M	N
.	.	.	.	.	.	.	.	.	.	.	.	.	.	.	.	.	.	.	.	.
.	.	.	.	.	.	.	.	.	.	.	.	.	.	.	.	.	.	.	.	.
.	.	.	.	.	.	.	.	.	.	.	.	.	.	.	.	.	.	.	.	.

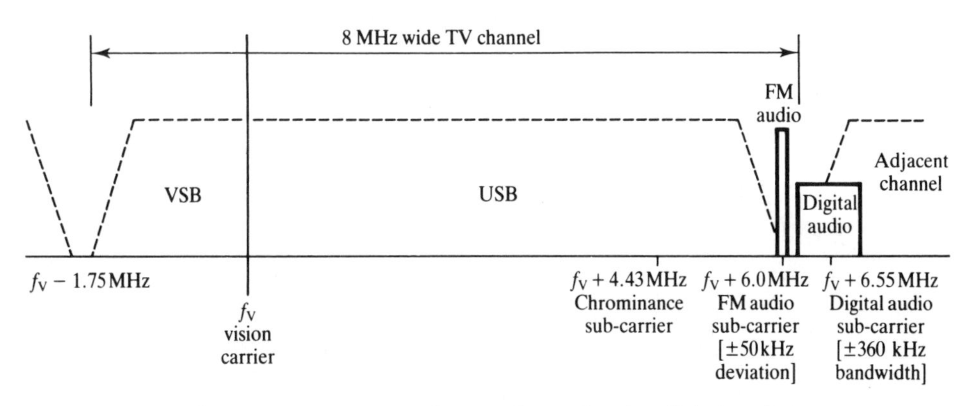

Figure 17.6 A UK 625-line TV band showing digital audio

signal has been digitized. This method is called **near instantaneous companding.** The basis of the technique is to digitize into 12, 13 or 14 bit samples and lose four of them in a controlled manner which compresses the signal (see Table 17.1). The resulting characteristic is not as close to the logarithmic ideal (to keep S/N_Q constant) as the analogue companding above, but is acceptable and easy to do. (See Coates' book for more details.)

A good example of its use is in the digitized sound now being added to UK TV broadcasts – called NICAM728 from its frame length and repetition rate. It uses 14 bit sampling, re-coded to 10 bits with 1 parity bit added giving 11 bit companded samples. 64 of these are time multiplexed, with 14 additional bits, to make frames exactly 1 ms long containing 728 bits at a rate of 728 kbits/s. This arrangement is possible because the sampling rate is 32 ksamples/s, hence two audio channels can be sent simultaneously allowing for stereo sound with the picture. The audio bandwidth is around 14.5 kHz. The companded signal is transmitted by QPSK-modulation onto a sub-carrier 6.552 MHz above the vision carrier in a nominal 720 Hz bandwidth, just above that of the f.m. sound (at 6.0 MHz, ±50 kHz deviation). Both sound signals are reduced in amplitude (20 dB for digital, 3 dB for analogue) to avoid interference with vision. Figure 17.6 illustrates this.

17.7 • Delta modulation

Another group of techniques aimed originally at reducing bandwidth whilst retaining digital transmission comes under the general name of **delta modulation** (DM or ΔM). Here the analogue baseband is sampled and quantized, as for PCM, but at a rate well above the Nyquist frequency. This sample is not digitized but a digital signal is sent which says whether the next sample is less than or greater than the one before, that is the sample **difference** (or Δ) only is sent. In fact these methods do not reduce transmission bandwidth very much because fast sampling is required to make the

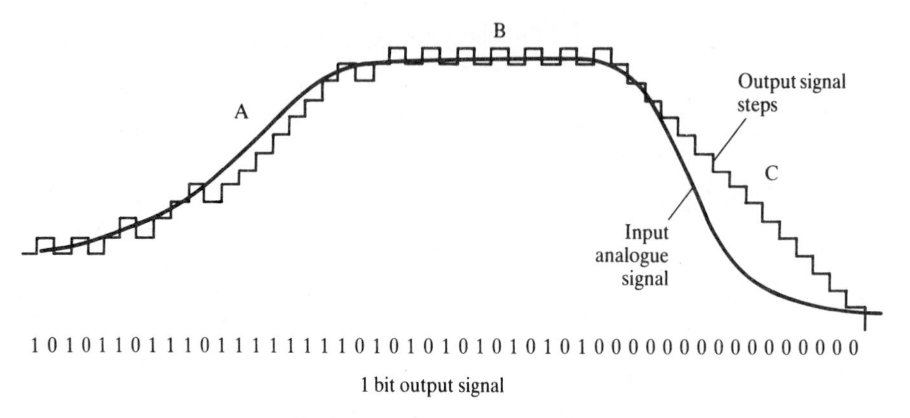

Figure 17.7 Basic delta modulation

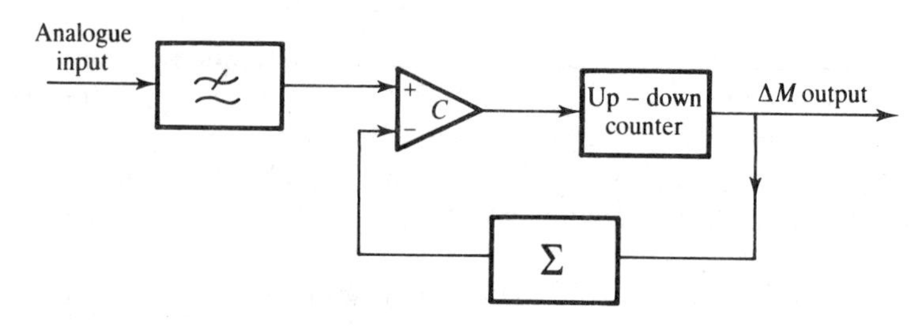

Figure 17.8 Basic delta modulator

quantized steps small enough. But they do have the greater advantage that they are easily realized in integrated circuit form as cheap **CODECs** (Coder/Decoder).

The basic delta modulation itself is quite easy to understand. The digital output merely follows the analogue input, going up or down one quantized level at a time. Figure 17.7 illustrates this process which is similar to a tracking A/D converter.

There are three features to note in this diagram:

A. The single step level changes can only just follow the rising input voltage.
B. The output has to "toggle" to follow a constant input voltage.
C. The single step level changes cannot follow a rapidly changing input voltage.

The actual digital signal transmitted will consist of a continuous series of binary digits in which the one and zero mean:

1 go up one quantized interval (or step)
0 go down one quantized interval (or step)

A **delta modulator** consists of an up–down counter whose output is fed back via an integrator and compared with the input signal. If the integrated output level is larger than the input then the next output digit is a one and, if smaller, a zero. An outline of a suitable circuit is shown in Figure 17.8.

Despite its simplicity this basic method has *serious disadvantages* for three reasons (two have been mentioned already):

1. Fast sampling needs a large digital bandwidth.
2. The output is unable to follow rapid changes of input level.
3. Only incremental voltage levels are sent.

The first disadvantage means that this simple ΔM method actually needs much more bandwidth than PCM for the same analogue signal bandwidth (e.g. 10 times for a 4 kHz telephone channel; see Coates, p. 222).

The second disadvantage is really the cause of the first (small steps need fast sampling) and can be overcome by the adaptive methods described later. The effect is to limit the slew rate of the output as shown in Figure 17.7 at C. The output bit rate is at the sampling rate so that a large transmission bandwidth will be needed if it is to be fast enough to allow the single step output to follow rapid input changes. Also, the slew rate problem will be worse for higher input frequencies and power levels, so some sort of compromise rate has to be used which is likely to be much higher than the 64 kbps of standard PCM.

The third disadvantage means that signals containing d.c. levels and very low frequencies cannot be sent satisfactorily (e.g. TV signals). For example, the standard voice frequency telephone channel avoids the d.c. problem by limiting its minimum frequency of transmission to 300 Hz, and the slew rate problem is contained by restricting the highest frequency to 3400 Hz. It also means that errors can only be corrected if the absolute value is sent occasionally.

Various forms of **adaptive delta modulation** (ADPM) have been devised to overcome the slew rate limitation. The most straightforward method is merely to

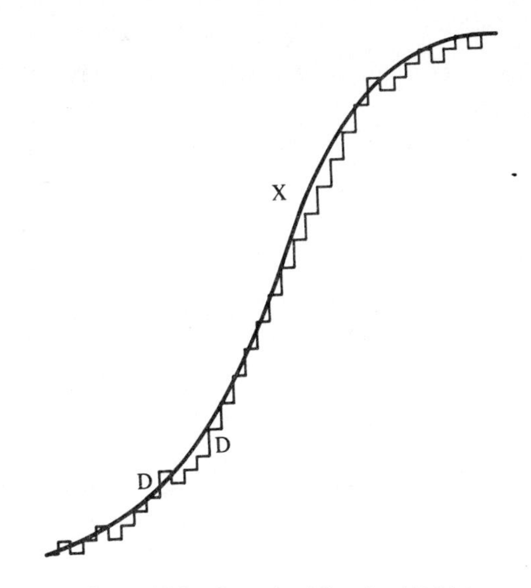

Figure 17.9 Step doubling in ADPM

increase the step size when necessary. If three identical digits are sent (one or zeros) it is assumed that the output is trying to follow an input which is changing more rapidly than it can keep up with. Then the step size is doubled and so on. This is illustrated in Figure 17.9.

Note how doubling occurs in two places (at D) at that, even then, the output is starting to be slew rate limited again (at X). To overcome this it can be arranged to double the level change again. Such a system would need 3 bits per sample to carry the full information.

Another method is called **Continuously variable slope delta modulation** (CVSDM) which merely increases the step size in proportion to the input signal power. The assumption is that larger powers have faster changes of signal level. It is a form of companding and CODECs using it can work at half the bit rate of PCM or less (see Coates, p. 225).

One use for the basic delta modulation method shown in Figure 17.7 is to provide a method of obtaining PCM at non-standard rates such as for NICAM above.

17.8 • Summary

PCM has three advantages (over analogue methods):

1. Time division multiplexing possible.
2. Sample values much less easily corruptible.
3. Signals easily processed.

And one main disadvantage:

4. Bandwidth much greater than original baseband bandwidth.

PCM is obtained by

S/H + A/D + P/S

and recovered by

S/P + D/A + LPF

where

S/H means sample and hold
A/D means analogue-to-digital conversion
P/S means parallel-to-serial conversion (PISO)
S/P means serial to parallel conversion (SIPO)
D/A means digital-to-analogue conversion
LPF means low-pass filter

A/D consists of two steps taken together – quantizing and encoding. For **quantizing** the range of sample heights is divided into a large number of equal voltage intervals separated by M quantized levels numbered from zero upwards sequentially, where M is always a power of 2. Actual sample heights are assigned to the nearest level.

In **coding** each level number is expressed as an n bit binary number where

$$M = 2^n$$

(for standard PCM, $n = 8$ and therefore $M = 256$, with transmission bandwidth $\text{B/W} = 2nB$, where B is the baseband bandwidth).

Quantization introduces **quantization noise** producing additional S/N of

$$(\text{S/N})_Q = 2^{2n}$$

To limit the errors it causes in low-level signals companding is used: either **analogue**

$$A\,\text{law (Europe)} \quad y = \frac{1 + \log(Ax)}{1 + \log A} \quad 1/A < x < 1$$

$$= \frac{Ax}{1 + \log A} \quad 0 < x < 1/A$$

$$\mu\,\text{law (USA)} \quad y = \frac{\log(1 + \mu x)}{\log(1 + \mu)}$$

where

$$x = v_{\text{IN}}/v_{\text{IN}}(\text{max}) \quad \text{and} \quad y = v_{\text{OUT}}/v_{\text{OUT}}(\text{max})$$

or **digital**, by 12 (or more) bit coding of samples (e.g. NICAM728).

Some bandwidth reduction and simpler realizations have been achieved by **delta modulation** which, at its simplest, uses 1 bit signals to indicate whether the next sample is smaller or larger than the last. There are several developed forms of this.

17.9 • Conclusion

PCM is now a well-developed process in common use worldwide especially for telephonic communication. It has great advantages over any other method of baseband coding, not least because it is very easy to combine many different signals by time division multiplexing as described in Chapter 19.

But first, in the next chapter, we shall look at some aspects of the two main causes of degradation in digital signals – noise and ISI.

17.10 • Problems

17.1 An analogue signal is limited to an amplitude range of 0–5 V.

(i) Calculate the quantization intervals when digitized with 4, 6, and 8 bits per sample.
(ii) Sketch the positions of the quantized levels for the 4 bit quantizations.
(iii) Sketch the form of the digital signal for the largest and smallest signal voltages and also for the middle 2.5 V one at 6 bit quantization.
(iv) What voltages do the following digital samples represent: 10000000, 01000000, 00000001, 01000000, 00100000, 00011010, 01111111, 11100101, etc. (set more for yourself)?
(v) For a 6 kHz baseband bandwidth calculate the minimum PCM bandwidth required for 8 and 12 bit quantization.

17.2 An analogue signal, limited to ± 5 V in amplitude, is converted to a PCM signal having n digits per sample. Calculate the quantization intervals when n is 3, 5, and 7.

17.3 An analogue signal, band limited to a standard voice frequency (telephone) channel, is sampled at the usual rate and coded into 8 bit samples.

(i) How many sample height levels are there?
(ii) What is the quantization interval, for a 10 V range?
(iii) How can negative signals be taken into account? (Two ways.)
(iv) What bandwidth is required for the coded samples?

Explain how you obtained your answers.

(v) It is generally thought better to send a PCM signal rather than just a sampled (PAM) one. Give reasons.
(vi) Why not use fewer bits per sample? Or perhaps more?

17.4 A signal, band limited to B Hz, is sampled at k times the Nyquist frequency using n bits per sample. What is the minimum bandwidth? Suggest your own, realistic, values for the algebraic symbols used in the question and work out a likely bandwidth.

17.5 UK TV needs at least 6.5 MHz baseband bandwidth. If it were sampled at 8 bits per sample what bandwidth would be needed? Communications satellites use TV transponders 41 MHz wide. See if you can work out how this is done (FM) and could be done (PCM).

17.6 Draw a block diagram showing the stages needed to convert an analogue signal to serial PCM.

Extend this diagram to show its transmission by ASK and subsequent demodulation and conversion back to the original baseband.

17.7 Straightforward PCM and PCM–TDM systems need modifying to improve

(a) fidelity for small-baseband signal levels
(b) bandwidth used
(c) access by many users – of satellites.

The following acronyms are relevant to the areas of improvement above. Work out (or look up) what they mean and discover how they apply above:

(i) Companding – A law and μ law
(ii) DPCM and ADPCM (sometimes ΔPCM)
(iii) QPSK and 8-PSK
(iv) 16-QAM and 64-QAM
(v) 9-QPRS (a hard one this)
(vi) TDMA (also FDMA)
(vii) CDM (i.e. spread spectrum)
(viii) M-ary PSK.

17.8 Explain how an analogue signal can be transmitted by coding it into a series of binary digits (PCM).

Discuss the advantages of coding an analogue signal in this way before modulating it on to an r.f. carrier rather than modulating it directly on to the carrier.

Determine the highest baseband frequency which can be transmitted using a PCM system having a sampling interval of 50 μs.

17.9 A PCM communication system time-multiplexes 30 audio channels and two control channels. The information rate is 64 kbps for each channel. The audio channels contain signals from a 256-level quantizer and one sample per channel is sent per frame.

(i) Calculate the total bit rate and the bit duration.
(ii) Calculate the number of bits per frame.
(iii) Sketch the frame structure, assigning the control channels to appropriate positions.
(iv) Give the highest audio frequency which can be transmitted and suggest the type of baseband signal used.
(v) Sketch the bit sequence for two samples at 40 % and 70 % of the maximum on one channel.
(vi) Draw a block diagram of the system needed to produce this multiplexed signal.

Explain all steps in your answers.

and show how their effect can be eliminated or reduced. Illustrate your answers by considering the effect of this corruption on a 1 kHz triangular voltage signal which is first encoded and subsequently decoded. You may take the equation of the triangular waveform as

$$v_m = E_c \cos \omega_m t + (E_c/9) \cos 3\omega_m t$$

Give quantitative answers, as far as you can, with relevant derivations.

17.10 Explain why an analogue signal requires a much larger transmission bandwidth once it has been binary pulse code modulated.

Describe the steps needed to carry out the BPCM process above, showing particularly the factors which decide the system clock frequency. Estimate the clock frequency needed to produce 8 bit PCM of a standard telephone channel.

Mention briefly methods which give digital modulations with smaller bandwidths but similar quality of recovered signal.

17.11 Define the term *quantization interval* for a PCM signal.

Explain the steps which occur when pulse-code-modulating an audio baseband, limited to B Hz and $\pm V$ volts peak-to-peak maximum amplitude, using n bit sampling. Illustrate your explanation by considering the effect on a signal for which $B = 4$ kHz, $V = 5$ V, and $n = 3$. Show clearly any causes of degradation of signal and obtain an expression for the PCM bandwidth which results.

See also Problems 19.12 and 19.14.

• 18 •
Errors in PCM transmission

18.1 • Introduction

In the development of PCM described in the previous chapter the emphasis was on two main things – **companding** as a way of overcoming quantization noise errors at low signal levels, and **delta modulation** to decrease bandwidth and allow i.c. realization. These are both techniques which are introduced at the sending end of the communication system.

Now we take a look at two effects which can degrade signals during transmission and on reception – **noise** and **ISI**. Systems need to be designed to minimize the effects of these and a complete description of all the techniques used is beyond the scope of this book. Here we shall merely look at the basic information and ideas needed to understand them.

18.2 • Errors in digital signal transmission

In the transmission of digital signals the aim is to recover each bit correctly at the receiving end but, inevitably, some bits will be wrongly detected (as a zero instead of a one and vice versa). Obviously, incorrect bits will occur at random intervals of time and may even occur in bursts rather than being spread out. Thus, for a given system, it is not possible to say whether any particular bit will be in error, but it is possible to say what the average rate of occurrence of errors is likely to be. This is the **bit error rate (BER)**, sometimes called the probability of error (p_e), or error probability, and is quoted as the average number of errors per bit to be expected. For example

for good quality audio BER $> 10^{-6}$ at least
for normal telephones BER $> 10^{-3}$ is adequate

These figures are sometimes quoted inversely with positive exponents of 10 (e.g. $< 10^6$). This means that, on average, the signal will send 10^6 bits before an error occurs. It means exactly the same as a BER of 10^{-6}.

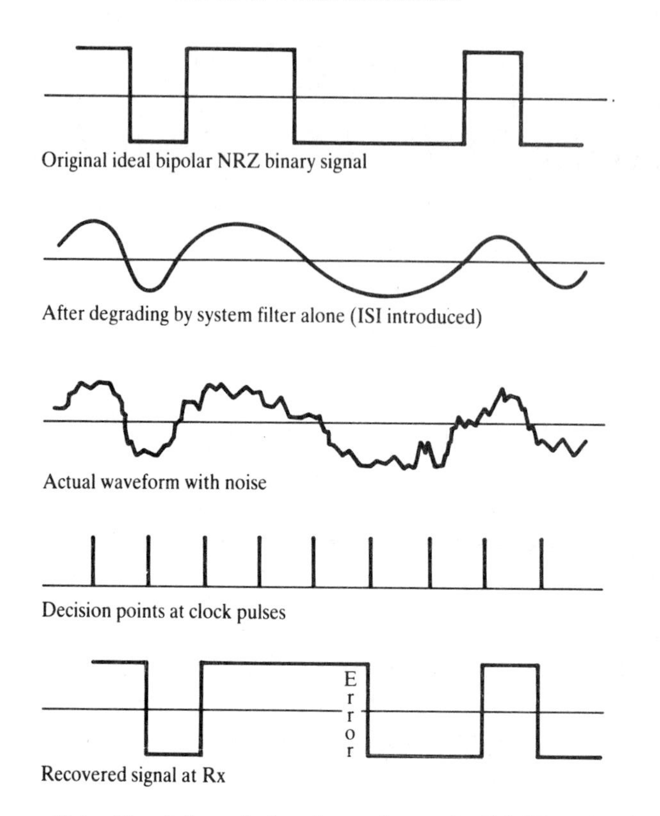

Figure 18.1 Signal degradation due to "smearing" (ISI) and noise

What causes these errors? There are two main ones: noise; and intersymbol interference (ISI). Figure 18.1 shows how these can cause errors in detected digits.

18.3 • The "language" of noise

Noise is mainly introduced into the signal by the transmission channel (e.g. the atmosphere for radio links) and at the front end of the receiving circuitry where the signal is smallest. For any given system the receiver detection circuitry requires a minimum signal-to-noise ratio into the receiver front end (e.g. 10 dB) which, in turn, requires a minimum received signal power.

Noise in communication systems is usually of the thermally generated type known as **additive white Gaussian noise** (AWGN) with rapidly and randomly changing noise voltages having a temperature-dependent mean square value independent of frequency. These noise voltages (n or v_n) have a **Gaussian** amplitude probability distribution as shown in Figure 18.2.

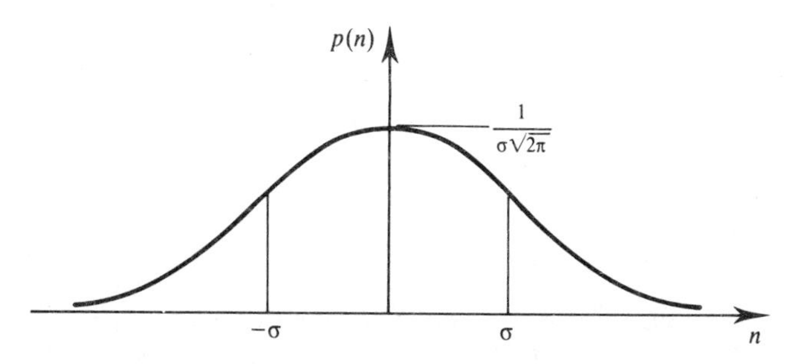

Figure 18.2 Gaussian noise voltage probability distribution

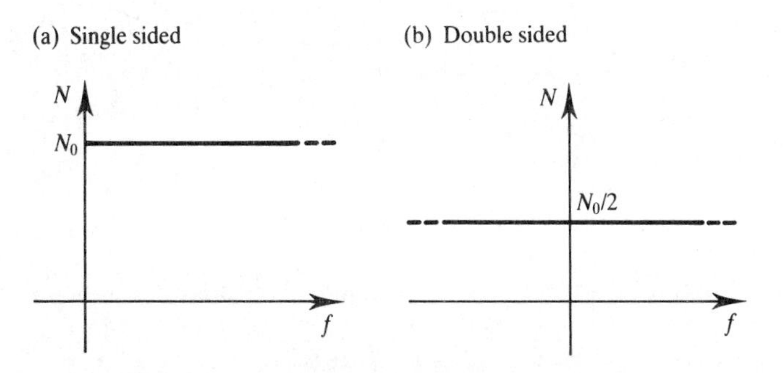

Figure 18.3 Normalized "white" noise power spectra

The expression for this curve is

$$p(n) = \frac{1}{\sqrt{2\pi\sigma^2}} \exp\left(\frac{-n^2}{2\sigma^2}\right)$$

where σ is the mean square value (the standard deviation or square root of the variance) often referred to as the "average" noise voltage.
These voltages produce noise power into a load with a spectral density independent of frequency as shown in Figure 18.3 – hence the term "white" by analogy with light.

N_0 is the normalized power spectral density (i.e. into 1Ω) and total power is

$$N = \sigma^2 = N_0 B$$

(into 1Ω load only). These powers are **additive** because, being randomly generated, their voltages add linearly.

The **thermal noise** produced by a resistor, $R\Omega$, at absolute temperature, T K, within a frequency bandwidth, BHz, has a mean square voltage given by

$$\overline{v_n^2} = 4kTBR$$

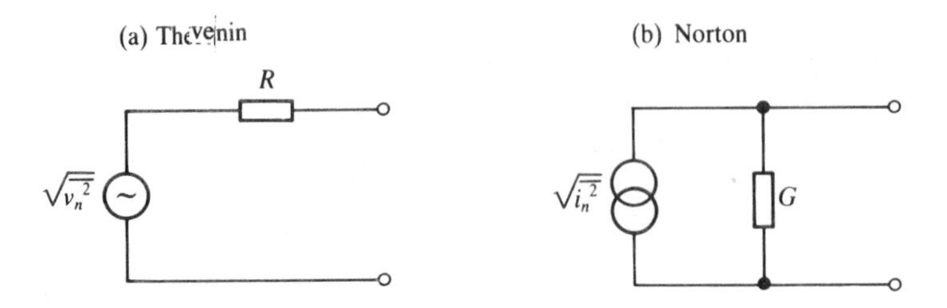

Figure 18.4 Noise source equivalent circuits

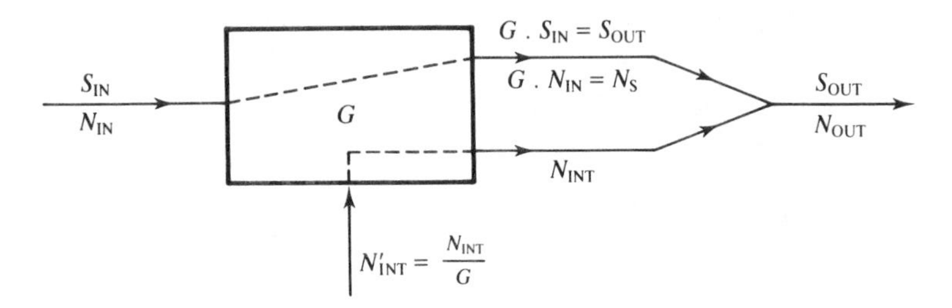

Figure 18.5 Noise and signal symbols for a network

where k is the Boltzmann constant of value $1.38 \times 10^{23}\ \mathrm{J\ K^{-1}}$ and noise power output

$$N = kTB \text{ watts}$$

(note that this is independent of the value of R). Here, v_n cause currents i_n where

$$\overline{i_n^2} = 4kTBG \qquad (G = 1/R)$$

and both v_n and i_n can be represented in noise-equivalent circuits as in Figure 18.4.

The **noise temperature** of a source is the apparent temperature which gives the noise power it actually produces. Often this is the actual temperature, especially the standard "room" temperature of 290 K (T_0) such as would apply for a passive component, but it may merely be an apparent "equivalent" noise temperature (T_e) which gives the noise power actually received. For example, **aerial temperature** (T_{AE}): an aerial receives noise of power spectral density as if it came from a source of temperature T_{AE}.

The **noise figure** of a system or network is a very important parameter in communications. It is a measure of the amount of extra noise power added by the network itself. It is defined and analysed using the terms in Figure 18.5.

The network has gain G (it can be attenuation), signal powers S, and noise powers N. Then F is **defined** as

$$F = \frac{\text{available noise power out}}{G(\text{available noise power in})}$$

"Available" means matched, that is $Z_{IN} = Z_S^*$ for S_{IN} and N_{IN}, and $Z_L = Z_{OUT}^*$ for the output powers. G is the "gain" when matched in this way. It is usually assumed, although often unknowingly, that a network is matched and then we can write F in terms of signal-to-noise ratio (S/N) as follows:

$$\begin{aligned} F &= N_{OUT}/N_S \\ &= N_{OUT}/G.N_{IN} \\ &= N_{OUT}/G.S_{IN}.S_{IN}/N_{IN} \\ &= N_{OUT}/S_{OUT}.S_{IN}/N_{IN} \end{aligned}$$

that is

$$\boxed{F = \frac{(S/N)_{IN}}{(S/N)_{OUT}}}$$

This is the usual way F is "defined" and used but remember that it tacitly assumes the matching conditions above. F can be expressed as a ratio or in decibels (being a ratio of power ratios). For an ideal completely noiseless network it would have value 1 (0 dB) as the two ratios would then be equal. Some useful values are:

$$F = 2 \equiv 3\,\text{dB}$$

$$F = 10 \equiv 10\,\text{dB}$$

It is also useful to express F in terms of the actual noise generated by the network, N_{INT}, and its input equivalent value, N'_{INT} $(= N_{INT}/G)$. That is

$$\begin{aligned} F &= \frac{N_{OUT}}{N_S} \\ &= \frac{N_S + N_{INT}}{GN_{IN}} \\ &= 1 + \frac{N_{INT}}{GkT_0B} \end{aligned}$$

or

$$\boxed{F = 1 + \frac{N'_{INT}}{kT_0B}}$$

F can also be expressed in terms of noise-equivalent temperature (T_e) by supposing that N'_{INT} comes from a source of equivalent noise temperature, T_e, so that:

$$N'_{INT} = kT_eB$$

therefore

$$F = 1 + T_e/T_0$$

or

$$\boxed{T_e = (F - 1)T_0}$$

When systems are arranged in cascade, as are the various units in a radio receiver, each unit will have a gain, G_n ("available" gain), and a noise figure, F_n. The question then is: what is the overall noise figure of the complete system? This is calculated by the de Friis equation which is expressed as

$$F = F_1 + \frac{F_2 - 1}{G_1} + \frac{F_3 - 1}{G_1 G_2} + \ldots + \frac{F_n - 1}{G_1 G_2 \ldots G_n}$$

Note how the overall noise figure is crucially dependent on both the gain and the noise figure, G_1 and F_1, of the first stage. If G_1 is large, $F \simeq F_1$ and the second and subsequent stages hardly matter. One practical example of the application of this idea is that of placing a high-gain low-noise front end of a microwave radio receiver as close to the aerial as possible.

For narrow-band signals such as occur on high-frequency radio links (e.g. microwave) the noise can be treated as if it were generated entirely at the carrier frequency, f_c, producing voltages which are random fluctuations of its phase and amplitude. This is **narrow-band noise** expressed as

$$\begin{aligned} n(t) &= x(t)\cos\omega_c t - y(t)\sin\omega_c t \\ &= r(t)\cos[\omega_c + \theta(t)] \end{aligned}$$

producing normalized noise power (into 1 Ω) of

$$N = \overline{n^2} = \overline{x^2} = \overline{y^2} = \overline{r^2}/2$$

Within a narrow band, in practice, the noise power is not usually uniform in density but peaks in the middle and tails off at the edges. To allow easy calculations in this ill-defined situation this power distribution is treated as if it were a rectangular

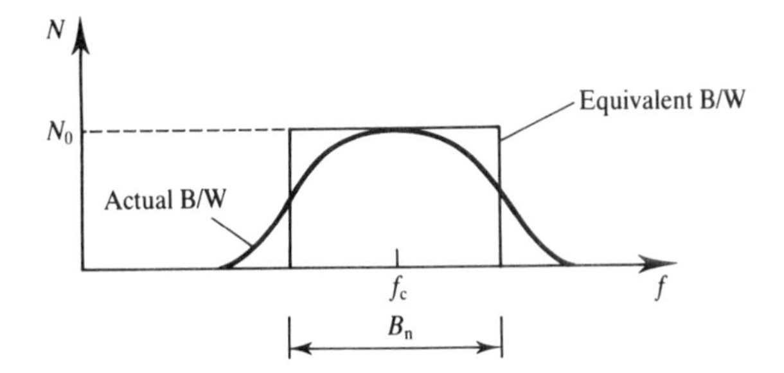

Figure 18.6 Noise-equivalent bandwidth $N = N_0 B_n$

distribution, at the peak value of the actual distribution, and within rectangular band limits, the **noise-equivalent bandwidth**, giving the same total noise power. This is all illustrated in Figure 18.6.

18.4 • Intersymbol interference

Intersymbol interference is caused by the unavoidable fact that no communication system has infinite bandwidth – it must act as some form of ill-defined bandpass filter which affects the digital baseband as if it had been passed through a low-pass filter. The effect of this on the bits in the signal is to spread them out in time so that they are not received in the pristine square form in which they were generated but in some smeared-out shape with a rounded top and tails fore and aft. This means that some of the bit energy occurs in time slots occupied by other adjacent bits and so changes their apparent voltage values. If this unwanted extra voltage is large enough these bits could be detected in error. This "smearing" process is illustrated in Figure 18.7.

Luckily it is possible completely to avoid errors due to ISI by two design factors:

1. The use of "decision" detection of bits.
2. By imposing an "anti-ISI" filter characteristic.

Decision detection is normally used in digital communication systems anyway. There is no need to recover any of the signal bits in their original clean square shapes because all that it is necessary to know is whether any particular bit is a one or a zero. The voltage of each received bit is then sampled at a single point on its waveform where it is most likely to be correct – usually at its mid-point. The receiver has to make a **decision** as to whether that bit is a one or zero and, to do it, some form of decision circuitry is used, probably a **threshold detector**. This technique is illustrated in block diagram form in Figure 18.8.

The threshold detector will be some form of **Schmitt trigger**. For bipolar input bits the threshold is set at 0 V so that, if greater than 0 V, the input will trigger a

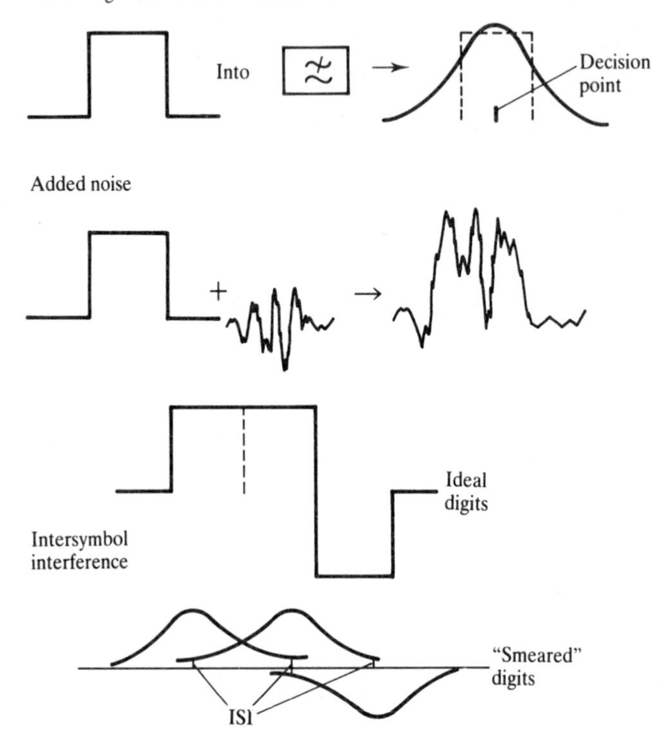

Figure 18.7 Bit error due to ISI

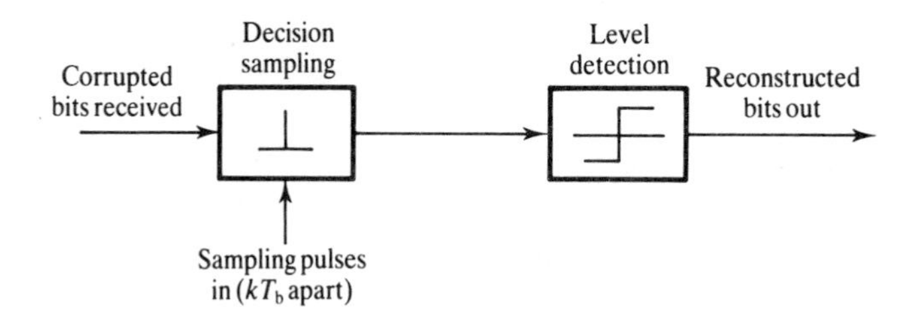

Figure 18.8 Decision detection of binary signals

one at the output and, if less than 0 V, a zero. The hysteresis inherent in the trigger will avoid unwanted extra output bits triggered by small noise spikes on the input.

Then ISI can be avoided by arranging that the smearing of the bits by the system filter occurs in such a way that their smeared waveform is always of zero voltage at the decision points for adjacent bits. This can be done by deliberately imposing on the whole system a filter of a spectral characteristic which will give an output bit shape having zero crossings at the required time intervals. Such a filter is called an

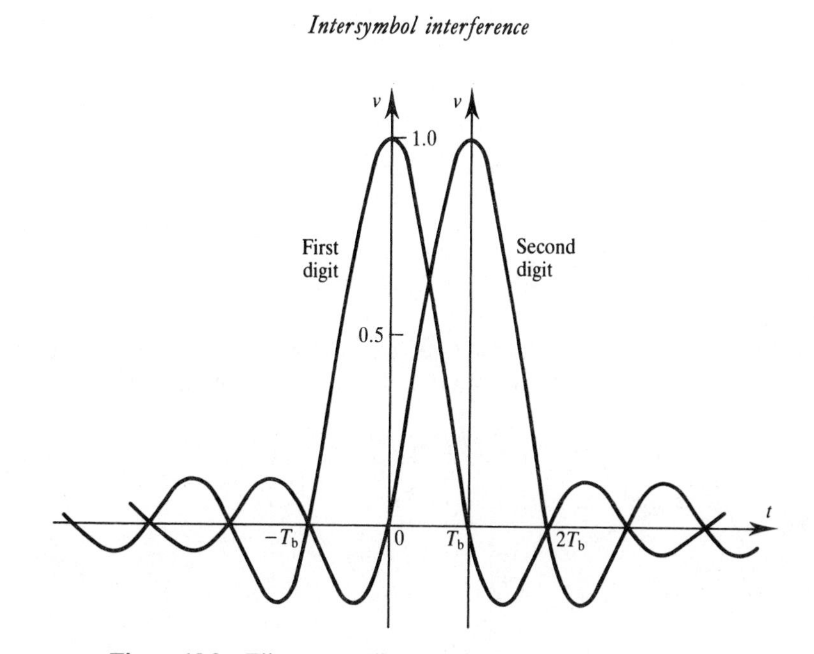

Figure 18.9 Filter-controlled output pulse shape avoiding ISI

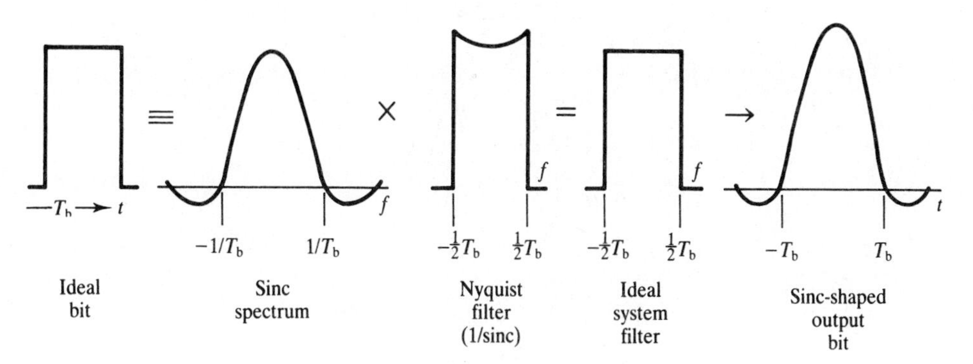

Figure 18.10 Action of Nyquist filter

anti-ISI filter and is usually included as two identical filters ($H_T = H_R$), one in the transmitter and one in the receiver, whose spectral product gives the overall filter transfer characteristic required. This is H_S (S for system) where

$$H_T . H_R = H_S$$

For reasons which will become clear later the anti-ISI filter is usually designed α for short and so H_T and H_R are designated $\sqrt{\alpha}$:

$$H_R = H_R \equiv \sqrt{\alpha}$$

For rectangular bits in, H_S gives an output pulse shape with zero values at intervals of T_b (the bit length) as shown in Figure 18.9.

The ideal shape for these output pulses is a sinc function which would occur if its spectrum were an ideal rectangular "brick-wall" shape. The ideal system filter (H_S) which would give this is something called the **Nyquist filter** as shown in Figure 18.10. The original rectangular pulse shape has a sinc spectrum and if the middle of its central peak is multiplied by the Nyquist filter the required output spectral shape results.

Of course, in practice a Nyquist filter cannot be made because of its abrupt cut-off. But if this filter is multiplied by a raised cosine filter (or any other symmetrical

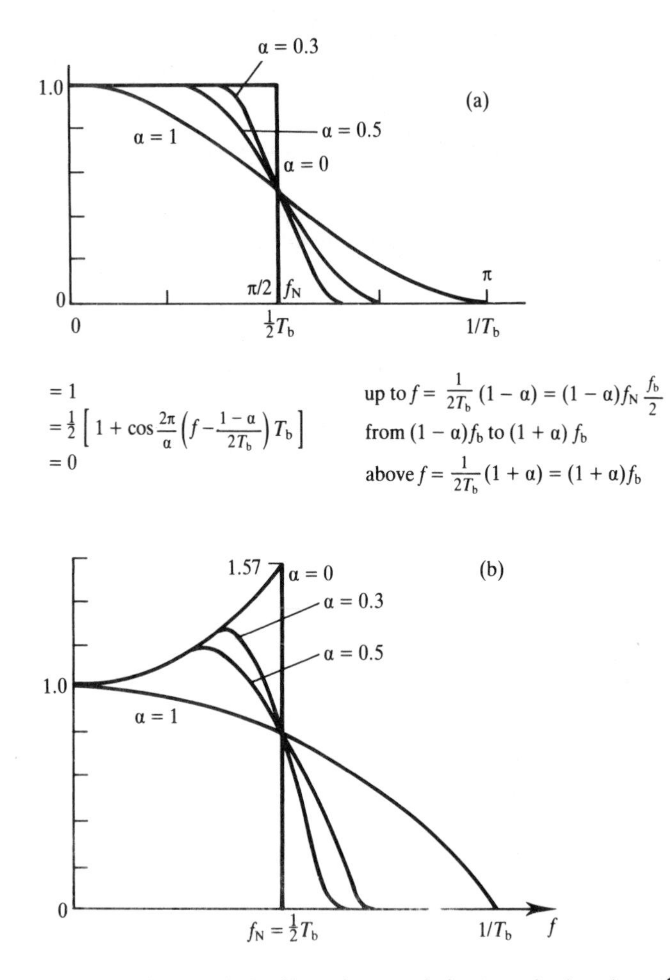

Figure 18.11 "Raised cosine" anti-ISI filter characteristic: (a) raised cosines ($\frac{1}{2}(1 + \cos\theta) = \cos^2\theta/2$); (b) Nyquist filter with various raised cosine roll-offs (linear scale); (c) as (b) but with log scale

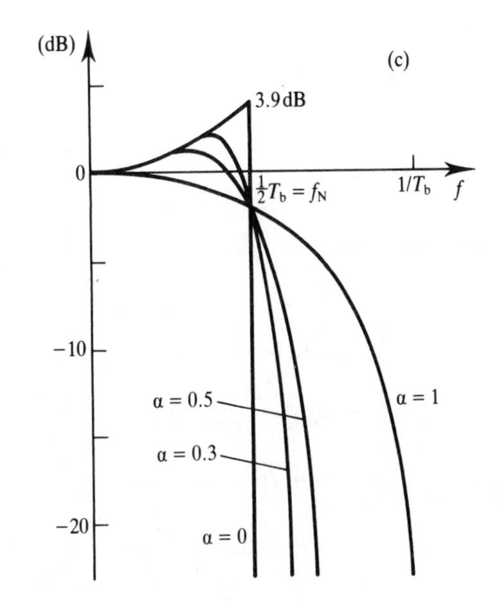

Figure 18.11 (*continued*)

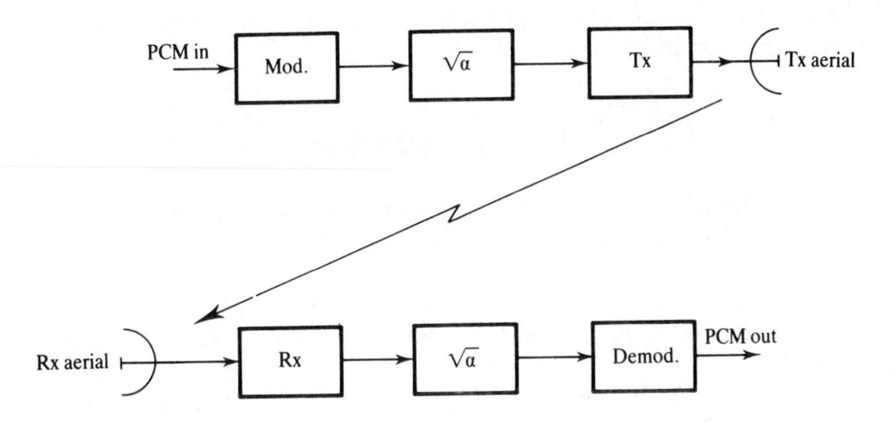

Figure 18.12 Communication system with anti-ISI filter

filter shape) the output waveform still has zero crossings T_b apart and the whole filter combination acts as an anti-ISI filter. Also it can now be realized in practice. This is shown in Figure 18.11.

The factor α introduced above is a constant of the raised cosine filter shape as given in Figure 18.11. Its effect is to increase the system bandwidth above that of the ideal Nyquist filter $(1/T_b)$ by a factor $(1 + \alpha)$. At **baseband** the Nyquist and raised cosine filters act as **low-pass filters** with bandwidths given in Figures 18.10 and 18.11 as

$$\text{ideal (Nyquist) B/W} = f_N = 1/2T_b = \tfrac{1}{2}\,\text{(bit rate)}$$

$$\text{practical (raised cosine) B/W} = f_N(1 + \alpha) = (1 + \alpha)\,\text{(bit rate)}$$

When PSK modulated for **transmission** onto a subcarrier these bandwidths are doubled (PSK is a form of DSBSC) to give

$$\text{Ideal B/W} = f_b = \text{Bit rate}$$

$$\text{Actual B/W} = f_b(1 + \alpha)$$

Hence the simple working rule already used that

$$\textit{Bandwidth} = \textit{Bit rate}$$

Of course, in a real system some additional bandwidth is needed, however small α is (it can be <0.3), to allow for realizable anti-ISI filtering. Hence signal degradation is avoided at the expense of increased bandwidth.

A complete radio communications system would have its $\sqrt{\alpha}$ filters either in the baseband or the i.f. parts of the circuit. Figure 18.12 shows this rather stylistically.

Thus noise becomes the main limitation on the performance of a digital communication system.

18.5 • Conclusion

This chapter has only looked at the basic ideas of causes of errors in digital signal reception. There is a great deal more to the design of digital systems to reduce these to a minimum – as well as taking into account other major design considerations such as the efficient use of both bandwidth and power. To find out more about these, readers should refer to more specialized textbooks some of which are listed in the Bibliography.

The next chapter is the final one in the book and really follows on directly from Chapters 7 and 17 rather than from this one. It looks at the multiplexing of signals for transmission.

• 19 •
Multiplexing

19.1 • Introduction

For obvious economic reasons it is often necessary to be able to send many signals simultaneously down the same transmission system. The process of combining signals for such transmission is known as **multiplexing**. The method used to do this will depend on both the type of signal and the transmission medium being used. There are two main methods:

1. Frequency division multiplexing FDM
2. Time division multiplexing TDM.

Other methods you may meet are code division multiplexing (CDM) used in spread spectrum transmission techniques, and wavelength division multiplexing (WDM) used for optical fibre transmission.

19.2 • Frequency division multiplexing

This is essentially an analogue method whereby each signal is allocated its own frequency slot within the overall transmission bandwidth of the system. The signals are then separated out again at the receiving end by filtering. The principle is best illustrated by looking at the formation of a standard telephone group which consists of 12 standard telephone channels in FDM as shown in Figure 19.1.

Each telephone channel starts off within the nominal baseband bandwidth of 0–4 kHz and is then SSB (LSB) modulated up on to one of a sequence of subcarriers spaced 4 kHz apart to put it in a 4 kHz frequency slot in the group transmission bandwidth from 60 to 108 kHz. These subcarriers will be at 64 kHz, 68 kHz, and so on up to 104 kHz and finally 108 kHz. The actual process is illustrated in Figure 19.2.

All this is now very standard technology, the main problem being in the filtering requirements which are quite stringent and require high-order crystal filters to achieve. These telephone channels provide 900 Hz of empty "guard band" between them, within which adequate filtering has to be achieved both in forming the group (removing

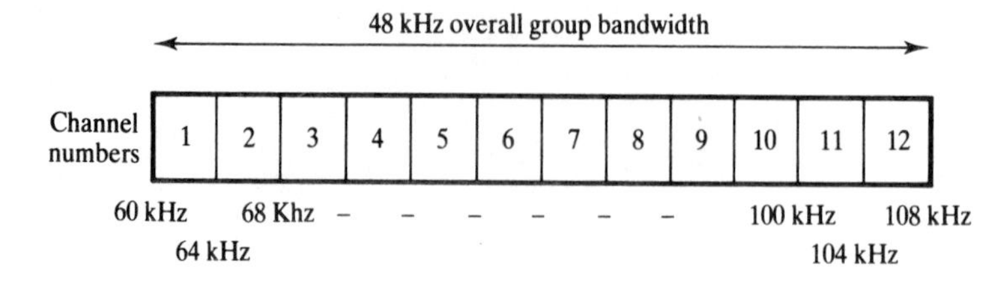

Figure 19.1 CCITT FDM telephone group structure

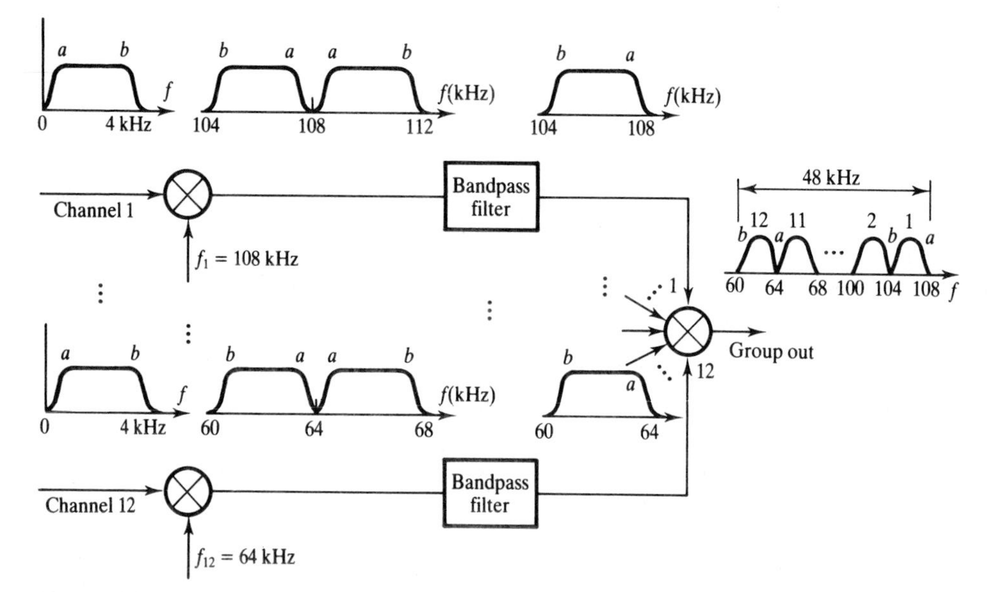

Figure 19.2 Formation of an FDM telephone group (after Schwartz, 1990). Reproduced with permission of McGraw-Hill, Inc

the other sideband) and in separating each signal from the others at the receiver. Luckily millions of such filters are needed in the whole telephone system, so the cost of each one is much reduced.

The total transmission bandwidth of several signals in FDM is obviously the sum of the bandwidths of the individual signals. In particular, if there are N signals of identical baseband bandwidth B Hz, then the total bandwidth is

$$\text{FDM B/W} = NB\,\text{Hz}$$

but note that, in general, this FDM bandwidth will not start at zero frequency but will be translated up to a higher band by the modulation involved in the multiplexing process. Thus, here it encompasses the range 60–108 kHz with a 48 kHz group bandwidth.

Telephone use of FDM has been enormous, and although much of it has now been replaced by TDM, it is still important to know about it, both as a basic idea, as here, and also in terms of the hierarchies described later.

19.3 • Time division multiplexing

This is essentially a digital technique applied to sampled signals in the form of PCM although it can be used with PAM samples. For both, the basic idea is that there are proportionately long gaps between each sample into which other samples can be placed. The principle is illustrated in Figure 19.3 for both PAM and PCM.

For PCM the interval between two samples of the same signal is unchanged, being just the sampling interval (T_s) of 125 μs (1/(8 kHz)). What has changed is the big length, eight of which would have occupied the whole sampling interval originally. Now each bit must be shortened to allow in the bits of the other signals. The amount of shortening is in exact inverse proportion to the number of signals being multiplexed. Each bit started off 15.6 μs long (125/8) so that two signals would mean bits 7.8 μs long (15.8/2) and so on – all within the same 125 μs **frame** as it is now called.

There seems to be no obvious limit to the number of signals which can be multiplexed into this frame and the limit is set in practice by bandwidth considerations and the convenience of traffic needs. Since this is partly a subjective decision it will come as no surprise to find that the world uses two different standards. In North America the AT&T standard of 24 channels per frame is used whilst in Europe there are 32 channels in one frame. Both are illustrated in Figure 19.4.

Note that the 24 channel frame has one extra bit "stuffed" in at the end and

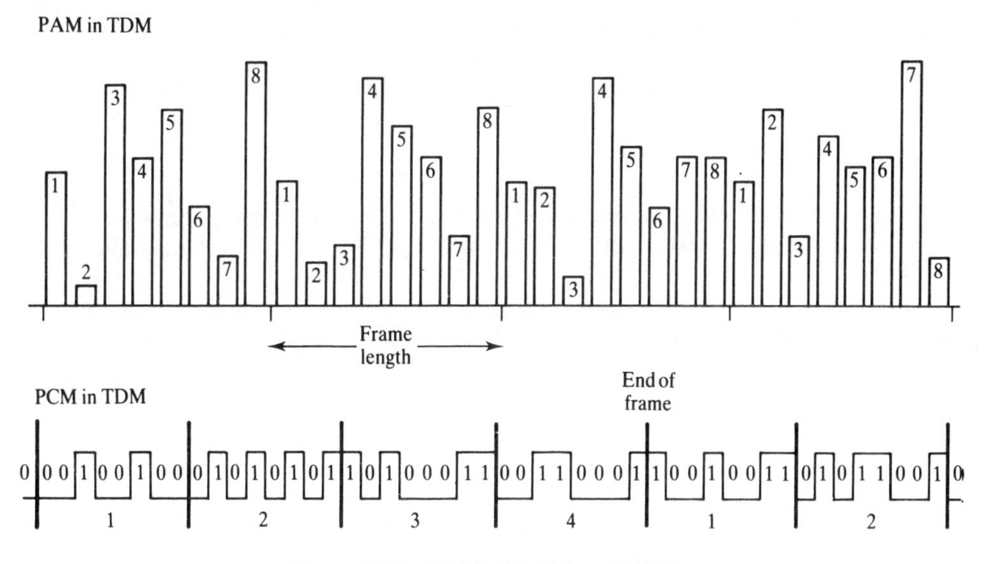

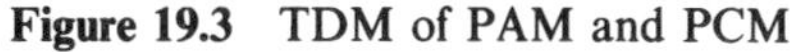

Figure 19.3 TDM of PAM and PCM

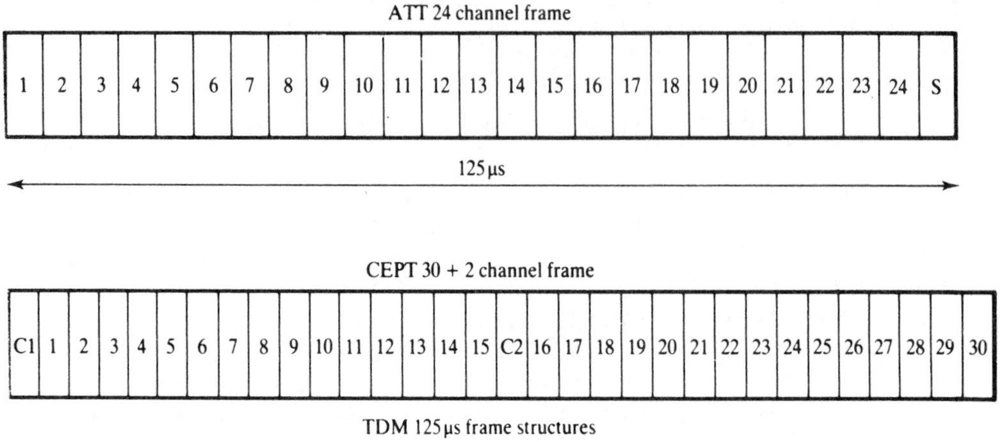

ATT 24 channel frame

1	2	3	4	5	6	7	8	9	10	11	12	13	14	15	16	17	18	19	20	21	22	23	24	S

125 μs

CEPT 30 + 2 channel frame

C1	1	2	3	4	5	6	7	8	9	10	11	12	13	14	15	C2	16	17	18	19	20	21	22	23	24	25	26	27	28	29	30

TDM 125 μs frame structures

Figure 19.4 CCITT PCM TDM frame structures (after Holdsworth and Martin, 1991)

that the 32 channel frame uses two of its channels (16 and 32) for signalling purposes. Although both frames have to be of the same length (the 125 μs sampling interval), they do not occupy the same bandwidth. In general, for N identical analogue signals of original baseband bandwidth B Hz, sampled at the Nyquist frequency and digitized using n bits per sample, the number of bits per frame will be

$$\text{bits per frame} = 2NnB$$

but with an extra one added for the 24 channel frame. Then the bandwidth occupied by the frame will be equal to this bit rate:

$$\text{TDM B/W} = \text{bit rate} = 2NnB$$

For the two frame types these give

24 channels: 193 bits, each 0.648 μs long; 1.544 MHz frame bandwidth
32 channels: 256 bits, each 0.488 μs long; 2.048 MHz frame bandwidth

The 32 channel frame is often called the 2Meg frame for short as that value is accurate enough for descriptive purposes. The 24 channel frame is similarly described for short as the D2 frame because of its designation by its AT&T originators (D for digital, version 2).

19.4 • Multiplexing hierarchies

As indicated in the figure captions, the FDM group and the two TDM frame structures are fixed by international agreement. This is done by treaty through the International

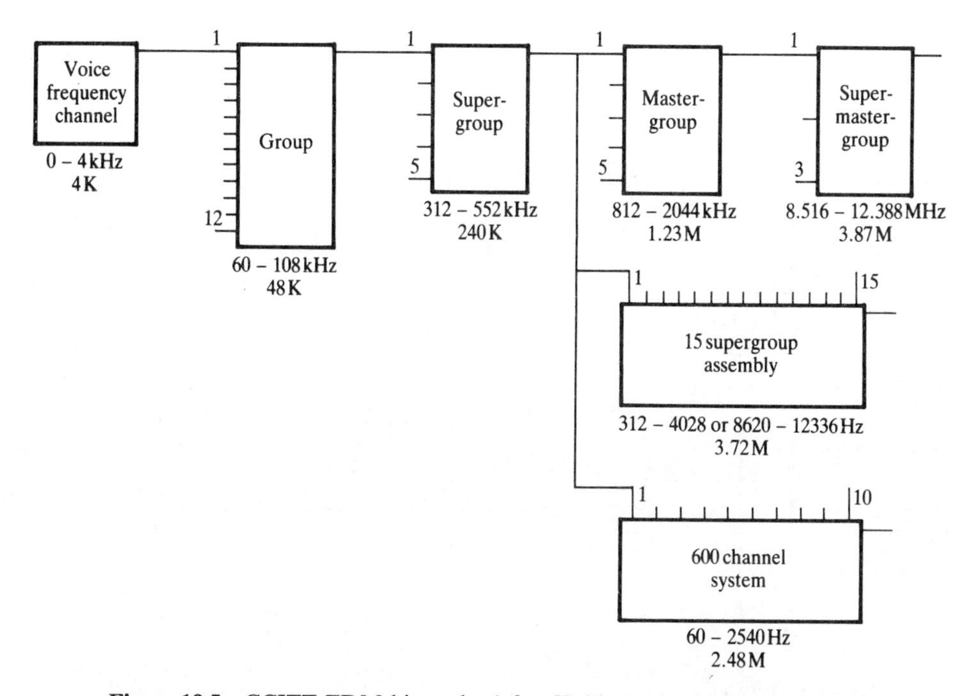

Figure 19.5 CCITT FDM hierarchy (after Holdsworth and Martin, 1991)

Telecommunications Union (ITU) using standards recommended by the International Consultative Committee on Telegraphy and Telephony (CCITT). The basic frames the latter recommended are as above, but there is a lot more to it because of the need to multiplex many more channels together for routes carrying heavier traffic loads – groups are gathered into supergroups; frames into superframes; and so on. The whole system is shown succinctly in Figures 19.5 and 19.6.

TDM has two standard hierarchies because of the two separate systems which grew up either side of the Atlantic – as for the basic frame structures above. Some steps in these hierarchies have particular uses. For example, in FDM the supermastergroup of bandwidth 3.9 MHz fits nicely into the two standard coaxial cable bandwidths of 4 and 12 MHz. When further frequency modulated it fits into the 36 (or 72) MHz bandwidth of a satellite transponder giving it its standard 900 channel capacity.

In TDM the 140 MHz channel, carrying nearly 1000 two-way conversations, is a common digital radio transmission standard whilst the nominal 565 MHz channel was a standard for optical fibre capacity.

19.5 • TDM developments

The hierarchies above worked very well for most of the 1970s and early 1980s. In particular, the TDM hierarchies were especially good when used to replace FDM in

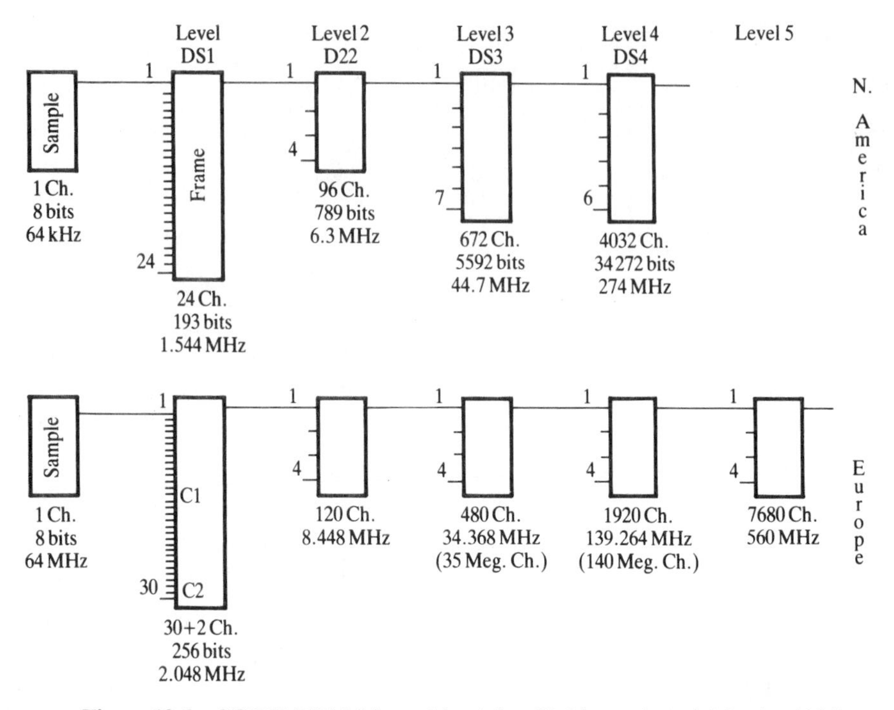

Figure 19.6 CCITT TDM hierarchies (after Holdsworth and Martin, 1991)

telephone networks where the requirements of accuracy and flexibility are fairly undemanding and the growth rate small (e.g. 5% a year).

Unfortunately (or perhaps fortunately) the rapid change over to digital transmission created, in effect, a new system where the signal transmitted is of the same type whatever the original source (merely a bit stream). This led to ease of processing (e.g. digital exchanges) and, more importantly, apparently unrestricted use by non-telephone traffic such as data, FAX, video, cellular, etc. All this became very attractive to commercial users, especially as the success of digital transmission made it very cheap. Thus a huge explosive non-telephone use began and is still going on.

This sounds marvellous, but problems soon became apparent because these new uses have requirements of accuracy, flexibility and availability not needed for telephones only. Then fresh problems arose when optical fibres came into general use with their own standards.

All this can be summarized into three areas of difficulty.

1. Synchronization
2. Interfacing
3. Management.

Synchronization has always been a problem in telephone use of TDM because different signal sources were allowed to have slightly differing clock rates – not much

($\pm$50 parts per million) but significant. This was fairly easily overcome by putting in extra bits during multiplexing to make all clock rates the same and then multiplexing bit-by-bit (rather than sample-by-sample) at levels above the 2Meg frame. This slightly new 'nearly' synchronous hierarchy became known as the **plesiochronous digital hierarchy** (or PDH). Satellite TDMA is a good example of a particular problem this technique overcame.

Interfacing had also been an obvious problem in TDM right from the start because CCITT was forced to agree to two international standard hierarchies. Whilst there were only two, and these used for telephones only, this was no real problem but interfaces from and between optical fibre systems and the need for unrestricted access by business users for new types of service (e.g. for cellular base stations) made an urgent solution necessary. The trouble is that, with the original hierarchies, the only way to cope with interfacing and access was to demultiplex to baseband and then remultiplex again – a very expensive method which also degraded all services whilst it was being done. Complete synchronization of the whole system throughout the world was an obvious idea but not a very practicable one. The answer was clearly a new system, a decision which was accelerated by the third area of difficulty.

Management via the control channels had, of necessity, been allowed for in TDM. The problem was that the new non-telephone uses require much more of it than can be fitted into the space allowed. Telephones are very undemanding because they can have large error rates without anyone minding much and the simple 4 kHz channels are very easy to route and monitor when only public telephone operators are involved. Once commercial use became widespread the performance required of the system became much more stringent. A much more error-free transmission guarantee was required with greater reliability of connection. Ease of access and flexibility of connection also became more important. Users need to have the facilities they pay for there; reliably; when they want them; without question – a tall order as anyone who has used FAX will know. Added pressure came because business use is growing rapidly in an increasingly competitive environment and is very lucrative. Yet further pressure came because individual operators were all putting their own solutions into effect so that prompt international action became urgently needed to avoid permanent confusion. The results were studies by CCITT which have led to international agremeent on an entirely new digital hierarchy, the **synchronous digital hierarchy** (or SDH).

SDH also uses a 125 μs frame structure (as for the older systems) but now it contains 2430 8 bit bytes arranged in 9 rows of 270 columns – a bit rate of 155.52 Mbps altogether. All inputs are synchronized (as PDH) and then multiplexed in hierarchies which give a built-in flexibility of operation able to cope with any type of digital traffic, whatever its requirements, both now and in the forseeable future. Naturally this makes it rather complex in detail so no attempt is made to describe it further here and the reader might like to start by looking at two excellent articles by Matthews and Newcombe in the *IEE Review* for May and June 1991. But, be in no doubt, it is the system of the future.

19.6 • Other methods

Code division multiplexing (CDM) is a method which allows many signals to be sent within the same large bandwidth. This is done by labelling each one with a distinctive code at the transmitter so that only the signal which matches the same code will be recovered at the receiver. The simplest way to do this is to modulate the digital signal with a higher rate pseudo-random bit stream generated in a way in which the receiver can reproduce it. The original signal bits are thus chopped up into several pieces called **chips**.

One effect of this treatment is to spread the signal over a wide bandwidth (set by the chip rate) at a low spectral power density which may even be below noise level. Then it goes under the name of **spread spectrum** (one type) and can be a clandestine communication method having military significance. It also allows the same large bandwidth to be used by many signals, each coded differently, which could lead to more efficient overall use of available spectrum (e.g. perhaps carrying voice channels on TV signals).

Wavelength division multiplexing (WDM) is really out of place in this book but included for completeness. It merely means using several light wavelengths (colours) down the same optical fibre, so carrying several messages at once – one per wavelength.

19.7 • Summary

Multiplexing means sending more than one (usually many) separate signals simultaneously down the same communication channel.

In **frequency division multiplexing** (FDM) each signal is assigned its own portion of the available transmission spectrum (e.g. 4 kHz in the range 60–108 kHz for a telephone group).

In **time division multiplexing** (TDM) each signal is sampled and (usually) digitally coded too. Then one sample from each signal is sent in time sequence, one at a time, during one sampling interval (e.g. 32 8 bit samples in 125 μs in a CCITT frame).

19.8 • Conclusion

The two main multiplexing methods enable enormous numbers of separate signals to be carried simultaneously down the same communication channel on high-density routes. At all traffic densities they enable groups of signals to be routed more easily as a unit, so facilitating the operation of complex systems such as national and international telephone systems. Digital multiplexing is increasingly becoming the more important and the new synchronous method represents state-of-the-art development in 1991 – a fitting point at which to end this book.

19.9 • Problems

19.1 Sketch time and frequency plots for the following PAM–TDM signals. Assume all sample pulses are the same length and each occupies half its time slot.

(i) Equal amplitude sine waves at f_0, $2f_0$, and $3f_0$.
(ii) As (i) but amplitude ratios 1, 0.5, and 0.3.
(iii) Equal amplitude square waves at fundamental frequencies of f_0, $2f_0$, and $3f_0$.
(iv) Sine and square waves of the same f_0 and amplitude.

19.2 For the CCITT voice frequency channel FDM hierarchy, answer the following questions:

(i) Why SSB?
(ii) Why does the first group start at 60 kHz?
(iii) How have coaxial cable bandwidths influenced the bandwidths set at the various levels of the hierarchy?
(iv) Why multiply by so much as 12 at the first level of the hierarchy?
(v) What is the unused (guard band) bandwidth between channels in a group? Estimate the order of filter needed to separate the channels of a group.
(vi) If you can, measure the actual useful bandwidth of your telephone and comment.
(vii) Examine the hierarchy and see where additional guard bands have been included. Why might this have been done?

19.3 For a.m. radio stations bandwidths of 6 kHz are used. Calculate the following:

(i) The baseband bandwidth.
(ii) The number of such basebands which could be frequency multiplexed into a standard telephone group bandwidth.
(iii) The bandwidth needed by such a group if the individual channels had been coded into 12 bit PCM and then frequency multiplexed.

Would method (iii) be a good way to transfer programmes from station to station? If not, what method would you suggest?

19.4 N binary coded signals, each having n digits per sample, are time multiplexed together. Calculate the overall signal bandwidths for the following conditions.

(i) $n = 8$, $N = 12$, all telephone channels.
(ii) $n = 6$, $N = 400$, all telephone channels.
(iii) $n = 8$, frame length $125 \mu s$, telephone channels B/W = 1.54 kHz. Find N.
(iv) Baud rates of 100, 300, 2400, and 9600.
(v) $n = 8$, $N = 24$, all telephone channels.

(vi) Bit lengths of 125, 16, and 104 μs.
(vii) UK 625 line TV, 8 bit quantization.

19.5 A 36 MHz wide satellite transponder carries 400 two-way telephone conversations using FM/FDM. Calculate the smallest value of modulation index (β).

If these voice frequency signals are instead converted to standard PCM and QPSK modulated into the same FDM channel widths, how many conversations can now be carried?

For PCM signals 72 MHz transponders are now preferred. Suggest a simple reason for this.

19.6 N signals each coded into 8 bit PCM are time multiplexed into one channel.

(i) Explain what time multiplexing (TDM) is.
(ii) Give the minimum transmission bandwidths needed for $N = 24$ and $N = 32$ for a standard voice frequency channel.
(iii) If each signal is B Hz wide and is sampled at k times the Nyquist frequency give the general expression for the minimum bandwidth for any N. What is k for (ii) above?
(iv) Why do you only send 30 signals in a 32 channel PCM frame?
(v) In North America 24 channel TDM frames are standard whilst Europe uses 32 channel frames. Find reasons why this is so. Also, think of problems introduced by this difference and how they might be overcome.

19.7 In the communications satellite Intelsat IV-A a standard transponder bandwidth was 36 MHz.

(i) Find out what a transponder is.
(ii) How many channels could be multiplexed on to it using

(a) SSB FDM
(b) 8 bit PCM TDM?

(iii) If the whole channel is used for one frequency-modulated TV signal, estimate the β value at the largest signal frequency for

(a) NTSC TV of 525 lines 60 Hz as in USA
(b) PAL TV of 625 lines 50 Hz as used in UK.

(iv) What bandwidth would be needed to send UK 625 line TV using 8 bit PCM?
(v) Recent (1986) tests have shown that acceptable digitized TV signals can be sent via a 36 MHz transponder. How might this be done? See what you can work out for yourself without having to look it up.

19.8 Show that the standard CCITT 32 channel TDM frame requires 2.048 MHz bandwidth. What bandwidth would be needed for the 24 channel frame? Recall that it is given one extra (stuffed) bit.

19.9 (a) Briefly explain the processes of time division multiplexing and frequency division multiplexing.

(b) A coaxial transmission line passes frequencies in the range 0–50 MHz. Determine the maximum number of speech channels, each of bandwidth limited to 4.0 kHz, that can be transmitted over the transmission line via

(i) a time division multiplexed PCM system. Assume the following: sampling frequency equals 1.2 times Nyquist frequency; 6 bit PCM words;

(ii) a frequency division multiplexing system employing single-sideband suppressed-carrier techniques. Assume that guard bands of 1 kHz exist between adjacent channels.

Explain all steps in your calculations.

19.10 Explain the reasons for sampling analogue signals before transmission. Show why the minimum sampling rate is twice the highest signal frequency and discuss practical considerations which modify this criterion.

A PAM–TDM system is required for telemetry transmission of six separate slowly varying quantities measured continuously. Values for two of these quantities are required twice as often as for the other four which are needed once every 0.1 seconds. Describe a suitable combined signal format and calculate the pulse repetition frequency.

19.11 Name the most common types of multiplexing used in communication systems. Explain how each method works in detail and give one example of each in common use today.

There are 60 telephone channels to be multiplexed by each of the two methods. Calculate the transmission bandwidths required, showing clearly how your values were obtained.

Discuss the relative advantages of using one or other of the multiplexing methods both in the light of your bandwidth values and of other considerations.

19.12 Twenty-four different analogue signals, each 5 kHz in bandwidth, are digitized using 7 bit PCM and transmitted as a continuous time-multiplexed binary signal. Answer the following questions:

(i) What is the minimum sampling rate for each signal?

(ii) If this signal rate is used, what is the bit rate for

(a) one signal

(b) the time-multiplexed output?

(iii) What is the bit rate in ii(b)?
(iv) What is the frame length?
(v) What is the minimum bandwidth for the time-multiplexed output?
(vi) How many quantized sampling levels are there?
(vii) Are the figures used in this question of standard usage? If not, what would standard values be?
(viii) What is the quantization signal-to-noise ratio introduced by the sampling?
(ix) How would you change the format to aid processing on reception?
(x) Say why it is adequate to sample only at the minimum rate in (i).

19.13 Ninety signals, each band limited to a telephone channel, are pulse code modulated using 8 bit samples and then frequency modulated in FDM on to a 36 MHz i.f. carrier using maximum deviation.

Calculate the smallest modulation index required, explaining how you obtained your value.

Sketch a block diagram circuit which would produce such modulation, making clear the purpose of each block.

Discuss how more efficient use could be made of the r.f. bandwidth available.

19.14 An audio baseband is limited to a nominal frequency range of 0–6 kHz. It is sampled at the Nyquist frequency for the highest signal frequency and coded into 8 bit PCM. It is then time division multiplexed with 24 other similar signals, limited to the same audio range, to form a TDM frame. It is then PSK modulated on to a 9.6 MHz i.f. subcarrier. Answer the following questions:

(i) What is the sampling frequency?
(ii) What is the bit length for one channel after PCM encoding?
(iii) What is the frame length?
(iv) Sketch the make-up of the frame.
(v) What bit rate does the PSK modulator receive?
(vi) Sketch the form of the bits as required by the modulator.
(vii) Sketch the form of these bits after modulation.
(viii) What is the minimum bandwidth needed for the modulated subcarrier?
(ix) Compare the answer to part (viii) with the bandwidth if the original audio baseband had been modulated for medium-wave broadcasting.
(x) What chief advantage does PSK modulation have over FSK modulation in this application?
(xi) Why would a carrier recovery circuit be required when the PSK signal is demodulated from the subcarrier at the receiver?

See also Problems 14.9 and 17.9.

Bibliography

The author has used most of the books below, sometimes in earlier editions, during the 13 years or so he has been teaching communications. They are listed roughly in order of similarity of coverage and/or treatment to this book with the ones near the end providing a much broader more general coverage of communications-related topics. There are, of course, many more useful books than those listed here and all of them will contain some material which will help your understanding. To dip into many books is a good way both of getting a preliminary idea of what is important in a subject and of finding the key explanation of those maddening details which just will not come clear. Older books are often especially useful in this way.

1. F. R. Connor (1982) *Modulation*, 2nd edn, Edward Arnold: London.
2. F. R. Connor (1982) *Signals*, 2nd edn, Edward Arnold: London.
3. F. R. Connor (1982) *Noise*, 2nd edn, Edward Arnold: London.
4. F. R. Connor (1982) *Networks*, 2nd edn, Edward Arnold: London.
5. F. G. Stremler (1990) *Introduction to Communication Systems*, 3rd edn, Addison-Wesley: Reading, MA.
6. K. S. Shanmugam (1985) *Digital and Analog Communication Systems*, 2nd edn, Wiley: Chichester.
7. M. Schwartz (1990) *Information Transmission, Modulation & Noise*, 3rd edn, McGraw-Hill: New York.
8. B. Rorabaugh (1990) *Communications Formulas and Algorithms*, McGraw-Hill: New York.
9. J. A. Betts (1970) *Signal Processing, Modulation & Noise*, Hodder & Stoughton: London.
10. R. F. W. Coates (1982) *Modern Communication Systems*, 2nd edn, Macmillan: London.
11. J. J. O'Reilly (1989) *Telecommunication Principles*, 2nd edn, Van Nostrand Reinhold: Wokingham.
12. R. L. Freeman (1981) *Telecommunication Transmission Handbook*, Wiley: Chichester.
13. J. K. Hardy (1986) *Electronic Communications Technology*, Prentice Hall: Englewood Cliffs, NJ.
14. M. S. Roden (1991) *Analog and Digital Communication Systems*, 3rd edn, Prentice Hall: Englewood Cliffs, NJ.
15. J. Dunlop and D. G. Smith (1989) *Telecommunications Engineering*, 2nd edn, Van Nostrand Reinhold: London.
16. J. Brown and E. V. D. Glazier (1964) *Telecommunications*, Chapman & Hall: London.
17. R. E. Ziemer, W. H. Tranter and D. R. Fannin (1989) *Principles of Communications*, 2nd edn, Macmillan: New York.

18. P. H. Young (1990) *Electronic Communication Techniques*, 2nd edn, Merrill: Columbus, OH.
19. H. Stark, F. B. Tuteur and J. B. Anderson (1989) *Modern Electrical Communications*, 2nd edn, Prentice Hall: Englewood Cliffs, NJ.
20. D. Roddy and J. Coole (1984) *Electronic Communications*, 3rd edn, Prentice Hall: Englewood Cliffs, NJ.
21. S. Haykin (1989) *Analog and Digital Communications*, Wiley: Chichester.
22. R. J. Schoenbeck (1988) *Electronic Communications*, Merrill: Columbus, OH.
23. G. Kennedy (1985) *Electronic Communication Systems*, McGraw-Hill: New York.
24. B. Holdsworth and G. R. Martin (1991) *Digital Systems Reference Book*, Butterworths: Oxford.

Index